# Modern TRIZ Modeling in Master Programs

Michael A. Orloff

# Modern TRIZ Modeling in Master Programs

Introduction to TRIZ Basics at University and Industry

 Springer

Michael A. Orloff ⓘ
Academy of Instrumental Modern TRIZ
Berlin, Germany

ISBN 978-3-030-37419-8          ISBN 978-3-030-37417-4   (eBook)
https://doi.org/10.1007/978-3-030-37417-4

This Springer imprint is published by the registered company Springer Nature Switzerland AG
The registered company address is: Gewerbestrasse 11, 6330 Cham, Switzerland

Michael A. Orloff is a founder of the Academy of Instrumental Modern TRIZ (AIMTRIZ, Berlin, Germany, 2000) and Global Enterprise for Mastery in TRIZ (GEMTRIZ, Minsk, Belarus, 2019). New company GEMTRIZ continues the activity of AiMTRIZ to develop educational Modern TRIZ society around the world.

Prof. Orloff is an author of several books published by SPRINGER and author of Modern TRIZ educational and project modeling with techniques "Extracting", "Reinventing" and "Meta-Algorithm of Invention T-R-I-Z".

His training and consulting are well known at many companies including SIEMENS (for *postautomation* and *electrotransportation* directions, Berlin, Germany), SAIT Samsung Advanced Institute of Technology (for *system development*, Suwon, Korea), TU Berlin (for several specialties), universities and dozens of companies in China (till 1000 trainees in auditorium at trainings), Republic of Korea, Egypt, Kazakhstan, Belarus, Russia, etc. Till now about 500 innovative and inventive solutions (many of them are patented by customers) were made with the author's consulting and problem solving (on the order of customer or together with Think Tank Teams of a company) at more than 120 enterprises.

The book is compiled using mainly the cases of international Master of Science Programs at TU Berlin "Global Production Engineering" and "Entrepreneurship and Innovation Management", and at Campus El Gouna, Egypt, for "Energy Engineering" (with attendees from faculties "Urban Development" and "Water Engineering"), and also in Russia, Belarus and Kazakhstan for the programs at more than 20 universities and industrial companies.

The learning material reproduces and represents the educational cases achieved also in co-operation with Institute of Aeronautics and Astronautics (ILR) and key partner-companies ECM Space Technologies GmbH (formerly) and ECM European Academy GmbH (generally for European programs TEMPUS and ERASMUS MUNDUS and others), S2M Science-to-Market GmbH (generally for Chinese aerospace institutions), COURSENTO UG and GEMTRIZ Global Enterprise for Mastery in TRIZ (LLC).

This textbook also includes the examples of pilot projects for teachers and students of high schools and universities in co-operation with National Research Nuclear University MEPhI (Moscow Engineering Physics Institute), UNESCO

Institute of Informational Technologies in Education (IITE), BSUIR Belarus State University of Informatics and Radio-electronics (Minsk, Belarus), Kaspian University (Almaty, Kazakhstan) and others.

The book is addressed, first and foremost, to the students of universities at Master Programs and postgraduate programs, to teachers and professors, interested in modernization of the subjects with Modern TRIZ modeling and inventing, as well as to any specialists interested in applying the modern systematic methods of creative design thinking.

El Gouna Center of Technical University Berlin, Egypt, August 2, 2017

*All my inventions were an overcoming of the "insoluble" contradictions between the requirements for functioning and available resources. My solutions were de-facto in strict accordance with TRIZ methodology. Today Modern TRIZ can become the basis of creative engineering education, including at our first in the world "SkyWay Academy"*

**Dr.-Eng. Anatoliy Unitsky,**
engineer, inventor of "SkyWay" and "SpaceWay" systems on the base of his string technology, Chairman of the Directors' Board, General Designer of "String Technologies" corporation

*I believe that together with SkyWay technologies, the methodology of Modern TRIZ will come to Sharjah Research, Technology and Innovation Park and to universities of UAE. MTRIZ can supply new tools of engineering thinking for new generations of Master of Science, Master of Engineering, Master of Business Administration, etc.*

**HE Hussain Al Mahmoudi,**
Chief Executive Officer (CEO) of the American University of Sharjah Enterprises (AUSE) and the Sharjah Research, Technology and Innovation Park (SRTI Park), UAE

Sketch of String system at SRTI Park, Sharjah, United Arab Emirates

For around 15 years, the students of Master's programs Global Production Engineering as well as Energy Engineering, Master Business Engineering, and Information Technologies for Energy have great interest and benefit from the course of Professor Michael Orloff on MTRIZ. We are glad to collaborate with the MTRIZ Academy, complementing our Master programs in Campus El Gouna with an extremely useful skill in systematic creative design.

**Prof. Dr. Tetyana Morozyuk,**
Dean of the Energy Engineering Department, Campus El Gouna TU Berlin, Professor at the Institute for Energy Engineering, TU Berlin

We are very satisfied and thankful to Michael for dozens of his MTRIZ lectures and seminars, which have been provided over more than 10 years of cooperation, for many hundreds of our Master students, engineers, professors, etc. at EU programs and customized trainings. And we believe that we will develop new effective Master-training programs for our customers both for engineering and pedagogical specialists.

**Dr.-Eng. Arnold Sterenharz,**
Founder of ECM European Academy, Berlin, project manager of European educational and research projects TEMPUS, Erasmus, FP7, etc.

Class by Prof. Michael Orloff at the El Gouna Center of TU Berlin, Egypt

# "TRIZ"

**is an acronym**
**from Russian title for the**

*"Theory of Inventive Problem Solving":*

Теория Решения Изобретательских Задач (ТРИЗ)
– Teoriya Resheniya Izobretatel'skih Zadach (TRIZ)

## *We cannot predict the future, but we can invent it.*

**Dennis Gabor**
(1900-1979)

Inventor of holography,
Nobel Prize in Physics, 1971

(citation from the book "Inventing the Future", 1963)

## *The best way to predict the future is to invent it.*

**Alan Kay**
(b. 1940)

Pioneer of object-oriented
programming and graphical interface

(citation from meeting of PARC, 1971)

ASSERTION OF NEW PARADIGM:

# "Creative" vs "Analytic"

**Three widely shared perspectives stand in relief:**

1) **complexity is escalating,**

2) **enterprises are not equipped to cope with this complexity,**

3) **creativity is now the single most important leadership competency.**

**IBM said that creativity is needed in all aspects of leadership, including strategic thinking and planning.**

**Capitalizing on Complexity: Insights from the Global Chief Executive Officer Study, July 2010**

*This study is based on face-to-face conversations with more than **1,500** CEOs worldwide.*

**Haydn Shaughnessy,** contributor

# What Makes Samsung Such An Innovative Company?

...it was TRIZ that became the bedrock of innovation at Samsung. And it was introduced at Samsung by Russian engineers whom Samsung had hired into its Seoul Labs in the early 2000s.

TRIZ is a methodology for systematic problem solving. Typical of its origins in Russia, it asks users to seek the contradictions in current technological conditions and customer needs and to imagine an ideal state that innovation should drive towards.

At Samsung even the subsidiary CEO has to take TRIZ training.

**TRIZ is now an obligatory skill set if you want to advance within Samsung.**

**3rd of March, 2013**

**Peter Fingar,**
one of the industry's noted experts on BPM

# Innovation as a Business Process

**Clayton Christensen, Harvard Business School's professor:** ...the creator of new value, innovation isn't hit-or-miss, trial-and-error lateral thinking, but a repeatable process. *What is innovative about innovation today is the realization that it can be achieved systematically, and that the innovator is an obsessive problem solver.*

**Howard Smith, CTO of CSC's "Innovation Center of Excellence in Europe":** ...is now added something that may be a way of thinking, a set of tools, a methodology, a process, a theory or even possibly a deep science, but which may be gradually shaping up

## as 'the next big thing.' It's called TRIZ.

...the antithesis of unreliable, hit-or-miss, trial-and-error, psychological means of lateral thinking. Its scientific, repeatable, procedural and algorithmic processes surprise all who first encounter them.

**September 24, 2015**

# BPM Institute.org ™
A Peer to Peer Exchange for Business Process Management Professionals

# Table of Contents

# On the Way Towards the Master of Modern TRIZ

*Author's Preface*

## ➢ Message to Readers and Gratitude

The first three books[1] of the author at Springer Publishing House form the necessary foundation for achieving a very significant understanding and skill in inventively solving to complex "unsolvable" problems.

Such problems are encountered in design practice and in production in large numbers. Some problems are being solved, but soon others will appear, not less complex, often requiring immediate solutions.

As the many years of teaching and consulting practice have shown, these books are enough to achieve the qualification that we have identified as Master of Modern TRIZ (M-MTRIZ) in our Academy of Instrumental Modern TRIZ (AiMTRIZ).

Also, over the years, an understanding of typical mistakes in training has developed, which must be eliminated before the use of TRIZ in practice. An understanding of the most important tactical approaches to solving problems has also formed. This experience was not previously widely covered in the TRIZ literature.

Successful TRIZ models, as well as typical errors, can be studied only with a variety of practical examples. Such examples arise from educational practice, from solving real design and production problems, from consulting practice.

Achieving the level of M-MTRIZ requires constant work on the application of MTRIZ modeling. Such modeling inherently occurs when solving real problems.

And self-improvement in solving problems through self-training is also possible on the basis of such educational and research methods as Extraction and Reinventing, developed by the author more than 20 years ago.

It is possible that some readers have studied these methods in the mentioned books of the author. I am sure that readers will agree with benefits of modeling based on the methods and formats of MTRIZ. Several thousand specialists, students, undergraduates, professors, schoolteachers and even high school students have successfully mastered the foundations of MTRIZ.

---

[1] Orloff, M. *ABC-TRIZ. Introduction to Creative Design Thinking with Modern TRIZ Modeling.* – SPRINGER: Switzerland, 2016. – 534 pp. – further in the text as ***ABC-TRIZ***
Orloff, M. *Modern TRIZ. A Practical Guide with EASyTRIZ Technology.* – SPRINGER: New York, 2012. – 465 pp. – further as ***Modern TRIZ***
Orloff, M. *Inventive Thinking through TRIZ. A Practical Guide.* – SPRINGER: New York, 2-nd edition, 2006 (2003). – 368 pp. – further as ***TRIZ Thinking***

It is especially useful when learning to study examples developed by other, previously trained colleagues. These can be examples from different specialties, of different levels of complexity, made both in an educational institution, and in production or at a design institute.

Two of the most important things that are a serious advantage of MTRIZ unify all of these examples:

1) the ability to standardly structure and model the process of creating (inventing) an effective solution, and

2) the ability to present the results (description) of the invention process during and after modeling in standardized formats.

The application of standardization, especially with the introduction of such a methodological tool as the Meta-Algorithm Invention T-R-IZ (MAI T-R-I-Z), made the learning process, as well as the process of solving real problems, systematic, simple in terms of logic, effective for communication of all participants of the multidisciplinary Think Tank Team, necessary to document the process of the invention, and to include such examples in sample databases for collective access.

A lot of students and readers informed me that it was exciting and useful for them to read the first three books on the foundations of MTRIZ.

They said that the study of examples is like a detective investigation, a story with an unexpected and wonderful outcome. I also received reviews in which users of MTRIZ reported that now, wherever they are, they are everywhere looking for objects to practice MTRIZ modeling, in order to extract effective inventive models from these objects, to understand, to discover, the logic of the invention, to see the beauty of unexpected ideas.

The author hopes that this book will also open to readers no less surprising and useful aspects of inventive creativity. The author also wishes readers to get useful skills from numerous examples, accompanied by comments by the author, and sometimes by a story of some inventions creation, whether patented or not, that appeared only as a training modeling or as a solution to a real urgent problem in production.

I am extremely grateful to everyone who actively worked on the examples provided in the book, as well as on many problems that I could not include in this book for various reasons (including due to the confidentiality of inventions and innovations).

I thank my faithful friends with whom I can share our successes and difficulties, who supported us in unbearably difficult situations. We survived also because these friends were with us and are with us.

And I am infinitely grateful to my sons Alexei and Nikolai, who support me in all my aspirations, and, of course, to my wife Valentina, with whom we have been traveling together already for 25 years in MTRIZ and 50 years in married life.

## ➢ Innovention and Innoventing

In higher education, much attention is paid to the processes of creating innovation. It is enough to see the extensive programs at universities on the so-called "Innovation Management". However, in these programs, almost everywhere, under the creative process is understood and offered Brainstorming only.

This means that the process of creating ideas for innovation remains outside the tradition of a systematic engineering approach and design.

And only in TRIZ, for the first time in the millennial history of engineering, it was proposed to apply systematic inventive design in order to create effective ideas – of high quality and in the shortest possible time.

Indeed (see figure[2] below), it is the invention of new ideas that precedes the subsequent design and implementation of these ideas as innovations in practice. There is no need to prove that if there is no effective idea, then there can be no innovation.

## COMPREHENSIVE PROCESS OF PROBLEM SOLVING FOR PRACTICE

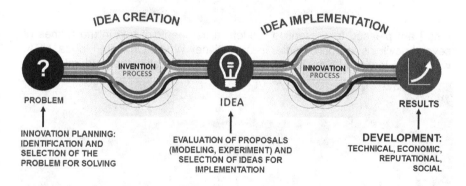

In our direction "Modern TRIZ", we call such a combined process "innovation + invention" in one word-purse **"innovention"**, or as "innovating with inventing" and shortly as **"innoventing"**. It is possible to see this word at our "coat of arms" (on the next page).

It also bears the call *"Improve yourself by transforming the mind (yours)"* in Latin *TRANSFOMAMIINI RENOVATIONE MENTIS*. This is very important message in MTRIZ mentality.

---

[2] Composition is based on https://www.ccl.org/wp-content/uploads/2017/03/Looking-Glass-Leadership-Program-Brochure-Eng.pdf

It should be noted that these combined words are not our invention, and they have been known at least since the 1940s, and were actively used, for example, by the Walt Disney Company, very successful in art-creativity.

It also means that at modern conditions of high competition and rapid technocratic development of society, engineering creativity methods are needed to be comparable or/and of equal worth to traditional engineering design in any applied areas.

Also I am pleased to present to readers a fragment[3] of text in the names of several pavilions at Epcot Center in Walt Disney World, Orlando, Florida, USA, which I like to visit with my family. This is my friendly call to the readers!

And I wish readers success in innoventing with this book. Good luck!

Michael Orloff                                               Berlin, September, 2019

---

[3]   the photos taken by author at *Epcot Center in Walt Disney World*, Orlando, Florida, USA. That is the fragment of photo at the stunning pavilion *Spaceship Earth* (and some others), constructed by *WED Enterprises* and based on the outstanding architectural invention of *Richard Buckminster Fuller*. I cannot but say that this pavilion where within fifteen minutes you travel with a "time machine" through the history of civilization and inventions. The "time flight" scenario was created by great futurist *Ray Bradbury*.

# Part I

# Key Concepts of
# MTRIZ Modeling

To effectively solve inventive problems of higher levels,
a heuristic program is needed, which allows us
to replace the search of variants
by purposeful advancement to the solution zone.

The contradiction inherent in this task
indicates the obstacle that must be overcome.

However ... the revealed contradiction
does not disappear by itself,
we have to find ways to eliminate it.

We need a rational tactics
that allows us to move step by step
towards the solution of the problem.

Such tactics are provided by the Algorithm
for Inventive Problems Solving (ARIZ). [4]

*Genrikh Altshuller*

---

[4]   Genrikh Saulovitch Altshuller (1926-1998) – founder of TRIZ, researcher, mentor, educator;
citation with abbreviation and re-arrangement according to the book Altshuller G. *Invention
Algorithm*. – Moscow, 1973 (Альтшуллер Г.С. *Алгоритм изобретения*. – М.: Московский
рабочий, 1973)

# 1  Basic Models and Methods of MTRIZ

## 1.1  Modern TRIZ

The figure (fig. 1.1), which opens this book, reflects the idea of a standard representation of the creative knowledge of the **Theory of Inventive Problem Solving (TRIZ)** in the format of a modern educational TRIZ-trend, called by the author of this book as **Modern TRIZ (MTRIZ).**

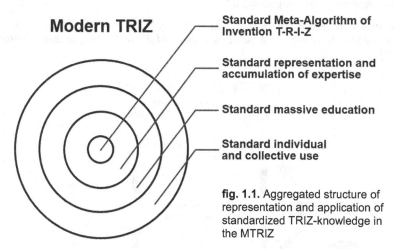

**Modern TRIZ**

Standard Meta-Algorithm of Invention T-R-I-Z

Standard representation and accumulation of expertise

Standard massive education

Standard individual and collective use

**fig. 1.1.** Aggregated structure of representation and application of standardized TRIZ-knowledge in the MTRIZ

Standardization of the representation of creative knowledge does not in the least reduce the freedom of creativity, but only enhances its effectiveness both as during the mastery of the theory and tools of creativity, and – especially! – during the creation of new ideas.

So the artist's art is only freed from routine and strengthened not only by a subtle understanding of the nature of drawing, color and composition, but also by the ability to wield masterfully the brush and paints.

At the same time the canvas, paints and brushes are prepared in advance. In moments of inspiration, there is no time to think about brushes and canvas – you have to create art.

After 55 years of my experience in TRIZ (since 1963 – I was 16 years old) I know exactly that the invention of ideas is the more free from routine, the more perfect **the mastery – knowledge and skill – in TRIZ.**

Yes, each of the methods of development of creativity is useful. However, the fact is that **TRIZ is the only methodology of purposeful, system-organized, creative design.**

© Springer Nature Switzerland AG 2020
M. A. Orloff, *Modern TRIZ Modeling in Master Programs,*
https://doi.org/10.1007/978-3-030-37417-4_1

## 1.2  Man Ingenious

**MAN INVENTED EVERYTHING BUT NATURE.**

Everything that surrounds us is created by man's thinking. At the same time, man has changed a lot in Nature, although, unfortunately, not always for the better.

The person continues to invent, and civilization continues its development. For the development of civilization, a man borrowed a lot from the Nature, and in some inventions even surpassed it

The man's invention appears (fig. 1.2) from three interrelated spheres (zones, noospheres) of thinking – the author's *model of 3Z (3 Zones, or also "three-zonal")*.

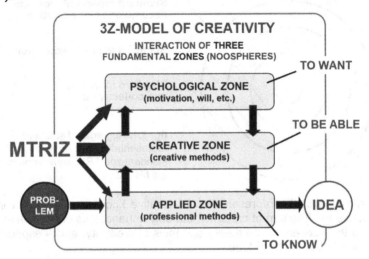

fig. 1.2. Creative knowledge and skills in the structure of the 3Z-model
– "three-zonal" representation of thinking

Next, we will keep in mind *engineering design thinking*.

However, with a completely uncomplicated interpretation, all the ideas and models presented here are completely adequate to any creative process, for example, in literature, theater, painting, and so on, even in music.

This has the deepest scientific confirmation from prominent artists, including musicians.

In the creation of an effective idea, all three "specialized" zones of thinking are always involved. The 3Z-model seems to be a metaphorical version:

**1) to know + 2) to be able + 3) to want!**

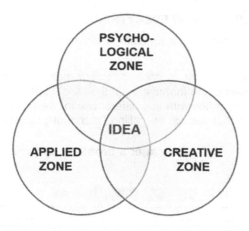

**fig. 1.3.** 3Z-model of interaction of three mental zones (noosphere) of thinking

The solution of any problem goes first in *the zone of applied knowledge* and skill, that is, in the main professional space.

If the solution is not obvious, *the creative zone* is included. In this case, the conventional brainstorming "works" on the basis of spontaneous ideas, accompanied by the question *"What if I do this so ...?"*.

If the brainstorming is unsuccessful, then the decision cannot continue in principle without activating *the psychological zone*. It is possible to reach a solution only through the presence of strong motivation, will, goal-striving, patience.

The strongest motivator is the desire to create a beautiful, effective, and, if possible, simple solution.

**ATTENTION:** we claim that at the "very-very" upper level (emotional – psychology!) the decision to recognize or reject the idea is accepted by any author, by any "judge," *according to a single "preference criterion"* –

*I like (this)!* **VS** *I don't like (this)!*

Here the symbol **VS** means **"against"** from English *versus* or Latin *adversus*.

TRIZ fundamentally changes the organization and character of creative thinking, *it provides systematic advancement to an effective solution*, although at the final stage it can also begin the presentation of the idea with the words *"What if we do this so ...?"*

But this is already a completely different process!

## 1.3  Why the Problem

**NOBODY IS ABLE TO KNOW EVERYTHING.**

Transformation of a task into a problem is forced by its increasing complexity (fig. 1.4).

Complexity increase, in turn, is defined by four main attributes. Two of those attributes are related to the formulation of the task, while the other two directly bear on its solution. In line with this differentiation, we recognize two levels of problem situations that we will be calling "manager problem" and "developer problem".

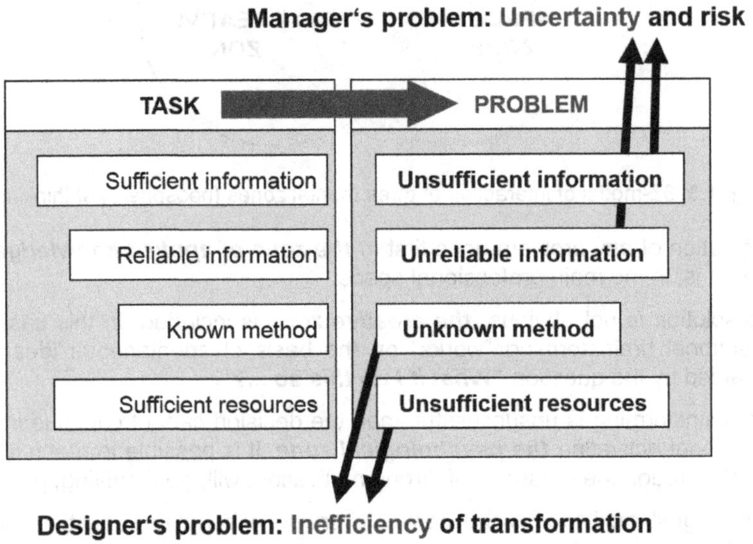

**Manager's problem: Uncertainty and risk**

| TASK | ➡ | PROBLEM |
|------|---|---------|
| Sufficient information | | Unsufficient information |
| Reliable information | | Unreliable information |
| Known method | | Unknown method |
| Sufficient resources | | Unsufficient resources |

**Designer's problem: Inefficiency of transformation**

fig. 1.4. Transition from TASK to PROBLEM = Increasing the complexity

**Definition of Manager's Problem**

| Manager's Problem | If the source data related to the task is incomplete or unreliable, the consequences may include the following: indistinct formulation of the purpose of development, imprecise determination of the causes and sources of the problem, insufficient or incorrect listing of existing and available resources that could be used to solve the problem, etc. |
|-------------------|------------------------------------------------------------------------------------------------------------|
| | These deficiencies in the formulation of the problem create, from the very beginning of the work aimed at improving the object, a risk that the desired objective will never be attained. |

This situation can be called a "Manager's Problem".

Indeed, the managers, including the chief designer, are responsible for the formulation of the problem and definition of the ultimate objective.

**Errors in this area are extremely costly.**

For example, wrong interpretation of competitor actions can directly lead to wasteful use of corporate resources to create products that will not be competitive a priori. The only solace is that eventually such products may find their niche in the market – but then again, this may never happen.

### Definition of Designer's Problem

| Designer's Problem | If developers do not know what method could be used to solve the problem – which often happens in stress situations, when the solution is required *"here-and-now"* and, naturally, no apparent resources are available – their actions may prove to be inefficient. |
| --- | --- |

This situation can be called a "Designer's Problem".

Indeed, the developers are responsible for specific modifications of the system under development.

**The cost of errors in this area falls within a broad range, but is normally lower than the cost of management errors.**

An ideal situation is one where managers are capable of providing well-substantiated and essentially correct system evolution projections, including those, which apply to competing systems, while designers are sufficiently proactive and creative to develop efficient conceptual frameworks and solutions giving the company a competitive edge in the market.

To solve the problems is a matter of high importance to forecast continuous system evolution based on TRIZ laws.

The absence of such ongoing forecasting modeling is the main cause of time shortages and stress situations experience by design teams.

It also creates conditions conducive to development of erroneous or inefficient solutions.

## 1.4  Poly-Screen Model of System Development

We represent an artifact as a continuously developing system (figure 1.5 in the center), consisting of subsystems and entering, for example, in the form of a subsystem into a system of a higher rank (super-system).

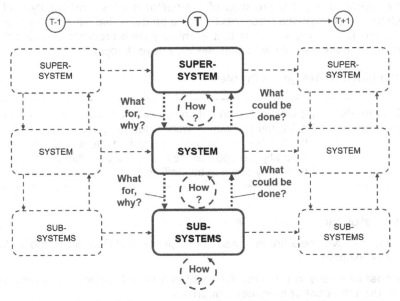

**fig. 1.5.** Poly-screen model of system development

The simultaneous evolution of all the components – *subsystems, system* and *super-system* – can be interpreted as a change in their state on all 9 "screens" when the position of the point **T – the "historical time"** – changes.

Higher-level systems set requirements (what for, why?) to lower systems, and lower-level systems meet these requirements with their realization variants (what and how?), and therefore the realization of the upper system, too.

Genrikh Altshuller, the founder of TRIZ, attached great importance to this model[5]: *"The purpose of TRIZ: relying on the study of the objective laws of the technical systems development, to give the rules for thinking organizing on a multi-screen scheme".*

Among the key **"rules of organization of thinking"** in TRIZ are:

- **implementation of the concept "Idealization" and**

- **construction and resolution of contradictions.**

---

[5]   Altshuller G. *To Find the Idea. Introduction to the Theory of Inventive Problems Solving.* – Novosibirsk: Science, 1986 (Альтшуллер Г.С. *Найти идею. Введение в теорию решения изобретательских задач.* – Новосибирск: Наука, 1986

## 1.5 System Development

The design object at the highest level of abstraction can be represented by the **"System Formula"**, which denotes fundamental aspects of the description and modeling of any complex system (fig. 1.6).

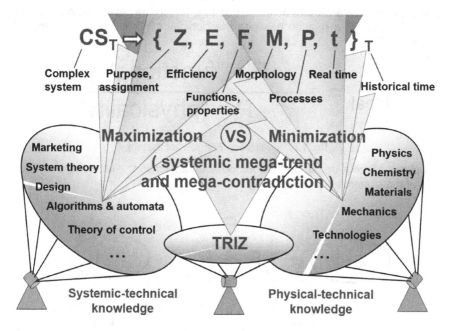

fig. 1.6. "System formula" and a complex of knowledge
providing system development

It is possible to divide the description of the system into two complexes: **the function models {Z, E, F} and the organization model {M, P}.**

Aspects **{Z, E, F}** represent properties related to the **"functionality"** of the system, for example, purpose, efficiency, functions.

Aspects **{M, P}** represent properties related to the **"physicality"** of the system, for example, dimensions, weight, energy consumption, waste, pollution.

The development of a system with the intended use **Z** is accompanied by a continuous increase in the efficiency **E**.

In turn, the evaluation of the effectiveness **E** is based on a set of quality indicators directly related to the composition of the functions **F**, the structure **M** and the functioning **P** (the change in state in real **t** and historical **T** time, that is, with processes) of the system.

## 1.6 Formulas of System "Idealization"

"Idealization" of the system means increasing "functionality" as a sum of positive properties and reducing "physicality" as a sum of all types of costs for the implementation of the system.

These two meta-trends can be combined (fig. 1.7) in the **"Idealization Formula"** in the form of a relation between the "functionality" of the system and its "physicality":

$$E_{ideal} = \frac{F = \text{"functionality"}}{P = \text{"physicality"}}$$

**fig. 1.7.** "Idealization Formula" of system

Directions of meta-trends on the methods of "idealization" – increasing efficiency – of the systems are shown in fig.1.8.

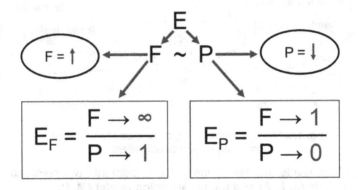

**fig. 1.8.** Directions of meta-trends in the "idealization" of the system

More constructively, the two meta-trends for increasing efficiency can be represented by the following two options, common in practice:

the increase in $E_F$ efficiency is mainly due to the growth of the functional **F** at fixed costs **P**;

increasing the efficiency of the $E_P$ mainly by reducing the cost **P** with a fixed functionality **F**.

# 1.7 TRIZ Laws of System Development

## 1.7.1. "Complete" Structure of System

Generalized and simplified principal scheme of any system is shown in fig 1.9.

We will repeatedly use the notations introduced in this scheme in the following examples.

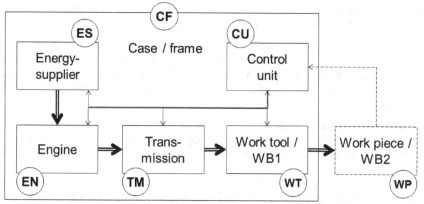

**fig. 1.9.** Generalized principal scheme of a system

This scheme often helps to clarify what changes need to be planned. This understanding helps to correctly identify the specific features of change.

In addition, the development of systems follows certain laws, which we will discuss in the next section.

Here we note one feature of this scheme, namely, the difference between two definitions of the "working body" of a system:

1) it is "work tool (**WT**)", "instrument", to process the "work piece (**WP**)", "product", belonging to another system from external environment; for example, a WT "hammer" processes a WP "nail";

2) it is one part of joint "tool", including the "work body 1 (**WB1**)", belonging to the system, and "work body 2 (**WB2**)", belonging to another system from external environment; for example, to move a car, the interaction of the WB1 "wheel" with the WB2 "road" is necessary.

Often many try to use the word "mover" or also "propeller", or even "propulsor" as a "working body"; this is possible if these words are understood to mean WB1. But it is necessary to see that the "propulsor" creates movement when there is an environment on which it acts. For example, a "propulsive screw" of a propeller-driven aircraft acts on air, a boat's "propulsive screw" acts on water, a wheel acts on a road. That is, in all these cases we are dealing with WB1 as "propulsor" and with WB2 as a component of external environment.

And only the "propulsors" on the reactive principle of motion act directly on the moving system, being specific WB1.

Here and further we will call and consider as a system any object if this does not contradict the content of the problem and if it is appropriate if we are interested in the development of this object. Because in these cases it is necessary to consider the object in the process of modernization, system changes, in the direction of the growth of "ideality".

And the last note: Consideration of the "complete" system scheme given here does not mean that we define the very concept of "system" in this way. We simply proceed from the necessary generalized idea of the structure of developing objects, such that in order to function, in any case, all the components and forces listed on the scheme are needed.

So, we use the word "system" as a systemic representation for the practical operation. Therefore, when studying an object, we can also not take into account the absence of some parts in the system scheme and still to consider this object as a system.

## 1.7.2. The TRIZ Laws of System Development and Evolution

The most important TRIZ laws of system development and evolution is shown in fig 1.10.

In the title, in the wording of the law, its content is reflected. The general principle of applying the law is to achieve the functional property that the name of the law speaks of. For example, an increase in dynamization means that the general development trend is the creation of controlled parameters, and this is possible only when the parameter has a range of changes. And so, changing a parameter value within this range of change means creating a "dynamized" parameter, which is what is needed for control.

Another example is resource coordination. Indeed, the development of the system is precisely related to how efficiently resources are used - space, time, material, energy, etc. The coordination of resources is a reflection of the law of continuous growth of "ideality." Moreover, at each stage of the development of the system, the problem of optimal coordination of resources is solved.

In practice, in any case, the sequence of application of key laws in a developed system is as follows: increased dynamization − increased coordination of resources − increased controllability − increased efficiency of the implementation of the main useful function in the direction of the growth of "ideality".

The action of some of the laws presented here we will see later in the examples of MTRIZ modeling. This will allow a better understanding of the content and application of TRIZ laws.

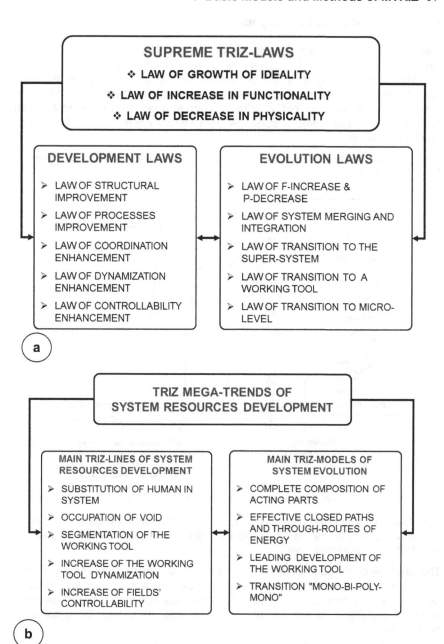

fig. 1.10. Generalized schemes of TRIZ laws

### 1.7.3. The Law of Spiral Progress

This law, which may complement the set of main TRIZ laws, corresponds to the dialectical principle of the spiral development of all systems. In accordance with this principle, progress is understood as the process of transition from one state to another (next) one, which is more perfect, the transition in time and history from the previous "old" qualitative state to a new qualitative state, from "simple" to "complex", and accordingly from lower to higher level of development and evolution.

The driving aspects and forces of such development as progress are shown in fig. 1.11 and fig.1.12.

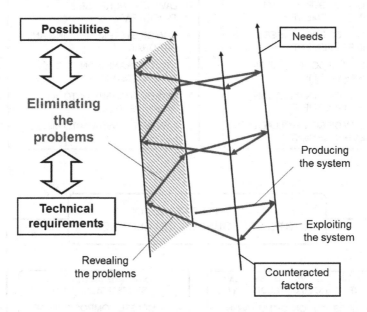

**fig. 1.11.** Generalized schemes of TRIZ laws

You can clearly see the development cycle, starting with the growth of needs and production of a system which has the main useful function that ensures the satisfaction of identified needs.

The exploiting of the system reveals the counteracting factors and disadvantages of the system. All this creates problems for system use and requires elimination of problems. As a result of the analysis of the causes of problems, technical requirements are formulated to upgrade the system or to create a new system.

Modernization design is limited by the possibilities typically associated with the limited resources available for use.

Nevertheless, a cyclical modernization is being carried out, and the progress continues until the moment when all the possibilities for development are exhausted. Usually it means a transition the system to new one with the same purpose but with a new principle of functioning. At this moment, the transition of the previous generation system to the new generation occurs, which means a step in evolutionary progress.

For the practical implementation of progress, another meta-cycle (fig. 1.12) is needed, including processes such as the evolution of the system itself, as well as the extraction of experience in designing and operating the system, and then refining of this integral experience in the form of scientific knowledge transferred to the new generation of designers in the educational process.

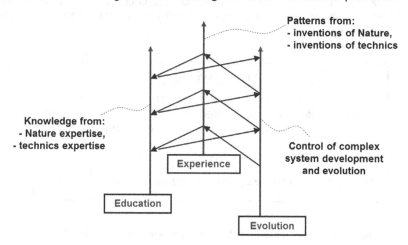

**fig. 1.12.** Spiral meta-cycles of system development and evolution

Symbolically, this meta-cycle is also reflected in one of the logos of our Modern TRIZ Academy (fig. 1.13).

**fig. 1.13.** Meta-cycle of spiral development and evolution of systems in the logo of Modern TRIZ Academy

## 1.8 General Contradiction

Let us consider very common situation when we only set a goal without knowing the way to reach this goal.

For example, we can require driving a car with speed of 250 km per hour. Of course, we should specify the additional condition for such driving. It is necessary to describe a road, whether condition, model of a car, day or night time, etc.

If some very necessary information is absent, we have typical General Contradiction because of we do not know how we could realize this requirement. Yes, it is an example only. Nevertheless, it demonstrates character of such contradictive situation when we have a goal only without understanding the method to reach this goal.

---

### GENERAL CONTRADICTION:
**THE GOAL IS INDICATED BUT
THE CIRCUMSTANCES OR-AND SOLUTION METHOD ARE
NOT SPECIFIED OR-AND NOT KNOWN**

Car▶

⊕     *speed should be 250 km per hour*   VS   **?**     ⊖

---

**fig. 1.14.** A brief practical definition of General Contradiction (GC)

**Definition of General Contradiction**

| General Contradiction | *General Contradiction* (in the classical TRIZ: *Administrative Contradiction*) is a system requirement which merely reflects the general need to attain a certain property (or state) or remove an obstacle preventing the system from operative as desired. |
|---|---|

Such a situation is typical for enterprises, when lists of problems for a solution are drawn up. In such lists, as a rule, only existing problems are indicated as targets for their elimination and nothing more.

Such General Contradictions in the transition to real design are transformed into the format of Standard Contradiction and Radical Contradiction.

## 1.9  Standard Contradiction

Let us consider two situations connected with very familiar circumstances. The first situation arises if we need to increase *the speed* of the car, but at the same time, *the safety* is reduced, since the traffic control is complicated. There is **a conflict between two different properties**, requirements, states, **for two different factors**. There is a conflict between *speed* and *safety* of traffic.

Such a conflict situation can be compactly represented in the form of a formula (fig. 1.15). This formula (the format of the representation of the initial conflict) will be called the **Standard Contradiction (SC).** A more complete formal definition is given below.

---

### STANDARD CONTRADICTION:
**TWO INCOMPATIBLE REQUIREMENTS
TO TWO DIFFERENT FACTORS**

Car▶

⊕     increase **the speed**   VS   **the safety** is reduced     ⊖

---

**fig. 1.15.** A brief practical definition of Standard Contradiction (SC)

**Definition of Standard Contradiction**

| Standard Contradiction | *Standard Contradiction* (in the classical TRIZ: *Technical Contradiction*) is a binary (two-component) model that reflects incompatible requirements for **two different functional properties (factors)** of the object. |
|---|---|

Usually, the requirement to change the first factor (positive **trend-factor**, or **plus-factor**) corresponds and promotes the **Main Positive Function[6] (MPF)** of the system, but the resulting change in the second factor (negative **problem-factor**, or **minus-factor**) does not match or counteracts the achievement of MPF. Most often, SC arises when one tries to improve one of the system factors, since another factor worsens.

**EXAMPLE OF SC.** Usually as the measurement speed increases, the measurement accuracy decreases. Here, two factors begin to clash (conflict) when the requirement for one of them changes (speed), because the requirement for the second factor – the accuracy, at least, is that the accuracy does not decrease. These requirements – to increase the speed and maintain the accuracy of the measurement – are incompatible in the new situation.

---

[6]  it is recommended to get acquainted with the MPF in more detail on the author's textbooks

## 1.10 Radical Contradiction

Analyzing the "wings" of SC for the example of movement on a car, you can formulate *requirements only for one of the factors.*

Therefore, you can consider the situation of "choice" of actions, based on the following conditions: it's necessary *to increase the speed of movement,* for example, because the driver is late for work, but the speed *cannot be increased,* as this leads to a decrease in traffic safety. Obviously, there are **opposite requirements for the same one factor (speed).**

Similarly, we can consider *the situation with the directly opposite requirements for the "security" factor:*

    **safety** *increases* with reducing the speed (thus time in a way will increase);
        **VS** but if the speed is increased, **security** *will go down.*

Both variants can be represented by the formula (fig. 1.16). A more complete formal definition is given below.

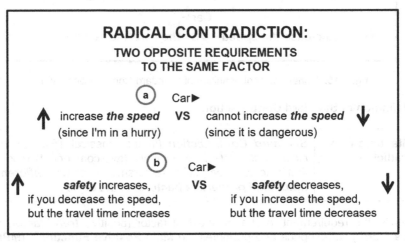

**Рис. 1.16.** A brief practical definition of Radical Contradiction (RC)

**ATTENTION.**   Only requirements to one factor (property, aspect) can be **opposite** !!!

For example, *to increase the **temperature** **VS** to reduce (not increase) the* **temperature**.

Generalized, metaphorically, we can assume that requirements in the RC **to the same factor** are of this nature:

        **a)** *increases* **VS** *decreases,*

        **b)** *to be* **VS** *not to be.*

**Definition of Radical Contradiction**

| Radical Contradiction (1) | **Radical Contradiction** (in the classical TRIZ: *Physical Contradiction*) is a set of two oppositely directed, mutually exclusive, system requirements **to the same one factor.** |
|---|---|
| Radical Contradiction (2) | **Radical Contradiction** is a binary (two-component) model that reflects a conflict situation in which, for one reason (for implementing one system function), a certain factor needs to be changed in a certain direction (conditionally: **plus-requirement**), and for another reason (to implement another system function) the same factor requires to be reversed in the opposite direction (conditionally: **minus-requirement**). |
| Addition | In a radical contradiction, the following conflict situations are often encountered:<br>a) both requirements for changing the target factor may be necessary for the implementation of the MPF;<br>b) the "minus-requirement" prevents the development of the MPF;<br>c) the "minus-requirement" is harmful, since it directly opposes the MPF, and should be excluded. |

**EXAMPLE OF RC.** On one revolution in orbit around the Earth, the satellite is one part of the time on the sunlit side, while its surface is unacceptably heated, and the other part of the time the satellite spends in the shadow of the Earth and is intensively cooled to unacceptably low temperatures.

In order for the satellite equipment to operate normally, the thermal stabilization system must satisfy the following RC:

Thermal stabilization system should ▶
***cool off*** *hardware compartments of the satellite* (on the sunny side) **VS**
***heat up*** *hardware compartments of the satellite* (on the shadow side)

Since the operation of the thermal stabilization system is associated with significant energy consumption, the problem arises to ensure an efficient operation of the system with minimal energy expenditure.

Note that in most cases this problem is solved strictly in the TRIZ style: the satellite *ITSELF* provides temperature stabilization for its internal space!

**By method 1**, the satellite rotates slowly (if possible).

***Method 2*** applies the principle of "heat pipe": the transfer of heat with a vapor of volatile liquid from the heated side to the cooled side and a return of the cooled condensed liquid, which has heated the shaded side, to the side heated by the sun.

## 1.11 What is Changed at the Invention Process

The invention according to TRIZ is the path (fig. 1.17) from the existing state of the artifact **"is"** to the future state **"should be"** or **"will be"** with the help of transformation models that play the role of navigators of thinking.

**fig. 1.17.** The process of invention as a "path" from the state "is" of the artifact-prototype to the state of "need" an artifact-aim of the same purpose

In more detail, the process of creating any invention can be presented in the form of four transitions to develop the initial artifact-prototype from the state "is" to the state "will be" (fig. 1.18):

**1) transition in time,** in which the artifact-prototype corresponding to the "is" state is changing in the target direction (in accordance with the given trend, aim) towards the state "will be"; the trend is set by growing needs and is limited by available resources, including the information resource – the amount of knowledge to create a new artifact;

**2) transition in construction,** in which the existing prototype construction in the "is" state fully defines all its properties (representation **"Construction →  Properties"** or **C-P**), and for the state "should be" or "will be" it is required to create a new, unknown construction-aim, which would meet the required properties "need" in accordance with the opposite direction of the mapping **P-C**, that is, **"Properties → Construction"**;

**3) transition in properties,** in which the contradictions always existing in the prototype state, must be resolved, and then in the "must be" or "will be" state, no one contradiction of the prototype should be;

**4) transition in idealization (the resolution of initial contradictions),** in which the contradictions inherent in the prototype are replaced first by the model and the requirements of the **Ideal Final Result**[7] **(IFR)** along with the **Functional Ideal Models**[8] **(FIM)**, and then also by the ideas of new principle of functioning of the artifact to obtain the construction with no initial contradictions.

The invention in terms of technical implementation is a transition in the construction. The invention from the point of view of system development is four transitions: in idealization, in properties, in construction and in time.

---

[7]   it is recommended to get acquainted with these aspects  in more detail on the author's textbooks
[8]   ditto

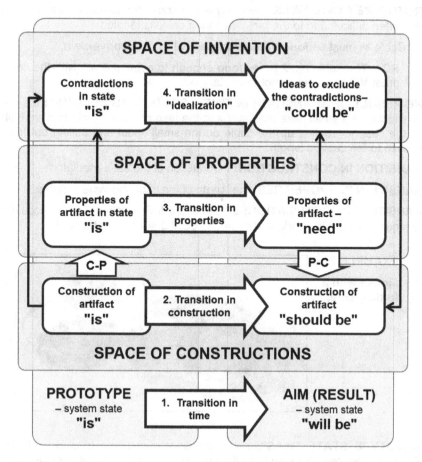

**fig. 1.18.** The invention process as a set of system transitions

**EXAMPLE 1.**

**Wired mouse.**

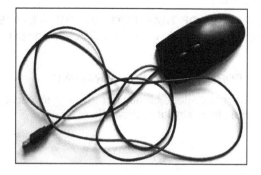

**PROTOTYPE STATE "WAS":** working with a computer mouse with a long cable is often difficult due to the fact that a long cable is tangled.

**SC:** Cable ▶ must be *long* **VS** gets tangled, which is *inconvenient*;

> **RC:**  The cable must be ▶ *long* enough to communicate with the computer **VS** *short*, so as not to get tangled.

**THE RESULT OF TRANSITION IN TIME – STATE "HAS BECOME":** the cable is hiding in a small case-mediator with the help of a spring mechanism built into the case for reeling up the cable on the small spool and pulling out from the case to the desired length.

**TRANSITION IN CONSTRUCTION:** introduction of the case-mediator.

**TRANSITION IN PROPERTIES:** the length of the cable became variable.

**TRANSITION IN IDEALIZATION:** implemented IFR – cable ITSELF acquires the required length for work, and when stored, it does not get tangled!

**EXAMPLE 2.**

**Mouse with a folding wire.**

**PROTOTYPE STATE "WAS":** it is necessary sometimes to work with the mouse at a different distance from the computer, for example, at a lecture, and it becomes inconvenient to constantly change the cable length according to the example 1.

**THE RESULT OF TRANSITION IN TIME – STATE "HAS BECOME":** a wireless mouse with radio communication with computer was invented.

*TASK FOR READER'S SELF-TRAINING:*

> Develop independently (for training) a description of each transition, similar to Example 1.

## 1.12 How to Invent by TRIZ

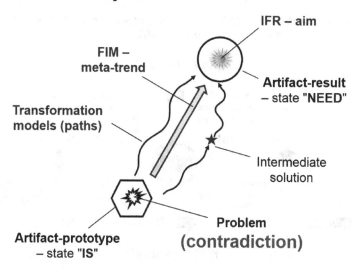

**fig. 1.19.** The process of creating an invention with TRIZ

The initial problematic situation is that it is required to make a developmental change in artifact-prototype, but a certain contradiction arises. **A contradiction is a model of the problem in the form of an extremely compressed (reduced) form.** For example, *it is necessary to build a large star and install it on a high tower, but the star has a large windage and can fall down from the tower in strong winds.*

Then it is necessary to set the aim – the **Ideal Final Result (IFR)** – and adopt a specific **Functional Ideal Model (FIM)**. FIM sets the mode of movement (meta-trend) towards the IFR. For example, IFR: *the star must be large AND reliable. Let's choose such a FIM: let the star be installed "sideways" to the wind! ITSELF!*

Then we have to choose the transformation models – "the paths to the aim". For different variants of the contradiction formulation mentioned above, TRIZ recommends such models of transformation as *dynamization, asymmetry, copying, turning harm into advantage,* and some others.

Now we have to "go through these paths" and invent specific solutions, changing the resources of the artifact-prototype towards a new artifact-result that meets the required IFR. Here come into play the imagination, the talent of associative thinking, the experience, the knowledge, the character of the person.

Now, following the models of TRIZ transformation as navigators of thinking, the appearance of *the idea of copying a weathercock* seems already quite logical and predictable!

## 1.13 Meta-Algorithm of Invention T-R-I-Z

The process of ideas generation for solving any problem can be represented by a constructive four-stage scheme (fig. 1.20).

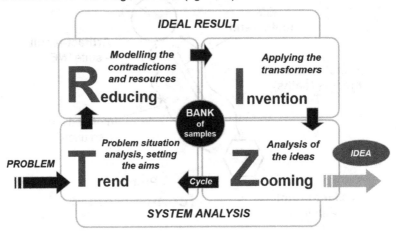

**fig. 1.20.** Meta-Algorithm of Invention T-R-I-Z

**The first letters of the stages name form the abbreviation "TRIZ":**

**TREND** – analysis of the problem situation, setting the goal and trend (strategic direction) of development;

**REDUCING** – construction of a problem model in form of contradiction, definition of the **Operational Zones[9] (OZ)** and **resource analysis**, formulation of IFR and FIM.

**INVENTING** – creation of ideas using transformation models; at this stage the TRIZ technique "meets" with the talent of problem solver – *TRIZ does not know your problem, only you know it, but TRIZ knows how to solve similar problems and gives you recommendations!*

**ZOOMING** – consideration of ideas on a different scale from different levels and with different focuses (from many perspectives and on different aspects) for evaluating effectiveness.

A cyclic transition (pointer *"Cycle"*) to the repeated passage of the algorithm is possible if a satisfactory solution has not been received on this completed cycle.

**Meta-Algorithm of Invention T-R-I-Z**, proposed by the author in 1995, is structurally and conceptually closest to the first **Algorithm of Inventive Problems Solving ARIZ-1956**, proposed by G.S. Altshuller and R.B. Shapiro.

---

9    see the author's textbooks

## 1.14 Algorithm "START"

In more detail, the MAI T-R-I-Z is presented in the MTRIZ toolkit in version **START: S**implest **TR**IZ-**A**lgorithm of **R**esourceful **T**hinking.

This scheme (fig. 1.21) shows *the minimum necessary aspects for understanding and practical work in the MAI T-R-I-Z*

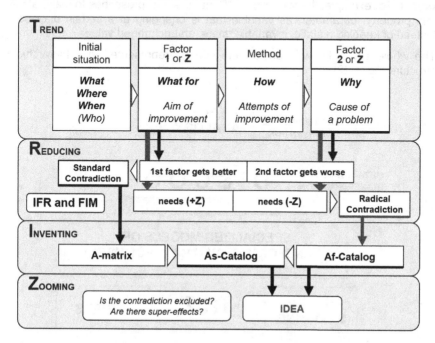

**fig. 1.21.** Algorithm of Invention START

**NARROW** arrows show the route of movement through the **Reducing** and **Inventing** stages to solve problems on the basis of **Standard Contradiction (SC)**. Solving the SC is based on the use of transformation models collected in the **As-Catalog**. The SCs with frequent use are represented in the A-matrix which gives statistically effective recommendations for resolving the SCs. These TRIZ tools are discussed in the following sections of the book.

**WIDE** arrows show the route through the **Reducing** and **Inventing** stages to solve problems on the basis of **Radical Contradiction (RC)**. The fundamental models for resolving the RCs are presented in the **Af-Catalog**, which is given below.

Both routes start at the **Trend** stage and end (the transition is indicated by arrows) in the **Zooming** stage.

## 1.15 Transformation Models

At the Inventing stage of the MAI T-R-I-Z, the instrumental transformation models (*transformers*, as well as *navigators*) are used as "thinking navigators".

Transformers are *metaphors of action, describing the way of changing the object.* For example, the transformer "Dynamization" prescribes to make a certain property, parameter, as variable that is changing at a certain diapason, instead of keeping a static, invariable, state, an unchanged value.

The whole set of known transformers can be represented in a hierarchical structure (fig. 1.22).

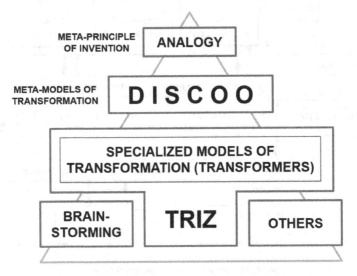

**fig. 1.22.** Hierarchical structure of constructive creative transformers

At the top level, there is not even a transformer in the sense of instrumentality of applying, but a principle that invariably and constructively enters into all creative transformations is an implicit foundation of creativity. This is a **meta-principle** called **ANALOGY**.

The principle of ANALOGY is uncovered through the invariant **meta-models DISCOO**, reflecting the nature of any instrumental transformation.

Instrumental executable models form a large list of **specialized transformers**. In the whole set of specialized transformers, it is possible to single out the models used most often in brainstorming, as well as in TRIZ. In practice, a wide variety of other non-systematized and unclassified models are used.

## 1.16 Meta-Principle of Invention

THE "ANALOGY" IS THE FUNDAMENTAL NATURAL META-PRINCIPLE AND META-METHOD OF HUMAN THINKING.

EVERYTHING WHAT IS INVENTED BY THE MAN, COULD HAVE BE INVENTED AND IS INVENTED ONLY ON THE BASIS OF THE ANALOGY, ONLY ON THE BASIS OF KNOWLEDGE ALREADY EXISTING, AND NO OTHERWISE.

But in every invention, necessarily, development of the principle, development of an invariant idea of artifact – constant, unchanged – plus change and substitution; formation of new functions, constructions, forms and properties that emanate only from existing knowledge; objective laws as natural science, and beauty as art at the same time; logic and fantasy, recombination and transformation together with invariable, invariant, are present inseparably.

All, without exception, elements, properties and attributes of the new have their predecessors, their prototypes and analogues.

In order to change and transform, one must have knowledge.

To know the profession, know the models of transformation, know the ways of psychological management of the process of creating effective ideas.

All these are analogies!

The question: actually, where are the creativity, inspiration, talent, intuition and even instinct?

The answer is in combination, connection and integration of art of fantasy with logic of changes, and also in individual preferences and feeling of beauty.

## 1.17  Meta-Models of Transformation

**THE ALL POSSIBLE MODELS OF TRANSFORMATION (TRANSFOR-MERS) ARE THE INSTRUMENTAL EXTENSION AND REALIZATION OF THE PRINCIPLE OF ANALOGY**

▶   *The following universal meta-models* are the nearest natural extension:

**DIFFERENTIATION** – change of properties,

**INTEGRATION** – fusion of properties, replacement with new common property,

**SEPARATION** – separation of properties,

**COMPOSITION** – combination, uniting the properties.

The totality of these transformers can be designated by one word **DISC**.

▶   One can add the following two models to the number of universal meta-models:

**ORIGINATION** which is the event of emergence of new idea, new solution, new system.

**OBLITERATION** which is the event of completion of the life cycle of object (system), discontinuation of production, termination of operation, disposal.

Adding these two models to the set DISC gives a new value of **DISCOO**, or, with combining the last two letters in one, **DISCO**.

▶   Knowing and understanding of the nature of the models DISCOO help determine **the strategy and tactics of changing the object** in the direction of increasing its effectiveness, "ideality".

In more detail these models are considered in the author's books in the sections devoted to **the management of system development.**

Numerous *specialized model-transformers* are instrumental extension and implementation of one of the universal meta-models.

## 1.18 Specialized Transformers

Specialized transformation models were accumulated in TRIZ in the so-called Altshuller's catalogs (As-Catalog, Part IV. Reference Materials) for solving to SC.

Specialized transformers were obtained by extracting from real patents, articles and other descriptions of the transformation of specific objects from the state of artifact-prototype to the state of artifact-result.

A metaphorical description of the action prescribed by the transformer is obtained by *generalizing the transformations on the basis of the principle of similarity of changes,* despite the difference in the objects themselves by purpose and design.

The "classical" As-Catalog includes 40 transformers (Table 1).

**Table 1. List of specialized transformers of As-Catalog**

| | |
|---|---|
| 01. Change in the aggregate state of an object | 20. Universality |
| 02. Preliminary action | 21. Transform damage into use |
| 03. Segmentation | 22. Spherical-shape |
| 04. Replacement of mechanical matter | 23. Use of inert media |
| 05. Separation | 24. Asymmetry |
| 06. Use of mechanical oscillations | 25. Use of flexible covers and thin films |
| 07. Dynamization | 26. Phase transitions |
| 08. Periodic action | 27. Use of thermal expansion |
| 09. Change in color | 28. Previously installed cushions |
| 10. Copying | 29. Self-servicing |
| 11. Inverse action | 30. Use of strong oxidants |
| 12. Local property | 31. Use of porous materials |
| 13. Inexpensive short-life object as a replacement for expensive long-life one | 32. Counter-weight |
| | 33. Quick jump |
| 14. Use of pneumatic or hydraulic constructions | 34. Matryoshka (nested doll) |
| | 35. Unite |
| 15. Discard and renewal of parts | 36. Feedback |
| 16. Partial or excess effect | 37. Equipotentiality |
| 17. Use of composite materials | 38. Homogeneity |
| 18. Mediator | 39. Preliminary counter-action |
| 19. Transition into another dimension | 40. Uninterrupted useful function |

## 1.19 Method of Extracting

The method of **"EXTRACTING"** is a fundamental method of training in MTRIZ. Each trainee will enjoy by this method, throughout his creative activity.

| Definition of method "Extracting-1" | ***Extracting-1** is retrieval of transformation models from any artifacts and from any information sources describing innovative ideas and objects.* |
|---|---|

Extracting is intended primarily for accelerated mastering of primary transformation models. The creative content of the method is that the method is oriented to the **discovery of the unusual and the "invisible" in the habitual and ordinary**. Learning and self-learning with the help of extracting is playful.

Briefly, the essence of the method is shown in the following examples.

**EXAMPLE 1. A mouse with a compact cable.** In the inventive solution, a spring mechanism is used by means of which the USB-cable is reeled up onto the spool 1. The spool together with the mechanism is mounted in the small case 2 (fig. 1.23, a).

During storage, the cable is completely reeled in into the case with the spool (fig. 1.23, b), and during operation the cable is pulled out to the desired length (fig. 1.23, c).

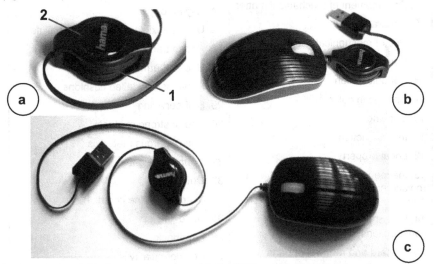

fig. 1.23. Computer mouse with the USB-cable of variable length

The prototype is a mouse with a long cable of unchanged size. The main technical difference between the new solution and the prototype is the use of a spool to reel up the cable.

It is possible **to single out (extract)** two dominant (main, most important) creative transformation models from the "classical" As-Catalog which are objectively present in the artifact-result:

**07. Dynamisation**: construction makes possible to change the length of the cable with a spool having a spring mechanism;

**34. Matryoshka**: a spool with a drive rotary mechanism is enclosed in a compact case.

These two models give *an idea of the creative way of inventing a technical solution*: indeed, with the help of Dynamization and Matryoshka, it is possible to eliminate the SC and RC of the initial prototype!

**EXAMPLE 2. Wireless mouse.** And yet the mouse with the reeling in cable continues to cause inconvenience when working with a long cable. There was a loss of time and a decrease in productivity to change the length of the cable!

A wireless mouse was proposed (fig. 1.24), in which the information connection between the mouse and the computer is carried out via radio communication!

**fig. 1.24.** Wireless mouse

Mouse 1 has a built-in receiver-transmitter, as well as a second receiver-transmitter 2 is hidden in its case, built into the construction of the USB-connector. For operation, the second receiver-transmitter 2 is taken out from the mouse case and inserted into the USB-slot in the computer.

Here it is possible to extract two dominant transformers:

**03. Segmentation**: detachment of the mouse and components of the communication system;

**04. Replacement of mechanical matter:** substitution of mechanical wire communication by radio communication.

A useful auxiliary model is the **34. Matryoshka**: before the application, the second communication device is hidden for storage in the mouse.

**Zooming:**

**1)** these models allow to completely eliminate SC and RC of wire prototypes!

**2)** there was a new drawback: the need to provide autonomous power to the mouse, which is now carried out with the help of relatively small batteries, and this leads to the fact that sometimes it is necessary to recharge the batteries at the most inopportune moment!

## 1.20 Solving to Standard Contradiction

To solve the SC in accordance with the START algorithm, one can use the A-matrix (Part IV. Reference Materials).

Usually SC is formulated in an arbitrary style using production vocabulary, although it has a "standard" structure:

### Object of transformation ► Plus-factor VS Minus-factor.

In order to use the A-matrix, it is necessary to select the formal factors (Table 2) from this A-matrix, which are as close as possible to the corresponding "informal" factors from the initial representation of the SC.

### Table 2. List of the formal factors of the A-matrix

| | |
|---|---|
| 01. Productivity | 21. Shape |
| 02. Universality, adaptability | 22. Speed |
| 03. Level of automation | 23. Functional time of the moveable object |
| 04. Reliability | |
| 05. Precision of manufacture | 24. Functional time of the fixed object |
| 06. Precision of measurement | 25. Loss of time |
| 07. Complexity of construction | 26. Quantity of material |
| 08. Complexity of inspection and measurement | 27. Loss of material |
| | 28. Strength |
| 09. Ease of manufacture | 29. Stabile structure of the object |
| 10. Ease of use | 30. Force |
| 11. Ease of repair | 31. Tension, pressure |
| 12. Loss of information | 32. Weight of the moveable object |
| 13. External damaging factors | 33. Weight of the fixed object |
| 14. Internal damaging factors | 34. Temperature |
| 15. Length of the moveable object | 35. Brightness of the lighting |
| 16. Length of the fixed object | 36. Power |
| 17. Surface of the moveable object | 37. Energy use of the moveable object |
| 18. Surface of the fixed object | 38. Energy use of the fixed object |
| 19. Volume of the moveable object | 39. Loss of energy |
| 20. Volume of the fixed object | |

The plus-factor corresponds to the row of the A-matrix, and the minus-factor corresponds to the column.

After that, you can select from the A-matrix a cell at the intersection of the row with the number of plus-factor and the column with the minus-factor number. In this cell you can see the numbers of transformers from the As-Catalog, such that were effective in thousands of problems by solving to SCs with such plus-factor and minus-factor.

Having the transformer numbers from the selected cell of the A-matrix, you need to carefully study the description of these transformers in the As-Catalog and try to interpret the recommendations of these transformers applicable to the problem being solved.

**EXAMPLE 1.** The informal SC for a mouse with a long cable (Example 1 in section 1.11) was formulated as follows:

The cable ▶ must be *long* **VS** *gets tangled*, which is *inconvenient*.

For the first (positive) factor, one can select a formal plus-factor from the A-matrix with the number 15. Length of the moveable object.

For the second (negative) factor, you can choose a formal minus-factor 10. Ease of use.

Then, from the cell of the A-matrix at the intersection of the row with the number 15 and the column with the number 10, we choose the following numbers of the transformers: 01, 07, 14 and 24.

In the As-Catalog we will see the following navigators:

01. Change in the aggregate state of an object;
07. Dynamization;
14. Use of pneumatic or hydraulic constructions;
24. Asymmetry.

We already know from the example of extracting that at least one of these transformers, namely, the navigator 07. Dynamization, was de facto applied in the invention of a mouse with a spool to reel in the cable.

This means that in this case, if we were to solve the initial problem, then for a given choice of formal factors for the initial SC, we could well have invented the idea of a reeling in cable in accordance with the transformer 07. Dynamization.

Indeed, the recommendations of the navigator 07. Dynamization are as follows:

a) the characteristics of an object or an environment are changed to optimize every work procedure;
b) disassemble an object into parts that are moveable among each other;
c) make an object moveable that is otherwise fixed.

With a certain skill in interpretation and application, a careful study of each of these three points and all of them together could well lead precisely to such an invention – a spool with a spring mechanism.

We will consider other examples in sections 1.23 and 1.24.

## 1.21 Resources of Operative Zone

Any decisions are made as a transformation of the resources of system. And more precisely: **for the solution it is necessary and sufficient to change the resources of the "Operative Zone"** that is the area in the system where the source of contradictions exists.

| Definition of "Operative Zone" | **Operative Zone (OZ)** is the set of components of the system and, sometimes, the system environment, directly related to the contradiction. |
|---|---|
| OZ is represented by some functional-structural model, reflecting the composition, communication and interaction of elements in a certain modeling system (language). | |

In addition, we must remember that usually a conflict occurs during a certain time, called the **Operative Time (OT).**

However, **the conditions for the emergence of a conflict had been "prepared" before the OT, and the consequences appeared after the OT.**

Consequently, the choice of a "place" for dealing with contradictions in the time may be related **with:**

> – **the prevention of conflict,** and therefore with the elimination of the conditions for its occurrence on the interval **"before OT";**
> – **the transformation of OZ resources at OT;**
> – **the elimination of the consequences of the conflict "after OT".**

**EXAMPLE.** The conflict with the cable of a wired mouse was that a long cable gets tangled. The place – OZ – of the conflict is the cable itself over its entire length. In fact, in order to resolve the conflict, in one case it was necessary to introduce a special element-mediator (coil with spring mechanism) into the structure of OZ.

In the second case, a completely radical solution was proposed: replacement of the mechanical matter, introduction of radio communication. The solution is achieved by changing the structural and energy resources of the OZ.

Pay attention to the system transition in the second solution. With the new transformation, *the ideal result* was realized: *the source of the conflict – the cable – disappeared,* and with it the original OZ also disappeared! That is, the idealization of the system for communication function has reached its limit: *there is NO cable, but a connection EXISTS!*

The formal definition of OZ gives only a minimal understanding of this notion, since considerable practical experience is required to master this concept. OZ resources are presented in Table 3.

**Table 3. Resources of system and OZ**

| SYSTEMIC-TECHNICAL RESOURCES | | | |
|---|---|---|---|
| **SYSTEMIC** | **INFORMATIONAL** | **FUNCTIONAL** | **STRUCTURAL** |
| general system properties | transmission of information-bearing messages | creation of functions | the composition of the object |
| Purpose of the system, its efficiency, productivity, reliability, safety, survivability, durability, etc. | Data integrity, accuracy, validity, interference immunity, methods and efficiency of measurement, management, encoding, etc. | Main useful function compliant with the purpose of the system, auxiliary functions, negative functions, description of the operating principle (functional model) | List of components and intercomponent relations, types of structures (linear, branching, parallel, closed, etc.) |
| PHYSICAL-TECHNICAL RESOURCES | | | |
| **SPATIAL** | **TEMPORAL** | **MATERIAL** | **ENERGETICAL** |
| geometric properties | temporal evaluations | properties of materials | properties of energy and its manifestations |
| Shape of the object, dimensions of the object – length, width, height, diameter, etc., shape features – presence of cavities, projections, etc. | Frequency of events, duration of time intervals, duration of time lags/leads. **Operating Time (OT):** interval of existence of problem situation | Chemical composition, physical properties, special engineering properties | Types of applied and measured energy, including mechanical, gravitational, thermal, electromagnetic and other forces; methods of energy utilization, etc. |

| **Properties of the resource** | **Value:** free → inexpensive → expensive |
|---|---|
| | **Quality:** harmful → neutral → useful |
| | **Quantity:** unlimited → sufficient → insufficient |
| | **Readiness for application:** ready → change in progress → creation in progress |

When solving new problem, selection of resources is carried out from the bottom part of the table **"Properties of the resource"** with decreasing resource preferences from the left to the right.

## 1.22 Solving to Radical Contradiction

To solve the RC, one should know that a radical conflict requires radical transformations that ensure **the separation of conflicting properties over a "critical" resource** (see also at Part IV. Reference Materials).

The most common solution is realized with transforming the following resources (fig. 1.25).

**fig. 1.25.** Key methods of resolving the RC

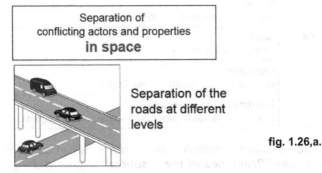

Separation of
conflicting actors and properties
**in space**

Separation of the
roads at different
levels

**fig. 1.26,a.**

Separation of
conflicting actors and properties
## in space

Separation of the
roads at different
levels

fig. 1.26,b.

Separation of
conflicting actors and properties
## in structure

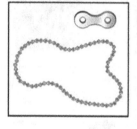

The chain link is
rigid, and the
whole chain is
flexible to circle
the sprocket
weels

fig. 1.26,c.

Separation of
conflicting actors and properties
## in material / energy

**PROTOTYPE:** sunglasses
with swivelling glasses

**INVENTION:** chromatic
glasses themselves change
its transparency depending
on the light brightness

fig. 1.26,d.

**fig. 1.26.** Classical fundamental models (simple A$_F$-Catalog)

## 1.23  Method of Standard Extracting-1 and Extracting-2

Initial training, as well as subsequent sustained self-training, is carried out using procedures for extracting transformers (Extracting-1) and contradictions (Extracting-2) from any artifacts of interest to the learner, for the researcher or for the designer.

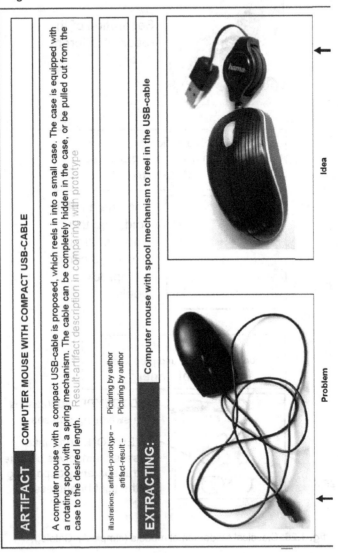

**fig. 1.27 - beg.** Example of extracting for a mouse with spooling USB-cable

The results of the extracting are recorded in a standard form, which allows you to store these results in a single format in a bank of effective samples, as well as exchange experience in training and design.

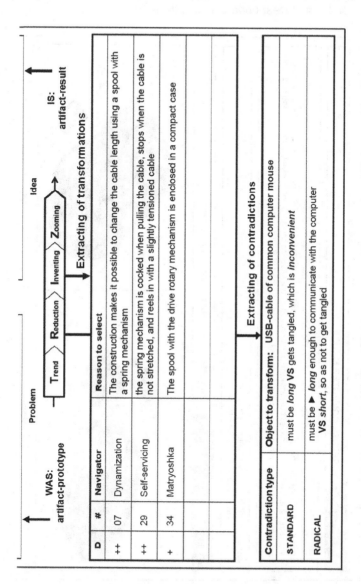

**fig. 1.27 - end.** Example of extracting for a mouse with spooling USB-cable

Results of extracting are used for reinventing that is modeling, reproducing the whole process of creating an invention.

When extracting, it should be remembered that correct modeling is possible only on condition of clear understanding that:

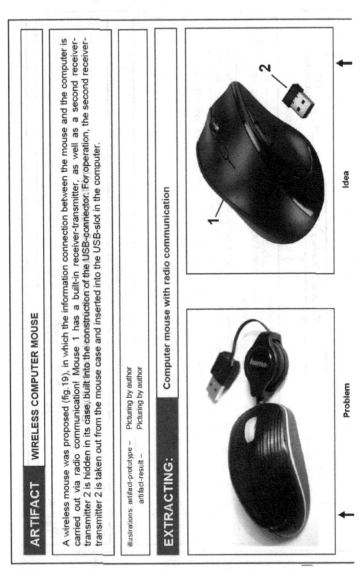

**ARTIFACT**   **WIRELESS COMPUTER MOUSE**

A wireless mouse was proposed (fig.19), in which the information connection between the mouse and the computer is carried out via radio communication! Mouse 1 has a built-in receiver-transmitter, as well as a second receiver-transmitter 2 is hidden in its case, built into the construction of the USB-connector. For operation, the second receiver-transmitter 2 is taken out from the mouse case and inserted into the USB-slot in the computer.

illustrations: artifact-prototype –   Picturing by author
             artifact-result –      Picturing by author

**EXTRACTING:**   **Computer mouse with radio communication**

Problem                                    Idea

**fig. 1.28 - beg.** Example of extracting for wireless mouse

a) the contradictions are extracted from the artifact-prototype,

b) transformation models (transformers) are extracted from the artifact-result!

c) the object of transformation can be only a part (subsystem) of the artifact-prototype.

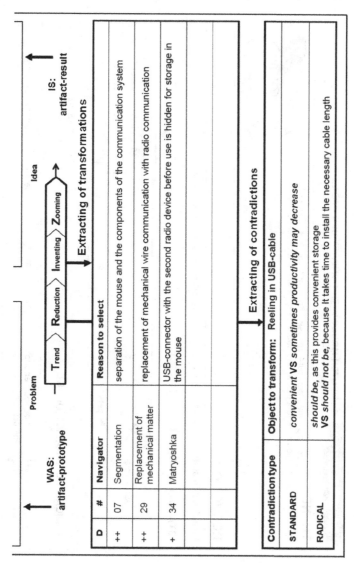

**fig. 1.28 - end.** Example of extracting for wireless mouse

## 1.24 Method of Standard Reinventing

Reinventing is the main method of training and self-training of each specialist to ensure continuous growth of mastery.

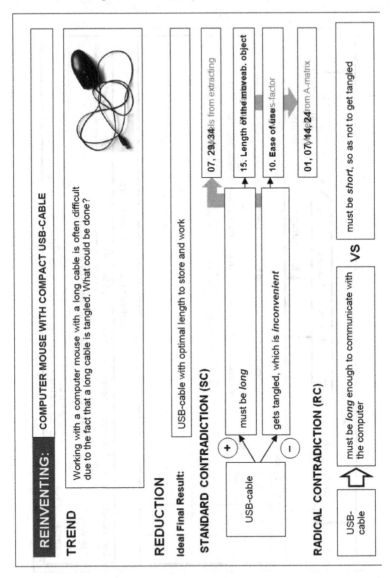

**fig. 1.29 - beg.** Example of reinventing for a mouse with spooling USB-cable

| Definition of method **"Reinventing"** | ***Reinventing*** *is modeling (reconstruction, reproduction, renewal) of the invention process as if invention was done with the help of TRIZ.* |
|---|---|

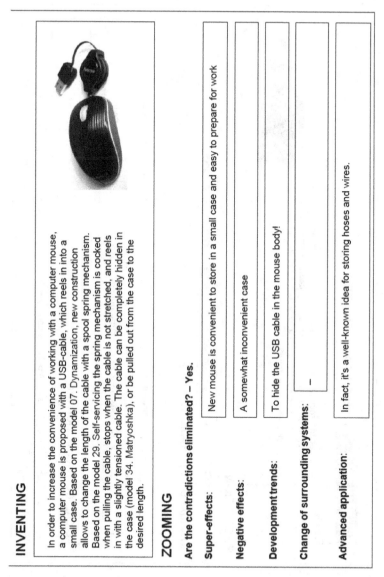

## INVENTING

In order to increase the convenience of working with a computer mouse, a computer mouse is proposed with a USB-cable, which reels in into a small case. Based on the model 07. Dynamization, new construction allows to change the length of the cable with a spool spring mechanism. Based on the model 29. Self-servicing the spring mechanism is cocked when pulling the cable, stops when the cable is not stretched, and reels in with a slightly tensioned cable. The cable can be completely hidden in the case (model 34. Matryoshka), or be pulled out from the case to the desired length.

## ZOOMING

**Are the contradictions eliminated? – Yes.**

**Super-effects:** New mouse is convenient to store in a small case and easy to prepare for work

**Negative effects:** A somewhat inconvenient case

**Development trends:** To hide the USB cable in the mouse body!

**Change of surrounding systems:** I

**Advanced application:** In fact, it's a well-known idea for storing hoses and wires.

**fig. 1.29 - end.** Example of reinventing for a mouse with spooling USB-cable

If the transformers chosen from the A-matrix do not coincide with those obtained during extracting, *preference is given to the extracting models provided that the plus- and minus-factors are adequately selected in A-matrix for the formal representation of the initial SC.*

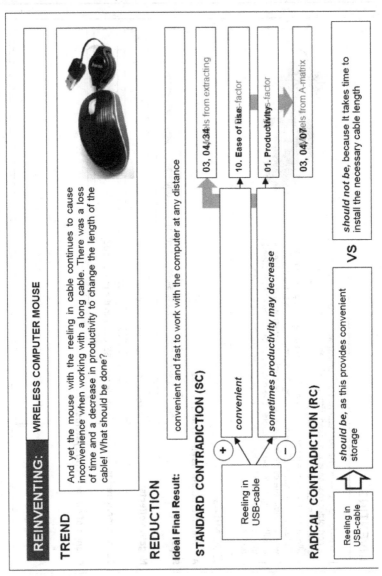

**fig. 1.30 - beg.** Example of reinventing for a wireless mouse

Continual practice in extracting and reinventing is the irreplaceable school for improving individual creative skills.

## INVENTING

In order to increase the productivity of working with a computer mouse 1, a wireless computer mouse with radio communication is invented.

The key transformer is model 04. Replacement of the mechanical matter: substitution of mechanical wire communication by radio. To implement the radio communication, the new mouse includes a detachable (model 03. Segmentation) USB-connector 2 equipped with a receiver-transmitter.

The mouse is equipped with a built-in receiver-transmitter (model 34. Matryoshka), and the USB-connector for storage is hidden in the mouse case (re-application of the model 34. Matryoshka).

## ZOOMING

**Are the contradictions eliminated? – Yes.**

**Super-effects:** This mouse is very convenient for the lecturer

**Negative effects:** Sometimes the USB-connector is lost, the need to change the batteries in the mouse

**Development trends:** USB-connector to be integrated into the computer! Do not change the batteries!

**Change of surrounding systems:** I

**Advanced application:** to apply this approach for remote object management systems

**fig. 1.30 - end.** Example of reinventing for a wireless mouse

# 2 Information Comprehension, Structuring and Formatting at MTRIZ Modeling

## 2.1. Crosshairs of Contradictions

To explain the key ideas of MTRIZ modeling the contradictions, let us to use a very understandable example with brewing tea.

A glass of tea is shown in fig. 2.1. To make tea, you need to pour the tea leaves with hot boiled water, after which the tea should be infused. You cannot take a hot glass with your fingers as shown in fig. 2.1. A glass will burn your fingers.

**fig. 2.1.** You cannot hold a glass of very hot water by your fingers when brewing tea.

In this situation, such conflicting relations are obvious:

> *Making tea* ▶ *needs boiling water in glass*
> *VS glass burns the fingers if try to take it*

This is a Standard Contradiction because there are **two different factors** in conflict:

---
### TWO DIFFERENT FACTORS:

***temperature*** (of water and glass surface) **VS *safety*** of fingers

---

You can also try to consider the cause of the conflict. Then we will see the conflict more deeply:

> *The glass* ▶ *should be **hot** (to make tea, boiled water was poured into a glass)*
> *VS should be **not hot** (to take the glass with your fingers)*

This is a Radical Contradiction because there are **two opposite requirements to the same factor** in conflict:

---
### TWO OPPOSITE REQUIREMENTS:

***to be hot*** VS ***to be not hot***, or also, ***to be hot*** VS ***not to be hot***

---

There are two opposite requirements to the same one factor – ***temperature*** (of glass surface in a point where the finger should touch the glass).

Next, let us examine the relationships between both types of contradictions in this conflict.

© Springer Nature Switzerland AG 2020
M. A. Orloff, *Modern TRIZ Modeling in Master Programs*,
https://doi.org/10.1007/978-3-030-37417-4_2

These relationships can be clear represented with a diagram named *"Cross-hairs of contradictions"* as shown in fig. 2.2.

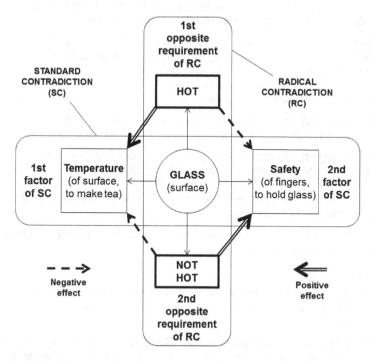

**fig. 2.2.** Crosshairs of contradictions

Such comprehension of information-functional relationship between Standard and Radical Contradictions gives the most exact and complete picture about the conflicting actors and features.

Some details: "hot" means "high temperature", "not hot" means "low temperature".

It also means, obviously, that in this conflicting situation, we need *at the same time (!)* to take a glass by our fingers **AND** to keep boiling water in a glass for making tea.

Conjunction "AND" here means that we cannot divide the processes "making tea" and "taking glass" in time.

**And this is precisely what becomes a conflict**, generated by the "interests" of different "systems" – glass (with hot water) and fingers, because glass "wants" to make tea and/but the fingers "want" to take a glass.

## 2.2 Modeling (Resolving) the Crosshairs of Contradictions

Sometimes the students are confused at the practical training without clear understanding what they should do to resolve the contradictions.

Often the reason for this lies in the fact that students do not understand *what and when* it is necessary (possible) to change in the objects during modeling and reinventing.

> There are two main training situations:
>
> **1) "here-and-now" training** with real objects in real time – it is necessary to propose effective solutions for existing objects;
>
> **2) "design-for-future" training** with any object to propose effective solutions for future modernization of the objects.

So, to continue the example with glass to make tea, we can arrange a training situation in the first or second style.

For the "here-and-now" situation, we may demand that you take a glass with your fingers and carry it to another table.

In this situation, you can:

- use the specialized TRIZ-model *18 Mediator* or/and "Separation of conflicting actors and properties *in structure*" and wrap the glass with a handkerchief or a sheet of paper;

- use fundamental transformation model "Separation of conflicting actors and properties *in space*" and try to hold the glass along the upper edge, or also support along the lower edge (fig. 2.3,a);

- use fundamental transformation model "Separation of conflicting actors and properties *in space*" and "Separation of conflicting actors and properties *in structure*" to support the glass with a saucer or a book (why not?! – fig. 2.3,b), or to carry it on a chair or even together with a table.

Did you imagine all these "inventions" already?

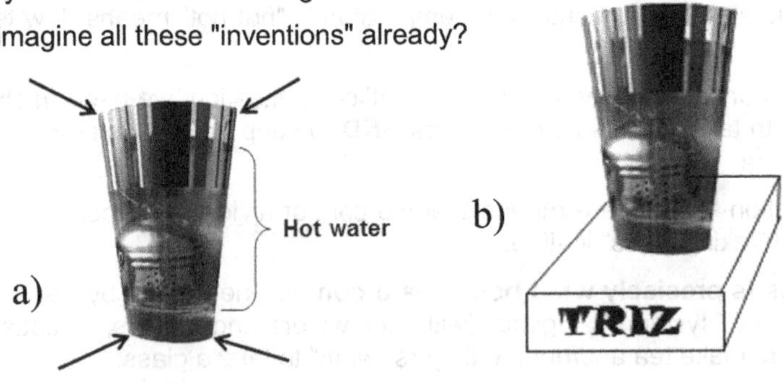

fig. 2.3. "Inventions" of methods to transport a glass with a hot water

In second situation "design-for-future" you can reinvent and/or model solutions as following:

- to add handle to a glass (dominating transformation in structure or as a model 05 Separation and 18 Mediator);

- to make the glass with more thick walls (dominating transformation in space);

- to make the body of wood (dominating transformation in material or as a model 01 Change in aggregate state);

- to make the body with double wall (dominating transformation in space and structure or as a model 10 Copying);

- etc.

And what is very important and interesting also that is to apply the approaches for resolving Standard and Radical Contradictions separately.

For the very beginners it is traditional to apply firstly resolving the Standard Contradiction and only after that resolving the Radical Contradiction.

---

What is useful to accent during modeling:

1) after solving the Standard Contradiction to explain *what resources are dominating in the solution*;

2) after solving the Radical Contradiction to explain *what specialized models could replace the fundamental transformations.*

---

Attention should also be paid to the following aspect of modeling: *the choice of the dominant factor in contradictions.* This factor exactly corresponds to the *Main Positive Function (MPF)* of the object (system). For example, in the example with a glass, MPF obviously consists in brewing tea that means using very hot water in a glass. And only after that we can think about safety for fingers during carrying the glass.

Together with this we can see that all our efforts were aimed to exclude negative factors, undesirable effects, namely, to get safety for the fingers.

---

This is very important moment in modeling (resolving) the contradictions – often to use changing the formulation of contradiction from form of opposition to the form *of non-contradictory co-operation, "win-win" coordination.* It means transition from opposition "**VS** (versus)" to coordinating conjunction "**AND**".

---

For example, we can require such non-contradictory coordination for the sample with a car speed/safety in the sections 1.8 and 1.9.

## 2.3  Spaces of MAI T-R-I-Z

> When solving problems, you must strictly remember
> and understand, to which "model spaces" belongs
> information at each stage of MAI T-R-I-Z.

Stages **Trend** and **Zooming** (fig. 2.4) belong primarily to the **"Space of Reality"**. Here, at the **Trend** stage, the real system is being exploited, the faults are identified, development tasks are set and the direction of modernization is selected, and the **Zooming** stage includes, in particular, mathematical modeling of functioning, as well as testing of prototypes.

The stages **Reducing** and **Inventing** (fig. 2.4) belong primarily to the abstract **"Space of invention"**. Here at the **Reducing** stage, in particular, the reduction of system conflicts to the models of contradictions occurs, as well as the creation (invention) of ideas at the stage of the **Inventing**.

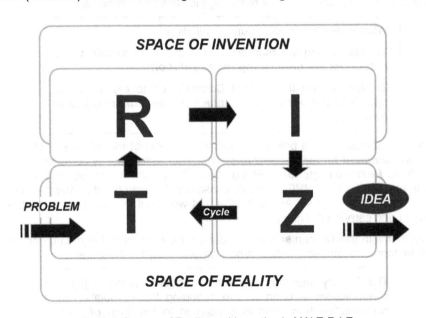

**fig. 2.4.** Spaces of Reality and Invention in MAI T-R-I-Z

The **Trend** and **Reducing** stages (fig. 2.5) belong to the **"Space of Artifact-Prototype"**.

**ATTENTION 1:** in the "Space of Artifact-Prototype" the description (information) includes *only that which relates to the prototype artifact!*

**For example:** any description of the original problem refers to the **Trend** stage, even if the indication of the trend of future changes is also placed there; all the wording of the contradictions, the description of resources, the definition of IFR and FIM – all this is given in view of the **NECESSARY FUNCTIONING**, but **not yet the SOLUTION**, at the **Reducing** stage.

The **Inventing** and **Zooming** stages (fig. 2.5) belong to the **"Space of Artifact-Result"**.

**ATTENTION 2:** in the "Space of Artifact-Result" the description (information) includes *only that which relates to the artifact-result!*

**For example:** any changes in the initial construction, branching of the solution process and cyclic repetition of MAI T-R-I-Z refer to the **Inventing** stage; all assessments of the quality of ideas are given in the **Zooming** stage.

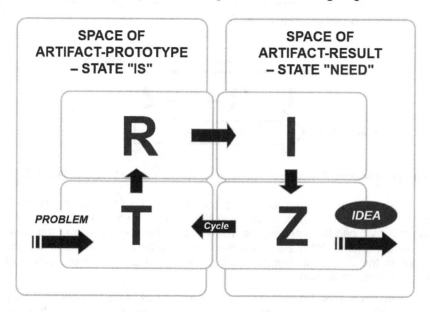

**fig. 2.5.** Spaces of Artifact-Prototype and Artifact-Result in MAI T-R-I-Z

**ATTENTION 3:** to study the above aspects, *it is necessary to carefully study all the examples presented in the following sections of this book.*

In these examples, which reproduce the processes of inventing the effective creative solutions (based on the author's *method of reinventing*), each stage of the MAI T-R-I-Z contains the very information which exactly corresponds to the contents and the required format the presentation of information at this stage.

## 2.4 Correct vs. Wrong Extracting & Reinventing

First of all it is necessary to consider correctly the process of Reinventing using the MAI T-R-I-Z.

What and when do we do at the each stage of MAI T-R-I-Z during Reinventing as opposed to problem solving process?

I hope that attentive readers keep in mind the methodical schemes for Inventing and Reinventing from "Part 6.1. Reinventing" of previous textbook[10]. It is useful to represent these schemes here with some new comments.

> When "inventing" we *work forward sequentially* through steps T-R-I-Z.
>
> When "reinventing" we *work in reverse order* through steps Z-I-R-T.

This situation is well illustrated in fig. 2.6.

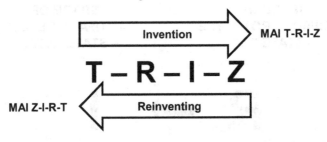

**fig. 2.6.** Schematic "Formula" of Cyclical Integration of MAI T-R-I-Z and MAI Z-I-R-T

Let us assume that we are studying a certain artifact – the product (result) of an invention. This product-artifact is in the state "is", i.e. already exists (fig. 2.7). It is matched to a certain prototype-artifact – the predecessor of the product-artifact in time. Both artifacts are vessels containing the experiences acquired in the course of their creation: technical, creative, and psychological (motivational, emotional).

The question is this: *How do you learn to invent?* How do you create a new target artifact? Can you use previous inventing experiences? If yes, how do you do that?

The answers to these questions are provided by the technology of reinventing developed by the author.

We need to organize – and then use on an ongoing basis – the "training–application" cycle as shown in fig. 2.7. In that cycle, each new invention, together with previously created artifacts, becomes the object of analysis and the material for subsequent teaching.

---

[10]  Orloff, M. *Modern TRIZ. A Practical Guide with EASyTRIZ Technology.* – SPRINGER: New York, 2012. – 465 pp.

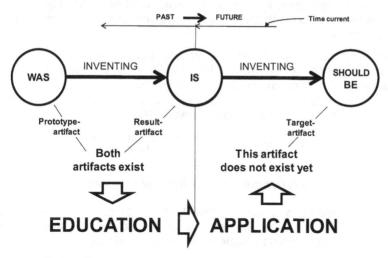

**fig. 2.7.** Training–Application Cycle in the Course of Time

Reinventing is the repetition (modeling, reproduction) of the process of known problems on the basis of the format and the schema of MAI by using the methods and models of TRIZ.

**Reinventing – fundamental teaching method in Modern TRIZ.**

When reinventing beginners are studying problems with known solutions which are illustrated in the MAI format in order to understand how these problems would have been solved on the basis of TRIZ and how in the future similar problems could be solved by using both, MAI as well as TRIZ.

| Definition "Reinventing" | **Reinventing** – *modeling (reconstruction, reproduction, renewal) of the invention process.* |
|---|---|
| Addition 1 | **TRIZ-Reinventing** – *modeling of the invention process* on the basis of TRIZ models. |
| Addition 2 | The initial reinventing has the first goal to allow students to *quickly and correctly acquire* the algorithm of inventive problems solving in the format of MAI T-R-I-Z. |
| Addition 3 | The second and supreme mission of reinventing is to reliably prepare students for an autonomous working on *any new practical problem.* |

MAI is only a general frame and a general navigator for inventive problem solving and it becomes a practical instrument only in connection with models and methods of TRIZ which equip the stages of MAI and thereby turn it into a certain variant of ARIZ – Algorithm of Inventive Problem Solving.

The content of reinventing is the reconstruction and description of all stages of an invention's creation with involvement of models and recommendations of TRIZ for each stage.

The complete reinventing process, as presented in its aggregated form in fig. 2.8 and 2.10, consists of two stages: the preliminary stage – *extracting*, and the main stage – *reinventing* on the basis of MAI T-R-I-Z.

fig. 2.8. Illustration to the Definition of Reinventing

**GOAL** We need to learn to model, on the basis of MAI T-R-I-Z, the process of creating any known product-artifacts, and then use that learning to create new inventions (fig. 2.9).

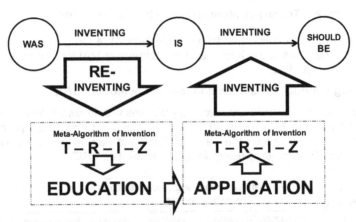

**fig. 2.9.** Reinventing on the basis of MAI T-R-I-Z
in the "education – application" cycle

We already know that when we study a product-artifact of interest we first and foremost, select a matching prototype-artifact, and then perform two extracting procedures:

**1) Extracting-1** – to identify the transformation models objectively realized in the product-artifact vis-à-vis the prototype-artifact;

**2) Extracting-2** – to identify the contradictions objectively eliminated in the product-artifact vis-à-vis the prototype-artifact. In other words, we are talking about contradictions that were **PRESENT** in the prototype-artifact, but are **ABSENT** in the product-artifact.

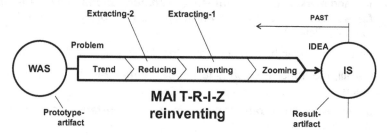

**fig. 2.10.** Key Extracting procedures for Reinventing

The Extracting-1 method is primarily designed for a faster acquirement of the basic models of transformation.

The creative content of this method is aimed at discovering the unusual and the "invisible" within the familiar and the usual. The studies when using the Extracting-1 are of a playful nature. Extracting-1 always leads to quite impressing results, to the discretion of unexpected creative ideas even in the apparently most simple objects and events.

The general description of the Extracting-1 process consists of the following[11]:

1) first we shall take any artifact and choose a real or an imaginary (virtual) prototype for it, assuming the prototype to be a predecessor of the artifact in the history of its constructional development. Ideally the prototype owns the same useful function (intent, purpose of use) as the artifact, but still we can only employ an object as a prototype which shares only a couple of the functions matching the ones of the artifact;

2) afterwards the positive changes inherent in the artifact have to be examined and compared to the properties of the prototype;

3) the identified changes are, of course, of a constructive nature. Still the essence of Extracting-1 is to find the creative transformation model, objectively taking part in the change, from every identified constructive change;

4) at first those Extracting-1 models are being carried out that can be found in the early edition of the TRIZ-catalogue of transformation models. If some changes cannot be convincingly described with the given transformation models then a new model can be described, named and included in your personal creative catalogue and further on be sent to the administrators of the course to get the definition evaluated.

Two basic variants of Extracting-1 are being differentiated:

1) objectively extracting of all transformation models **which are partly or completely participating** in the transition from the prototype to the artifact;

---

[11] Orloff, M. *ABC-TRIZ. Introduction to Creative Design Thinking with Modern TRIZ Modeling.* – SPRINGER: Switzerland, 2016. – 534 pp.

2) selecting and marking of the **dominating transformations, which are necessary for such a transition** and which should be closely connected with eliminated contradictions.

Three characteristics can help to determine the dominating transformation:

> First of all, the transformation is taking part **entirely** in the integral transformation.
>
> Secondly, every such transformation is **very important** to describe the basic inventive idea.
>
> **Third, when tentative eliminating such a transition, the description of the transition from prototype to artifact is impossible or absolutely incomplete.**
>
> Dominating and supporting transformers are marked with two pluses and accordingly with one plus showing the "Level of Concordance" in the tables of Extracting description.

It should also be noted that **the main aim of Extracting-2 is to define the contradictions** removed from artifact-prototype by applying the key transformations.

ATTENTION 1.   **The contradiction is being stated concerning the artifact-prototype accordingly in state "WAS" and not the studied artifact-result accordingly in state "IS"!**

ATTENTION 2.   **The extracted navigators were not applied to the artifact-result but to the artifact-prototype to exclude the contradictions in it!**

**ATTENTION 3 – AS RESUME:**

> The **artifact-prototype** is an object of the initial problem situation and the carrier of the conflicting properties studied at the stages of **Trend** and **Reducing** in MAI T-R-I-Z.
>
> The **artifact-result** is the "product" generated/designed by creative and applied transformations at the stages **Inventing** and **Zooming** in MAI T-R-I-Z.
>
> That is why the transformation models are taking part in the transition from the artifact-prototype to the artifact-result and **their assignment was to exclude the contradictions of the prototype.**

So, an experienced TRIZ-modeler can perform reinventing for relatively simple artifacts even easier and faster by completing the extracting directly during main reinventing. It is necessary to remember only that reinventing is based on extracting creative transformations from the product-artifact, and extracting contradictions from the prototype-artifact.

So, let us assume that we are studying a certain artifact – the fruit (product) of an invention.

> **ATTENTION: Critically Important Methodological Recommendations**
>
> When doing extracting and reinventing, it is absolutely necessary to re-member the following (fig. 2.11):
>
> 1) transformation models are formulated for the **Product-artifact** in com-parison with the prototype-artifact;
>
> 2) contradictions are formulated for the **Prototype-artifact** – in the product-artifact, the initial contradictions are already removed;
>
> 3) MAI T-R-I-Z stages Trend and Reducing are "linked" to the prototype, while stages Inventing and Zooming are "linked" to the product.

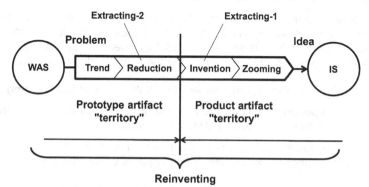

**fig. 2.11.** "Territories" of Correct (!) Formulation of
Extracting and Reinventing Models

### Additional Recommendations for Meta-Algorithm of Reinventing

Attentive readers must have already noticed that during the reinventing exer-cise artifact analysis proceeds strictly from the PRODUCT to the PROTOTYPE in accordance with the following "algorithm":

> **Stage 1:** Review the properties and construction of the product-artifact under analysis, and compare them to the properties and construction of the selected prototype-artifact – this corresponds to the *Zooming* stage in "direct" MAI T-R-I-Z;
>
> **Stage 2:** Extract transformation models objectively participating and realized in the product-artifact – this corresponds to the *Inventing* stage in "direct" MAI T-R-I-Z;
>
> **Stage 3:** Extract contradictions – this corresponds to the *Reducing* stage in "direct" MAI T-R-I-Z;
>
> **Stage 4:** Describe the original problem situation for the prototype-artifact – this corresponds to the *Trend* stage in "direct" MAI T-R-I-Z.

*Meta-Algorithm of Reinventing*
*Z-I-R-T*

In fact, all reinventing stages are performed in reverse order compared to MAI T-R-I-Z; accordingly, we can define and call the Meta-Algorithm of Reinventing either "MARI T-R-I-Z" or better as **"MAI Z-I-R-T"**.

The opposition of MAI T-R-I-Z and MAI Z-I-R-T can be schematically presented as follows (fig. 2.12).

Here it is also important to note that any object that we consider a product-artifact for reinventing purposes becomes the prototype-artifact as soon as we decide to enhance it, invent a target prototype.

This scheme clearly shows that in Modern TRIZ any artifacts and inventions can become the objects of study and analysis.

*To research and discover the world of TRIZ models is an entertaining, exciting and useful pastime! And it is also EASY!*

To do that, you do not need a special laboratory. On the contrary, you need to "build up" a habit of seeing TRIZ models in surrounding artifacts – extracting such models from such artifacts and to model its invention! *So your laboratory is wherever you are.* In other words, it is the entire world around you.

### Special Recommendations to work with A-matrix at Reinventing

**Case 1.      Special important case of mismatch of A-matrix recommendations with your extracting models**

If you have chosen suitable plus- and minus-factors in the A-matrix, but the relevant matrix cell does not contain the navigators which you selected at extracting stage,

> **YOU MUST USE ONLY NAVIGATORS SELECTED BY YOU AT EXTRACTING!**

You can describe the navigators and their influence in the "Inventing" field of "paper form" (the windows "Idea" in software EASyTRIZ). These navigators could be recorded also in the additional fields under the field "Navigators".

**ATTENTION:**   The recommendation above is extremely important!

**If you stick to "your" navigators as you identified it at the extracting stage, you will avoid "FICTITIOUS" or even "RASCALLY" representation of data!**

**ONCE MORE:**  do not use inaccurately defined plus- and minus-factors from some A-matrix cell containing the "required" navigator! In other words, relying exclusively on Extracting-1 and Extracting-2 results, you should model the entire process correctly and objectively using formal factors from the A-matrix *only when they truly correspond to the factors that were selected by you at extractings!*

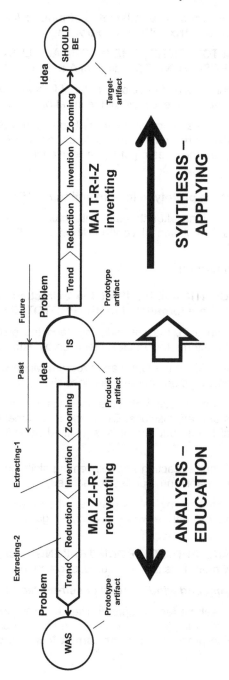

fig. 2.12. "Reverse" MAI Z-I-R-T: Reinventing as Analysis-Training

"Direct" MAI T-R-I-Z: Invention as Synthesis-Application

**Case 2.** **Special important case of mismatch of plus- or/and minus-factors from A-matrix with your extracting**

It is absolutely important first **TO IDENTIFY THE INFORMAL PLUS- AND MINUS-FACTORS WITHOUT RESORTING TO** factors of A-matrix.

This will help you to hone your skills and develop a habit of fully and accurately defining the nature (content, meaning, essence) of the initial problem.

If you have a doubt to select some factor, you can try two or three other variants. Maybe you can get different navigators from different cells of A-matrix. If some navigators can open you other ideas, put these reinventings in separate forms.

**Case 3.** **On the role of the prototype in reinventing AGAIN!**

One of the typical mistakes that students make in the course of reinventing, is formulation of contradictions… for the invention (result-artifact).

**This is totally wrong!**

**ATTENTION AGAIN:** the CONTRADICTIONS at the extracting and reinventing are formulated for the PROTOTYPE only !!!

In the result-artifact (new, invented), the contradictions and the problem are REMOVED with the help of transformation models identified by extracting!

**LAST BUT NOT LEAST:** All fields in the forms and program windows must be completed as concisely as possible.

In particular, it is necessary to avoid using obscure specialist terminology, abbreviations and slang words and expressions. Of course it is sometimes difficult to model without special terms, but then this is justified by indicating the effects.

It is advisable that you train your extracting and reinventing skills on an ongoing basis. This must become a habit and a professional need.

This is routinely done, for example, by chess players: when they are alone, they often do chess problems or analyze well-known chess games to pinpoint novel ideas.

Similar training (**THIS IS HARD AND STUBBORN TRAINING!** – or how else would you call it?) is done by musicians and, of course, sportsmen.

*All these are steps in the long and winding road of self-perfection.*

Modern TRIZ opens up new self-perfection opportunities by providing a technology for exchanging efficient examples. This is supported by standardization of the information presentation process **using the format Meta-Algorithm of Invention T-R-I-Z**.

# Part II

# Samples of MTRIZ Modeling at MSc Programs

Finding the new in the old –
the lot of genius.[12]

Jacob Perelman

We have considered further in this book dozens of illustrative reinventing examples developed by various users of Modern TRIZ who were the students of Master programs. These examples mainly are very simple – but this, after all, is our ultimate objective: we want to give universal examples, which are as simple as possible, which do not require special knowledge.

Master graduates have completed their training and have been certified (in most cases, within Master programs at TU Berlin) to use also the EASyTRIZ software package.

The purpose of this book is also to show you that all readers of this book can master Modern TRIZ and create dozens of similar examples using both artifacts that surround them in their daily life or other items making up their professional environment. This book can also help the readers who have embarked on a study TRIZ-basics **to more quickly acquire practical skills required to use fundamental TRIZ notions and models**.

Of course, it is not enough to just read this book.

**To learn to invent with TRIZ**, you need continuously to keep training in extracting and reinventing.

*Let us constantly improve ourselves!*

---

[12] *Jacob Isidorovich Perelman* (Russian: *Яков Исидорович Перельман*; November 22, 1882, Bialystok, Grodno Province, Russian Empire – March 16, 1942, Leningrad, USSR) – Russian and Soviet mathematician, physicist and world scholar, journalist and educator, populariser of exact sciences, founder of the genre of *entertaining science*, the author of the concept of *science fiction*; created dozens of books including *Physics Can Be Fun* (1913; under the title *Physics for Entertainment* look at https://archive.org/details/physicsforentert035428mbp) and *Mathematics Can Be Fun* (1927).

# 3  GLOBAL PRODUCTION ENGINEERING[13]

### 3.1. Engineering and Art of "Velocipede" Inventing[14]

The topic of research on creative and engineering content in the evolution of the bicycle is one of the most beloved for the author. Unfortunately, the format of this book does not allow me to develop MTRIZ-modeling of the evolution of a bicycle at least in the volume as it was done in the first publication[15].

I selected only a few works made by my students, who were also keen on the bicycle theme. I saw how much for many years at GPE Master Program they were enthusiastic over discovering creative space in engineering solutions for bicycles, for scooters, for other designs moved by pedals or simply by pushing away with feet.

The development of compact and efficient electric motors and electric batteries leads to a new generation of these devices. Therefore, despite the creation of new versions of traditional "non-electric" machines, my undergraduates were happy to model mainly "electrical" devices.

After collecting many of these works, I have made a small "classification" of the currently popular areas of development of *"velocipede"* systems (fig. 3.1).

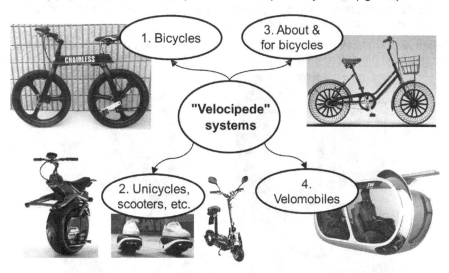

**fig. 3.1.** Some interesting directions of *"velocipede"* systems today

---

[13] international Master of Science Program at TU Berlin "Global Production Engineering (GPE)"
[14] these works were basically made by many students' teams at the GPE Master program trainings before certification stage which usually was professionally specialized
[15] Orloff, M. *Modern TRIZ. A Practical Guide with EASyTRIZ Technology.* – SPRINGER: New York, 2012. – 465 pp.

© Springer Nature Switzerland AG 2020
M. A. Orloff, *Modern TRIZ Modeling in Master Programs*,
https://doi.org/10.1007/978-3-030-37417-4_3

Today, so many directions are developing in personal vehicles such as bicycles or scooters. We can confidently say that today the engineering design of new models of bicycles has become a real beautiful art!

So, it is very difficult to choose particularly interesting examples – there are too many of them! Therefore, we give examples of modeling that attracted the attention of the undergraduates. We will show examples in the order of numbering the development directions of the **velocipedes**[16] in fig. 3.1.

### 3.1.1. Bicycles

The development of the bicycle exactly corresponds to the laws of system development formulated in TRIZ on the examples of hundreds and thousands of systems for various purposes. So, one of the most important laws is the law of the spiral progress of systems, corresponding also to the well-known dialectical principle of ascending spiral progress.

This pattern is appeared in technical systems in a fact that many systems repeat the previous principle of organization at a new level of historical development, achieving higher efficiency. For example, the principle of sail today has found its reappearance in new types of sails – made from metal, or made rotating (using the Magnus effect), or made from artificial fabric equipped with solar cells for generation of electricity over a huge surface of the sail, elongating from the mast and retracting into the mast, etc.

Here we can recall the constructions of some bike-oldies using pedaling on the front wheel (fig. 3.2 and fig. 3.3).

**fig. 3.2.** So called *"penny-farthing"*, *"high wheeler"* or *"spider"* etc. from the 1870s
(pictures made by author at PEUGEOT exhibition in Berlin and at Vienna Technical Museum)

---

16   look in the Oxford dictionary at *https://www.lexico.com/en/definition/velocipede* as from early 19th century French *vélocipède*, from Latin *velox, veloc* – *"swift"* + *pes, ped* – *"foot"*

**fig. 3.3.** Examples of the "spider" – "boneshaker" from 1860s and
very nice "penny-farthing" from 1880s at the German Museum of Technology
(Deutsches Technikmuseum), Berlin

And now look please at modern newest bicycles with pedals on the front or
rear wheel in fig. 3.4 – 3.9.

First this is *Bicymple Nuvo,* the brainchild of Josh Bechtel of *Scalyfish Designs,*
Bellingham, Washington, USA, – just a *"bicymple penny-farthing"* model (fig. 3.4;
all the pictures and fragments of some texts are cited from the web-sites of the
innoventors ***kickstarter.com*** and ***bicymple.com***).

https://www.kickstarter.com/projects/joshbechtel/bicymple-nuvo

**fig. 3.4.** *Bicymple Nuvo* – bicymple *penny-farthing*; for comparison: on the left is the
outline of a modern bicycle, on the right is the outline of a prototype – *penny-farthing*.

What is especially remarkable in this figure is a clear comparison with the sizes
and even the mechanism of a regular bicycle, a penny-farthing and a bicymple.

It is easy to see that the new bike is a modern replica of the old concept of a bicycle with a pedal drive on the front wheel. And this is the reproduction of a well-known principle in a new turn, new round, of the spiral of development!

What is really valuable and interesting here? Answer goes from the developers of this *Bicymple Nuvo*: **"bicymple. a bicycle, simplified"**

Josh Bechtel says: *"Is it possible to evolve from the established bicycle design while adhering to the basic principles of simplicity, functionality, style, and excitement? This is the question that gave birth to bicymple. By removing the chain, the number of moving parts and the overall complexity is significantly reduced. A unique direct-drive, freewheeling hub helps minimize the design. More than just a stylish concept bike, each bicymple model is comfortable, fun to ride, and brilliantly simple to maintain."*

It's hard to say better! And I like one definition-metaphor[17] more:

### Bicymple is Functional Minimalist Design on Wheels

And now we can extract the contradiction of an ordinary bike with a chain, and also see which transformation models quite adequately describe the transition from a bike with a chain to a bicymple. We will use a form from me previous book[18] (fig. 3.5).

| | Radical contradiction | |
|---|---|---|
| **Chain problem:** it becomes dirty, falls off, breaks; multiple-speed gear is too heavy | (+)-factor | (−)-factor |
| | The chain *must be* to transfer energy to the drive sprocket | The chain *must not be* as it becomes dirty and easily falls off |
| **Solution (2011):** to install pedal drive directly to the front wheel! | | |
| **Specialized TRIZ models:** *05 Separation, 10 Copying, 11 Inverse action, 12 Local property, 21 Transform damage into use, 34 Matryoshka (nested doll)* | | |
| **Fundamental TRIZ models:** *Separation in structure, in space, in functioning, in energy (path and point to apply)* | | |

**fig. 3.5.** Extracting the contradictions and transformation models
from construction *Bicymple Nuvo*

---

17  https://mashable.com/2012/12/30/bicymple-kickstarter
18  Orloff, M. *Modern TRIZ. A Practical Guide with EASyTRIZ Technology.* – SPRINGER: New York, 2012. – 465 pp. – see Part "3.7 History of Bicycle Evolution in Pictures, Historical Facts, Contradictions and Inventive Solutions"

With next bike we jump into the Future through the Past (fig. 3.6)!

*And we may consider this, according to TRIZ, as one next step in spiral evolution of bicycle concepts!*

Design company *"RSW Rudolph Schelling Webermann"*, Hannover, Germany, has developed breakthrough *Concept 1865* as *integration of the art and engineering*[19] using the newest materials of world leading chemical concern BASF!

**fig. 3.6.** *Concept 1865 Bike – effective jumping to the Future through the Past!*

«Conspicuous[20] with its wheels of different sizes, the velocipede was the first pedal-powered cycle in history. Together with BASF we have rebuilt the 19th century bike as a modern e-bike. *But why?*

*With the "Concept 1865", we are taking a trip back to the year 1865, when BASF was founded.*

This was also the point in time when *Karl Drais'* wooden *"Dandy horse"* was given its first pedals, which launched the bicycle on the road to global success.

As a tribute to this era of enthusiasm for technology and invention, we have embarked on an unparalleled thought experiment and asked:

*How would the first pedal cycle have looked if the pioneers of the bike had had today's advanced materials to work with?*

In cooperation with BASF, we have developed the e-velocipede *"Concept 1865"*. It is a ready-to-ride prototype with an electric drive (built in the rear wheel with battery inside the saddle!) and 24 polymer applications, some of which are highly innovative like the bearingless all-plastic pedals made of *Ultrason®* or the light and punctureproof tires made of *Infinergy®*.

By implementing this design study RSW obviously does not intend to reinvent the bicycle, let alone the wheel. Under the slogan *"Rethinking Materials"*, the unusual e-bike is in fact an invitation to customers to join the company in developing new applications and product ideas utilizing advanced plastics.

*It is an invitation to question the status quo and create something new – just as the pioneers of cycling did in their time. »*

---

[19] https://newatlas.com/concept-penny-farthing-plastic/29547/
[20] composed and cited from https://rudolphschellingwebermann.com/en/projects/concept-1865

And now we can show another minimalist solution similar to *Bicymple Nuvo* – with a pedal drive to the rear wheel (fig. 3.7). This is not *penny-farthing* but effective constructive step forwards!

**fig. 3.7.** *Bicymple Go* with a pedal drive to the rear wheel

The idea of innovation is the same as for *Bicymple Nuvo* – to exclude a chain. But the point to apply energy and to get route for energy pass is just opposite, namely not to the front wheel but to the rear!

Created in late 2011, the bicymple was an idea that *Josh Bechtel* had been developing for some time and needed to bring to reality. *Bechtel* partnered with co-conspirator *Gabe Starbuck* in 2012 and together they launched a successful Kickstarter campaign (look at *bicymple.com*).

We repeat hear that the concept began with the question:

> *Is it possible to evolve from the established bicycle design
> while adhering to the basic principles of simplicity,
> functionality, style, and excitement?"*

They did it! And now, not detracting the wonderful and beautiful decision of the authors, we note that this solution corresponds to model 21 Copying, as well as the law of spiral development of systems according to TRIZ.

Is not it? Given the analogy with *penny-farthing*?

Here we do not discuss many issues of the effectiveness of new bikes in general, we only study the possibilities of modeling creative content in engineering projects. Our goal is to master such design creative models in order to successfully apply them in the future to solve problems in our engineering activities.

Further, it will be fair to show a few more similar solutions that meet the idea of a spiral renewal of some pioneering ideas in the history.

So similar solutions could be found from other inventors and innovators (fig. 3.8 – 3.10).

**fig. 3.8.** *The Inner City Bike* by *Joey Ruiter*

Project "The Inner City Bike" was started by "JRuiter + Studio" in Grand Rapids, Michigan, in 2009.

*Joey Ruiter,* founder and principal designer, said[21]: *"Our project, simplicity in inner city bicycling, was at first glance a fun aesthetic opportunity in new trends, color, and materials. Our target lived / worked in an inner city environment with minimal space. Bicycling at this level can be more about fashion and culture than speed and performance ...We knew there were bigger opportunities. The project rethought what a "frame" meant, getting rid of basic key components, and creating a new type of compact bicycling... The final design came down to a frame system and a free-wheeling unicycle rear hub."*

> It is worth looking at two more examples of minimalist bike designs, which are outstanding works of design art and engineering art, the same as the two previous examples.

This is what Joey Ruiter said[22]:

"The Inner City Bike is how I'm starting a conversation about new products and how they change the world.

The bicycle is iconic. Throughout its history, its design has evolved. Big wheels, small wheels, even the number of wheels. It's been made of wood, metal, and plastic.

Is there room for another take on the bike? Can we re-define classic objects?

I think so.

---

[21] from https://jruiter.com
[22] cited from http://www.delood.com/design/inner-city-bike-36er

It is about simplicity in design. The *Inner City Bike* is the ultimate stripped away piece. So stripped even the chain is gone. It's a statement on bare essential transportation and new ways of thinking about materials, scale, manufacturing processes and function.

For me, the art of design happens when you change the way things are perceived, when a new word is coined to express what you've done. It challenges conventionality and creates new stories, interactions and rarity we strive for."

hinge to rotate frame for folding

**fig. 3.9.** *The Chainless Bike* by Sean Chan

Creator of the bizarre-looking bike, Sean Chan, explained that he was sick of bicycle chain maintenance, which led him to develop the new transportation product. Mr. Chan had several years of experimenting and engineering with his father to produce the prototype[23].

According to Chan[24], the Chainless Bike can be a commuter, trick bike, road, cruiser, or swing bike even though it doesn't fall into any category of bike types.

"I wanted to create a bike that was very basic and has folding capabilities and is collapsible," says Chan. "It was also important to build a bike that was not a lot of work and not a lot of maintenance for riders."

And to finish this art of the bikes, let us to look at one very interesting model with some kind of pedals – levers – joined to the hub[25] of rear wheel (fig. 3.10).

Here we will immediately find two fantastic features that make this artifact a work of design and engineering.

---

[23]   http://www.itechpost.com/articles/95802/20170414/would-you-ride-this-chainless-bike-bizarre-new-bicycle-ditches-the-chains.htm
[24]   https://finance.yahoo.com/news/bike-ditches-chain-140000946.html
[25]   https://newatlas.com/bygen-hank-direct-bike/33558/

The first is the abandonment of rotating pedals, instead of which swinging levers are used, which transfer pressure from the leg into the rotation of the rear wheel. It has three available gear ratios, thanks to a hub transmission.

> Please, note that the model of the **Ideal Final Result** is effectively implemented here: *the bicycle chain no longer exists,* **BUT** *the function of transferring energy to the drive wheel exists.*

And the second feature is reducing the length of the bike not by rotating it, but by shifting along the frame to the smaller size. The rear end slides forward, along the rail-like front section. The handlebars can also be folded back, for easier carrying and storage.

**fig. 3.10.** *The Hank Direct Bike* by Korean manufacturer *Bygen*

And to finish this first part about the designing the bikes, here is another statement by the outstanding designer Joey Ruiter about his understanding the process of creating the new, the process of inventing the new, which he also implemented in his *The Inner City Bike*[26]:

> *"I soon learned that the criteria, the constraints, are what drive good design. When I'm constrained, I am more creative. I want to see how good I can make something despite the constraints. I take complex problems and make the solution as simple as possible."*
>
> And we can say that this is always overcoming the contradiction between known and unknown, between "visible-and-existing" and "invisible-and-nonexistent" still.
>
> These statements both are extremely important for each designer, each student, each engineer, each manager and teacher or professor.

[26] cited from https://www.hermanmiller.com/designers/ruiter/

And before to continue the students' works, I would like to remind you of such absolutely minimalist designing construction of the folding and chainless bikes naming as of *X-bike* (fig. 3.11[27] and 3.12[28]).

**fig. 3.11.** *X-bike* by
Mark Sander, 1990

And although next two X-shaped bikes do not satisfy the minimalist and folding concepts, they quite satisfy a concept of chainless machine (fig. 3.13[29], 3.14 and 3.15[30]). These concepts belong to the company Cannondale.

**fig. 3.12.** One more model of minimalist *X-bike*, 2010

After the first publication, many questions remained, for example, how exactly the front wheel is rotated when the bicycle handle bar is used, and how the transmission from pedals to the rear wheel is carried out.

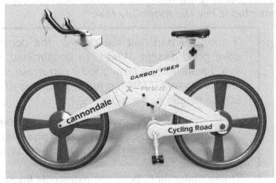

**fig. 3.13.** *The CERV Bike* by German manufacturer Cannondale in cooperation with *Priority Designs,* 2012

27   https://mas-design.com/timeline/
28   https://www.designboom.com/project/x-bike/
29   https://www.yankodesign.com/2015/08/13/a-bike-without-excess/
30   https://www.prioritydesigns.com/content/bicycle-design-engineering-and-development-consulting

**fig. 3.14.** *The CERV Bike* with automatically adjustable saddle depending on the road inclination – two dominating TRIZ-models: *07 Dynamization* and *12 Local property*

**fig. 3.15.** *The CERV Bike* with a single-sided swing arm instead of a traditional fork

It was impossible to do adjustable saddle with a traditional wheel fork, so a single-sided swing arm (TRIZ-model *24 Asymmetry*) plus a multi-axis and multi-link (TRIZ-models *03 Segmentation* and *18 Mediator*) turning gear were proposed. You can see that a chain is substituted here with a shaft to transmit energy from pedals to rear wheel hub.

And to continue this part of students' works and vision, supported with my comments and corrections, I would like to bring here, as some newest trend, a direction combined the ideas of chainless bike and electric bike.

This trend is in concordance at once with a group of TRIZ laws and trends. Here we have selected the following ones (from fig. 1.10):

*1) Leading development of the working tool,*

*2) Increase of fields' controllability,*

*3) Effective closed paths and through routes of energy,*

*4) Substitution of human in system.*

We start express-studying and creative MTRIZ-modeling the chainless e-bike from fantastic construction[31] of JIVR (fig. 3.16).

**fig. 3.16.** *The JIVR* by designer and manufacturer *Marcin Piatkowski*

Power from pedaling is transferred to the rear wheel via a shaft with two bevel gears and then a durable carbon belt (fig. 3.17) and other custom-designed components, all protected from exposure to moisture – and from any contact with your leg. The single-speed drivetrain produces at 3:1 rotation ratio – for every one rotation of the pedals, the rear wheel turns three times. When extra speed is needed, an electric boost is always one click away.

**fig. 3.17.** Pedaling possibility with transmission of power from pedals to rear hub with shaft and carbon belt

---

[31] https://jivr.co/our-story/

We can see here two main creative tricks: 1) using the shaft and belt instead of metallic chain as a chainless transmission[32] hided in the frame tubes according to TRIZ-model *34 Matryoshka (nested doll)*; 2) using the electromotor as it requires in the newest trend according to model *04 Replacement of mechanical matter* in points of *b) use of electrical, magnetic, or electromagnetic fields for the interaction of objects,* and *c) replacement of static fields with dynamic ones,* to implement the TRIZ-trends *"Leading development of the working tool"* and *"Increase of fields' controllability"*.

If all the power in the battery is used[33], or if you just want an old-fashioned bike trip through a park or along quiet streets, *JIVR* can always function as a conventional bike, powered by nothing more than your own two feet. Old school or high-tech – you can ride either way or alternate between them.

Starting in 2011, founder Marcin Piatkowski, then studying at University College London and the London Business School, began searching for ways to find solutions to this problem through the use of clever technology and attractive design. He says[34]: *"We are a team of designers, engineers and two-wheeled artists working together to bring that vision to life every day."*

Next step in evolution of chainless bike is an automatic e-bicycle[35] (fig. 3.18).

**fig. 3.18.** *The Bike2 by Danish manufacturer Bike2*

The *Bike2* system has no mechanical transmission by chain. It operates by pedals pushing a generator located in the crank-box. According to *Bike2* its system *"Excels by giving the cyclists a natural cycling experience with a feeling just like riding a conventional or pedal-assisted electric bicycle."*

As pointed (here and below cited from bike-eu.com[36]) by the producer, *Bike2* also claims that its *Hybrid Drive System* is the first e-bike system to have a completely step-less digital gearing and an automatic gear mode. ***This keeps the cadence at a fixed level despite different loads from the cyclist.***

---

[32]  https://jivr.co/chainless/
[33]  https://jivr.co/electric/
[34]  https://jivr.co/our-story/
[35]  http://bike2.dk/index.html
[36]  https://www.bike-eu.com/home/nieuws/2017/06/bike2-starts-shipping-chainless-hybrid-drive-e-bike-system-10130336

From TRIZ point of view it means implementation of the *"Law of transition to system joining up and integration"* and the trend *"Substitution of human in system"* mentioned above. *"Substitution"* means here excluding the bicyclist from the function and process of the load following and adapting the pedal powering.

Other benefits of the *Bike2* system are the exclusion of many mechanical parts, simple and easy assembly and maintenance, as well as a smoother level of assistance compared to that of conventional e-bike systems.

*Nils Sveje* and *Jesper Allan Hansen,* founders of *Bike2,* claim that tests by interested e-bike makers helped them to develop a system that can be fine-tuned to multiple segments and users: *"Next to the tests done by producers of regular bicycles, we also have had a lot of interest from makers of other amazing types of pedal assisted vehicles."*

And here now is a moment to demonstrate one earlier design (fig. 3.19[37]) by *Kyoko Inoda, Nils Sveje, Gustavo Messias, Peter Anderson* (2010) with fantastic idea of intelligent cadence leveling (stepless automatic gearing)!

**fig. 3.19.** *The Bike 2.0* (2010) with
automatic support of comfortable pedaling and super-minimalistic design

There is influence of X-bike style (frame in relatively similar to minimalistic concept in fig. 3.11 and 3.12) and development of newest concept for "full-electric" bicycle without mechanical transmission from the pedals to hub of rear wheel!

---

[37] http://www.tuvie.com/bike-2-0-next-generation-bicycle-with-chainless-transmission/

It was great step to excluding the mechanical transmission from the bike! Despite the certain controversy of such an idea, it nevertheless meets the TRIZ trend of *excluding a man from the energy supply unit and the system control unit.*

This solution uses rotating the pedals to produce electrical energy with alternator built into the pedal hub. Also it was also noted that with a lack of energy, a super-capacitor can be used, which was a fairly new word 10 years ago. This creative "trick" is in accordance with TRIZ-model *28 Previously installed cushion – "increase the relatively low security of an object with safety measures in advance."*

And now we can see one more of the most interesting and promising examples – folding e-bicycle[38] *Mando Footloose* (fig. 3.20).

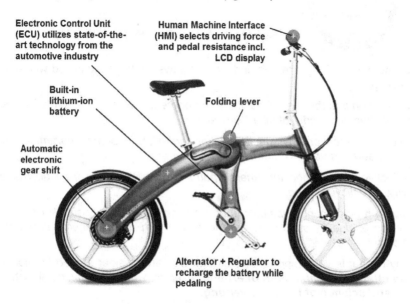

**Electronic Control Unit (ECU)** utilizes state-of-the-art technology from the automotive industry

**Human Machine Interface (HMI)** selects driving force and pedal resistance incl. LCD display

**Built-in lithium-ion battery**

**Folding lever**

**Automatic electronic gear shift**

**Alternator + Regulator** to recharge the battery while pedaling

**fig. 3.20.** *The Mando Footloose* by Korean manufacturer *Mando*

"All products of *Mando Fooloose* removed the chain, a defining feature of bicycles, and replaced it with the innovative *Series Hybrid System* in order to enable it to move as effortlessly as a car. This first-of-its-kind drive system propels the bike forward without any mechanical connection between pedals and wheels.

---

[38] as a global auto parts leader, *Mando* has been supplying state-of-the-art automotive brake, steering, and suspension systems to car manufacturers, such as Hyundai-Kia, General Motors, Audi, BMW and Volkswagen, since its establishment in 1962 (https://www.mandofootloose.com/en/pages/01about/companyinfo.jsp)

You can choose whether to accelerate using the pedals or the *CityBoost* function. Our unique drive system allows you to adjust the riding mode and pedal resistance to your personal preferences.

By pushing down the throttle and pedaling at the same time you activate the *CityBoost* function, which effortlessly propels you to about 15 km/h.

All products[39] of *Mando Fooloose* transforms the mechanical energy you generate while pedaling into electrical energy, which is then used to recharge the battery and to extend the range."

**fig. 3.21.** *The Mando Footloose in folded state*

Implementation of TRIZ trends and relevant laws could be explained with next extractings:

*1) integration into the rear wheel the e-motor means accelerated evolution just the main "working body" of bike – a wheel;*

*2) use of electrical energy supplies higher controllability of main parameters of bike – speed and riding comfort;*

*3) effective combination the energy paths from battery and from alternator powered by bicyclist' pedaling;*

*4) partial replacement (substitution) of a person as an energy supplier and also as an engine by supporting the movement with electric energy and electromotor.*

In any case it is very important for us that we can consider *Mando Footloose* as **an effective combination of art and science**, we can say as combination of **art and science of bicycle inventing.**

And I want to finalize this chapter with the words of *Mark Sanders*, the chief-designer[40] of the project *Mando Footloose*:

> **"An ideal combination of art and science gives birth to undeniably elegant products.**
>
> **I love the word 'elegant' because it makes sense both with respect to the sophisticated technology and the artful design."**

---

[39] https://www.mandofootloose.com/en/pages/02product/technology.jsp
[40] https://www.mandofootloose.com/en/pages/01about/brandinfo.jsp

### 3.1.2. Unicycles, Scooters, etc.

We will classify the "Unicycle" as a bike with a seat or without it and one wheel. That's enough for definition. Welcome to the art of unicycle inventing!

It is interesting to pay attention to the fact that the unicycle can be considered *as a penny-farthing*, provided that its rear wheel becomes smaller and smaller and can be considered only as a supporting foot[41] supplied with a small free rolling wheel, one or two (fig. 3.22).

Starting this chapter, we will not follow a strict classification, which could be encouraged by picture 3.1. And we will also limit the modeling to only one or two interesting artifacts representing constructions of a certain type according to picture 3.1.

First of all it is critically important to take in account that all the most effective and interesting unicycles are equipped with two things: 1) electromotor and 2) electronic control of motion and stabilization.

Really, motion and stabilization are of interest to us in the first place, and issues such as measuring and fixing parameters or the route of movement will be left unattended here.

Thus, we will focus on creative modeling of key technical ideas and solutions that set the device's development trend.

So, *YikeFusion* (fig. 3.23) is a next generation of folding "e-velocipede" *YikeBike* from Christchurch, New Zealand, described in my previous book[42].

**fig. 3.22.** Modernized (2014) "penny-farthing" *Halfbike 3* by *Mihail Klenov* and *Martin Angelov* from *Kolelinia LTD*, Sofia, Bulgaria

saddle          handlebar

**fig. 3.23.** Folding "e-velocipede" *YikeFusion* as modern successor of *Penny-Farthing* but without the pedals *(www.yikebike.com)*

---

[41] https://newatlas.com/the-halfbike-minimalist-bicycle/31238/ and https://halfbikes.com
[42] Orloff, M. *Modern TRIZ. A Practical Guide with EASyTRIZ Technology.* – SPRINGER: New York, 2012. – 465 pp.

**fig. 3.24.** *YikeFusion* and *penny-farthing* together at Tweed Run London 2013.

**fig. 3.25.** Unicycle *One Point Five* by Japanese company *Outre*.

Because the handlebar is located behind the cyclist at the saddle level, riding style on *YikeBike* is a little bit different from driving on *penny-farthing* as shown[43] in the photo (fig. 3.24) taken at Tweed Run London[44] 2013. Designers say[45]: "To design a personal transportation device that was safe, manoeuvrable and as easy to ride as a bike but specifically designed to be smaller so that it can be easily taken anywhere in a congested city. Rather than take a normal bike and crunch it up like most folding bike designs we took a step back to see if there was another safe stable configuration that is vastly smaller when folded.

We started from the assumption that you need a decent sized front wheel so you can go through pot holes, up curbs and over bumps in a safe comfortable way. You can see from the development history that it took a lot of trying to find a stable easy to ride design.

Although we started with pedal only versions we found that we could make a smaller lighter more useful version using latest battery, motor and controller technologies. "

And here (fig. 3.25) we can see the *penny-farthing*[46] ***"back to front".*** It features what could be considered one and a half wheels ... hence the name.

Really, it seems pretty likely that the designers of the *One Point Five* have at least seen the German-made ***"Halbrad"*** (or ***"Halfbike"*** in English). With this, tiny pivoting front wheel located under the crankset, so we got a *penny-farthing* "back to front".

---

[43]  fragment of picture cited from *thecyclehub.net/yike-bike,* posted on 15 April, 2013 by *doug*

[44]  The Tweed Run is a group bicycle ride through the center of London, first held on 24 January, 2009, in which the cyclists are expected to dress in traditional British cycling attire, particularly *tweed plus four suits.* Any bicycle is acceptable on the Tweed Run, but *classic vintage bicycles* are encouraged. Now the format of Tweed Run is copied in many countries (https://en.wikipedia.org/wiki/Tweed_Run)

[45]  http://www.yikebike.com/our-dna/

[46]  https://newatlas.com/bicycles/outre-one-point-five-bike/ and https://www.outre.jp/bicycle.html

**fig. 3.26.** An ultra-portable commuter *SBU V3* by *FocusDesigns*, USA

Next idea coming from TRIZ is a trend *"Transition the system into working body"!*

Really what we see in fig.3.26? – There is only one wheel and saddle seat as a personal one-wheel vehicle. **No frame, no chain and even no handle bar!** But the cyclist can seat and drive!

The "working body one" (WB1) here is a single wheel. The WB2 here is a road.

But of course all is not so simple! So let us to consider some unicycles.

We will not select the type or prefer one type or exemplar to another one. I use the examples simply supplied by my students mainly as a result of group study.

And so, the modern e-unicycles do not have the pedals (fig. 3.25)! And what does it mean from TRIZ point of view? Yes, really it is the displacement of a person from a transmission, not to mention substitution in the supply of energy and the role of an engine!

What producer[47] about the first of the considering machines: *"No handlebars, no steering wheel, no need for any of that! Control your SBU with natural leaning motions (similar to the well-known Segway) and experience a new way to travel. Lean forward to go, lean back to slow down/stop. "*

What else is here especially important according to TRIZ? It is *exclusion a person also from the control unit – one of the most effective modern TRIZ-trend of system development –* due the sensitive sensors and high-torque motor providing to the SBU's superior movement capabilities. Multiple 3-axis accelerometers and gyros provide superior inertial measurement enabling for a ride like safe and fun.

*"It's clean. It's green. It's an elegant machine!* The SBU weighs only 27lbs. and for many people worldwide the SBU has been an integral part of their daily commute. The SBU's portability makes it the perfect in-between commuter as it fits on the bus, the subway, the train, and underneath your desk at work. "

Today we can find a lot of such machines for sale with growing competition between manufacturers whose names are increasingly impossible to reliably determine. But still, we are only interested in the technical side of development. And many key functions, features, parameters, principles and technical solutions we can see quite definitely from pictures and marketing descriptions. And our conclusion is univocal: *the most of technical solutions strictly meet the TRIZ-trends mentioned in chapter "1.7 TRIZ Laws of System Development."* Several additional examples (fig. 3.24 – 3.28) can confirm the main TRIZ-trends if to analyze these examples using previous samples and explanations.

---

[47]  http://focusdesigns.com/sbuv3/

I think that the examples, studied by readers earlier in this book, have demonstrated main trends and key models for development and even evolution of such technical systems as "velocipede".

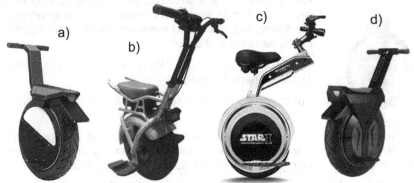

**fig. 3.27.** Examples of self-balancing e-unicycles of different brands:
**a)** *GORILLA WHEEL*; **b)** *UNO BOLT*; **c)** *Apex Star I SP800* (all the models are from http://selfbalancingelectricunicycles.com); **d)** *MACWHEEL D2* self-balancing electric unicycle (https://www.banggood.com/de/17-Inch-One-Wheel-Motorcycle-Waterproof-500W-Self-Balancing-Electric-Unicycle-Scooter-Monowheel)

It is easy to note that handle in these models is not a handlebar to control the turns! It is only the support for the cyclist to grip on it. And next support is two folding or fixed steps to lean on them with legs.

Control of turns is carried out by tilting the cyclist's body in the necessary direction. Start and braking are carried out by tilting the cyclist's body forward or backward.

**fig. 3.28.** Self-balancing e-scooter *Airwheel A3*

We will complete a brief description of unicycles and make transition to scooters using the examples of "e-velocipedes" by company *Airwheel*.

One of the most effective products of the *Airwheel* is self-balancing e-scooter *Airwheel A3* (fig. 3.28).

The company sets the rules[48]:

*"Abide by ergonomic principle. Adhere to the habit of manipulation and mindset. Control the speed and whizz about comfortably. Independent buttons and brake guarantee safe ride."*

---

[48]  https://www.airwheel.net/home/product/a3

a) *Airwheel R6* firstly intro-duces the automatic folding system to make the folding and unfolding process is super easy.

b) *Airwheel E6,* as the first e-bike in Airwheel, made many breakthroughs, like the unique frame and saddle design and USB port, etc.

c) *Airwheel E3* backpack has multi-functional handlebar, 300W hub motor, OO-frame design and folds up easily to backpack size.

d) *Airwheel S8MINI* is self-balancing scooter with dual ride modes and App remote control.

e) *Airwheel Q5* twin-wheeled electric scooter is easy to learn, with LED headlight and taillight, all-terrain rub-ber tire, led battery.

f) *Airwheel SE3* is not only a storage travel equipment, but also a personal equip-ment that can be used for transportation.

**fig. 3.29.** Products under the brand *Airwheel*:
(https://www.airwheel.net/home/products)

*These machines perfectly confirm our thesis that today the creation of the "ve-locipedes" has become a real integration of engineering and design art!*

The *Airwheel* designers say[49]:

> *"World-class industry design: Pay manic attention to details.*
> *Interpret craftsmanship and fashionable beauty*
> *of industry design."*

---

[49]  https://www.airwheel.net/home/product/a3

The variety of scooters and especially mono-wheelers continues to increase. The models are continuously changed in form and size (fig. 3.30 – 3.38).

**fig. 3.30.** Examples of scooters of different brands:
**a)** *eFlux Street 40*; **b)** *CityCoco*; **c)** *Dapang Electric Scooter*; **d)** *eFlux Harley Two*

One of the effective trends is making the wheels of scooters smaller and smaller (fig. 3.31) but wider, without significant loss of speed and comfort, in particular, some mono-wheelers can have a seat (fig. 3.31, d).

A fundamental modern TRIZ-trend is clearly visible on all e-devices considered: *the growth of automatic controllability of devices.* This trend is ensured by the organization of *a through passage of information and energy flows* across all components of the "velocipedes". The units for stabilizing the bikes and scooters in space, including the possibility of stable "standing" without additional mechanical support, are particularly important. *New materials* are used in sensors and processors, making them *compact and high-performance*, as well as in tires and frames, making them *durable and light.*

**fig. 3.31.** Examples of scooter/mono-wheelers: **a-b)** *Kiwano KO1*;
(https://electrek.co/2019/03/04/kiwano-ko1-plus-electric-scooter)
**c-e)** *Uno Bolt mini* (https://newatlas.com/uno-bolt-mini-one-wheeled-scooter/55443/
and http://www.unobolt.com/mini-by-uno-bolt.html)

In parallel, new e-devices appear, as simple e-scooters without seat, hoverboards, hovercarts or hovershoes (fig. 3.29, c,e; fig. 3.32 – fig. 3.38).

It is very difficult to choose 2-3 examples of each e-devices because of their very great variety. This choice should not be a promotion or even just a preference for some model. This is not our goal. Therefore, the examples presented here are a completely random choice from the Internet.

**fig. 3.32.** Simple folding e-scooters:
a) *Bluewheel IX7 E-SCOOTER* (https://www.bluewheel.de);
b) *Urban Scooter* (https://www.wheelheels.de/p/scooter/whus/)

**fig. 3.33.** Hoveboards:
a) *6.5" Premium Hoverboard Bluewheel HX310s* (https://www.bluewheel.de);
b) *City Cruiser* (https://www.wheelheels.de/p/balance-scooter/whcc/)

**fig. 3.34.** E-monowheel Segway One S1
(https://transportationevolved.com/segway-one-s1-review/#prettyPhoto)

fig. 3.35. Segway
S-PLUS
(http://de-
de.segway.com/products
/segway-miniplus)

fig. 3.36. Segway
NINEBOT Elite
(https://www.de-
bedienungsanlei-
tung.de/)

fig. 3.37. Segway Drift
W1 E-Skate
(https://www.youtube.com/
watch?v=F2JCVcWvhOc)

And here we can see two hovercarts as the newest fashionable trend to connect the hoverboard with special chair and control levers (fig. 3.38). They have different methods to control turns. First one is implemented by cart of Wheelheels (fig. 3.38, a) to control a rear hoverboard under a seat, and second one is a cart of Robway to control a front hoverboard (fig. 3.38, b).

In any case we see an adaptation of the known device hoverboard to new functionality if connect with also known components from carting. It looks as implementation of TRIZ-law *to integrate of different and alternative (for staying or seating movement) technical systems* to increase functionality of hoverboard in two interpretations: 1) to equip a hoverboard with a chair and control levers to get cart functionality, or 2) to exclude a motor from a traditional cart and motorize it with a hoverboard. As a result you can get hoverboard separately or as a component of a self-made cart.

**fig. 3.38.** Hovercarts: a) *Wheelheels Cart for Hama Balance Scooter* (https://www.wheelheels.de/p/balance-scooter/zubehoer/whhc/); b) *Robway Original DRIFTKART 360*

So, as a short summary we can reveal several tendencies demonstrating the objective existence of TRIZ-laws in modern innovations and inventions (innoventions!) of "velocipede" technics.

First, we see motorization as a common, fundamental modern trend in personal vehicles of real or conditional class "velocipede". Motorization is a phenomenon that accompanies the development of bicycles since the end of the 19th century. Moreover, what is important, these were attempts to install on a bicycle not only gasoline engines, but also electric (!) motors. These attempts were not very successful and could not turn into mass production, but continued until the end of the 20th century.

But from the beginning of the 21st century, we see the rapid mass development of electric "velocipedes" as a successful "spiral repetition" and achievement after previously made attempts for more than a hundred years. This is obviously realization of *TRIZ-law of spiral evolution*.

Second, in connection with electric motorization we can see a *TRIZ-law of displacement of a person from the function and units of energy supplying and transmission*. And, example by example, we see the death of even such traditional and indispensable elements of a bicycle as chain and pedals! Really, supplying the electric energy from battery to directly into the electric motor requires only thin electric wires hided into the frame or case of electric "velocipede" no matter it is a big e-bicycle or small e-skates.

Third, we can clear comprehend implementation of TRIZ-law of *excluding a person from control function and unit(s)*! Really, the human reaction rate is clearly insufficient to control e.g. such motorized "velocipede" as Segway. And then we understand that this became possible only with emergence of miniature electronic components – processors, sensors, as well as algorithms and software for controlling all processes of movement, stabilization, and now navigation, of "velocipede". It means implementation of such basic TRIZ-laws as *"complete composition of acting parts", "controllability improvement"* and *"transition to micro-level"*.

Here we also have a *"through passage of energy and information"* – according to TRIZ-law!

And next example shows the unusual and important realization of *"integration of alternative systems"* as combining the system with hand drive and automatized system[50] (fig. 3.39).

**fig. 3.39.** Scooter *Pendel* by *Huka*, Netherland
left picture cited from https://www.youtube.com/watch?v=AY7s90ZBt0Y

The wheelchair user can easily drive directly into the scooter *Pendel* through built-in ramp without assistance. Once the wheelchair is in place and firmly locked, user may lift the ramp and scooter also goes up itself for the ride.

Most of the available mobility solutions for manual wheelchair users require one form of transfer or another. The *Pendel* gives greater freedom of movement and maximum comfort for the user. And here we also see the opportunity to separately use the chair and scooter as in previous example.

To complete this part I would like to give my friendly advice to the readers: in order to really understand TRIZ-laws, it will be very useful to review the above examples repeatedly, perhaps more than once, and then you can still enjoy the rediscovery of these laws and the meanings of the development of technology.

---

50   https://www.huka.nl/en/product/pendel/

### 3.1.3. ...for Bicycles

Here we will consider only three interesting examples of components for bicycle, remembering that incredibly many components and accessories are under improvement and proposing at the market! And we will extract the TRIZ-models and laws from each engineering solution and idea.

**Example[51] 3.1. How to convert almost any bike into an e-bike.**

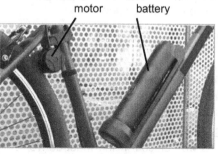

motor       battery

**fig. 3.40.** *"Revos"* kit supplies electric assistance to you on your bike

Whether you're cycling to the office and don't want to arrive too sweaty and out of puff, or you just need some help getting up a steep hill, e-bikes can be a great way to get around. For many though, buying a brand new bicycle when you already have a perfectly good one just doesn't make sense. That's where add-ons like *"Revos"* kit from UK-based *Revolution Works* come in, transforming an existing two-wheeler into an e-bike. *Revolution Works* reckons that riders will be able to go from standard non-powered bicycle to pedal-assist e-bike in under 10 minutes with no special tools. The kit will come as a drive unit, a pedal-assist sensor and a battery.

The 250 W aluminum alloy drive unit is clamped on the seat tube, between the seat stays, with the unit's roller resting on the tire of the rear wheel. A battery holder is mounted on the down tube, in place of a bottle holder perhaps. Then one of two available Li-ion battery packs (100 Wh or 209 Wh) is slotted into the holder and cabled up to the drive unit. Finally the magnetic pedal sensor is zip-tied to the chain stay and directed toward the cassette at the rear. The sensor is also cabled to the drive unit.

Once fitted, all a rider needs to do is start pedaling, the sensor will detect movement of the cassette and the *"Revos"* drive unit will kick in and help out until a speed of 15.5 mph (25km/h) is reached. The drive unit will only engage while a rider is pedaling, so will stop while coasting. And if you want to turn the unit off completely during a trip, you simply back pedal by half a turn. Then if you notice a particularly threatening incline approaching, the system can be reactivated with another half turn back pedal.

---

[51]   pictures and fragments of text cited from https://newatlas.com/revos-add-on-ebike-kit/54876/

"The drive unit (patent applied for) and motor controller ensure that pedal assist is provided extremely efficiently," company director *Mark Palmer* told *Newatlas*[52]. "This gives the rider a very positive experience as well as maximizing the available energy stored in the battery. When people see *"Revos"*, their first reaction is that it's too small give significant help. Once around the car park and they've changed their minds."

How can this example be interpreted in term of TRIZ? The dominant models will be considered first.

There are two dominant models *05. Separation: to include the only really necessary part (necessary property) into the object* and *12. Local property: a) change the structure of the object (the external environment, external influences) to a different one; b) different parts of an object have different functions; c) every object should exist under conditions that correspond best to its functions.*

So, the according to model 05 the electromotor was introduced into the system and was placed in the most convenient place according to model 12 to influence effectively onto the rear wheel.

The transformations *04. Replacement of mechanical matter (to introduce electrical power), 10. Copying (to use electromotor as in other machines),   11. Inverse action (to rotate a wheel instead a hub)* and *18. Mediator (to use temporary installed device to support efforts of driver)* could be taking in account as auxiliary models.

And now I would like to introduce incredible invention[53] by *John Schnepf*, New York, filed May 6, 1898. Excerption from a patent (fig. 3.41): "The chief object of my invention is to provide a means which may be attached to or detached from a bicycle of any ordinary type, which means may be utilized as a primary or auxiliary driving means.

$\mathcal{A}$ is a battery of any suitable portable type. Thus battery is detachably supported in the forward inside portion of the frame of a bicycle by suitable festening-clamps $a$ $a$ $a$. $\mathcal{B}$ is a motor. The motor $\mathcal{B}$ is mounted in a suitable frame carrying forwardly-projecting arms $\mathcal{D}$ $\mathcal{D}$, having semicircular depressions did near their outer ends. $\mathcal{F}$ is a pulley, preferably grooved and carried by the armature of the motor $\mathcal{B}$. The motor $\mathcal{B}$, through the medium of the frame and $\mathcal{D}$, may be clamped to the bicycle-frame so as to cause the pulley $\mathcal{F}$ to bear upon one of the bicycle- wheels $\mathcal{J}$, preferably the rear or driving wheel. "

We can see in fig. 3.41 a fragment of a patent description as an example of similar historical inventions **to confirm the cyclic spiral character of technical development** to formulate the TRIZ-law of **spiral evolution**.

---

[52]   according to *Paul Ridden*, June 01, 2018; https://newatlas.com/revos-add-on-ebike-kit/54876/
[53]   Patent USA No. 627,066, June 13, 1899.

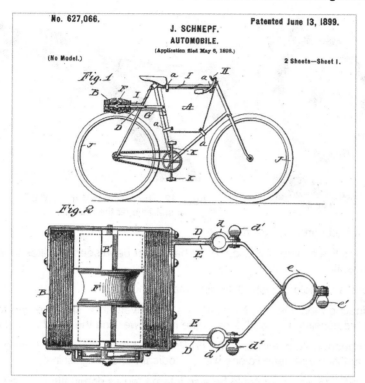

**fig. 3.41.** E-motorized bicycle by John Schnepf, NY, 1899,
according to patent USA No. 627,066

## Example[54] 3.2. *reTyre* bike tires feature interchangeable skins.

Serious cyclists will often swap back and forth between different tires, depending on the type of riding they plan on doing ... and yes, getting those tires on and off of the rims is a hassle. That's why the *reTyre* system was developed, with its zip-on tire skins.

At the heart of the Norwegian-designed system is a set of 568-gram smooth-tread base tires that mount on the rims in the usual way, and stay on the bike full-time – they're not tubeless, although a tubeless version is in the works. These can be used on their own, for commuting or touring.

There are also, however, a variety of rubber "skins" that can be zippered on over top of the tires, each skin featuring a different type of tread. These can reportedly be installed or removed within seconds, they won't slide back and forth against the underlying base tire, and the zipper's tab gets locked in place once it's done up.

---

[54] fragments of pictures and text according to Ben Coxworth, June 01, 2018, cited from
https://newatlas.com/retyre-zip-on-tire-skins/54877/

**fig. 3.42.** The *reTyre* winter "skin" is installed over the base tire

What is it according to TRIZ-modeling?

This seems like a simple solution, but it has never been seen on bikes before – it's an invention!

And there is a whole "bunch" of transformation models:

1) dominating model is *03. Segmentation* – to invent this "skin" we had to imagine a tire consisted of two "skins" – the main lower and the attached upper;

2) next important transformation is *08. Periodic action* – to transit from a continuous function to a periodic one with periodically changing "skins";

3) the model *18. Mediator* could be also interpreted as dominating but together with *10. Copying* – temporary to attach additional "skin" with copying the winter tread pattern.

### Example[55] 3.3. Bridgestone's Airless Tires Will Soon Let Cyclists Abandon Their Bike Pumps.

First revealed way back in 2011, Bridgestone's airless tires (fig. 3.43) use a series of rigid plastic resin spokes to help a wheel keep its shape as it rolls, instead of an inflatable inner tube that can puncture and leak. Military vehicles have been some of the first vehicles to adopt the unorthodox design, but Bridgestone will soon be making a version of its airless tires for bicycles.

Airless bike tires aren't a new idea (look below), you can already get wheels made from a solid rubber composite if you'll be riding on terrain where the risk of punctures and flats is high. But Bridgestone's approach, which replaces the inner tube and a portion of a bicycle wheel's spokes with thermoplastic resin supports, is better engineered to absorb bumps and provide an overall smoother ride, without ever requiring the rider to have to adjust the air pressure in their tires.

---

[55] fragments of pictures and text cited from https://gizmodo.com/bridgestones-airless-tires-will-soon-let-cyclists-aband-1794492775

So, cyclists would no longer have to carry a pump and spare inner tube(s).

And again we can recall the very old inventions[56] of airless wheels to reveal *the spiral law of evolution.* Working for this book with a purpose to compliment the students' essays I have found new information about such bikes (fig. 3.44) at wonderful website[57]. The *1905 Herrenrad Victoria "Model 12"* with Spring Wheels was invented to exclude inflatable tires from bike used at army. This bike was produced by one of the most traditional German bicycle brands *"Victoria"* founded in 1886 in Nuremberg by two bicycle enthusiasts *Max Frankenburger and Max Ottenstein.*

Nevertheless the first invention in this kind of airless tire (fig. 3.45) *Tweel* was made by the French tire company[58] *Michelin* in 2006. The *Tweel* is a portmanteau[59] of *tire* and *wheel.* Its significant advantage over pneumatic tires is that the *Tweel* does not use a bladder full of compressed air, and therefore cannot burst, leak pressure, or become flat.

**fig. 3.43.** *Bridgestone's Airless Tires*

**fig. 3.44.** *1905 Herrenrad Victoria "Model 12" with Spring Wheels*

---

[56]  Orloff, M. *Modern TRIZ. A Practical Guide with EASyTRIZ Technology.* – SPRINGER: New York, 2012. – 465 pp.

[57]  many thanks for samples from https://oldbike.eu

[58]  https://www.michelintweel.com/aboutTweel.html

[59]  https://en.wikipedia.org/wiki/Tweel

Instead, the *Tweel's* hub[60] is connected to the rim via flexible polyurethane spokes which fulfill the shock-absorbing role provided by the compressed air in a traditional tire. All these airless wheels became possible only in our time, at a new turn of development of initial idea during almost 115 years, thanks to new materials. Together with this, of course, we can see essential contribution of inventive designing into structure and shape.

**fig. 3.45.** *Michelin Tweel at work*

### 3.1.4. Velomobiles

A *velomobile*, or *bicycle car*, is a *human-powered vehicle (HPV)* enclosed for aerodynamic advantage and protection from weather and collisions.

They are similar to *recumbent bicycles, tricycles* and *quadricycles*, but with a full fairing (aerodynamic shell). A fairing may be added to a nonfaired cycle, or the fairing may be an integral part of the structure, monocoque like that of an airplane. First velomobile[61] was proposed by *Charles Mochet* who built in early 1930s many models of small vehicles called *"Velocar"*. Some models had two seats, most were pedal powered with steering using bevel gears, but as the years went by, many were fitted with small engines.

And today we can see not only the pictures[62] (fig. 3.46) but also a nice replica, carefully reproduced by enthusiasts (look website).

**fig. 3.46.** a) Early 1930s *Velocar* Serie and b) 1931 *Velocar Confort-Camionnette pickup truck*

---

[60] https://www.youtube.com/watch?v=27iJWJXBmnc
[61] https://en.wikipedia.org/wiki/Mochet
[62] cited thankfully from very informative website https://horseonline.site/mochet.htm

There are so many modern models of velomobiles ready to buy and use! This is **the spiral evolution** using modern materials and components! You can verify this by looking at several models e.g. from Norway[63], Netherlands[64] and Germany. Today we can see models of cycle cars of three classes: 1) *classic* only with pedals; 2) *hybrid* – pedals plus an electric motor, and 3) *modern e-car* – with an electric motor and without pedals – it is a fully electrified car, only relatively small in size, so you can remember that it is developed from a bicycle.

**fig. 3.47.** Hybrid E-quadricycles from Norway:
left – commuter *CityQ* and right – cargo *CityQ*

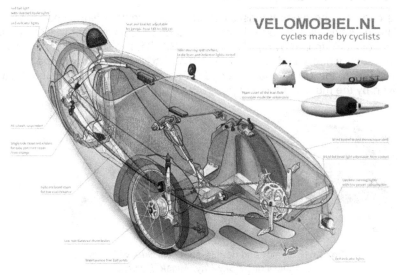

**fig. 3.48.** Layout of the main components of the classic
velomobile-recumbent *Quest* by *Ymte*

---

[63] CityQ - German Flyer 2018_final.pdf and https://www.cityq.biz
[64] http://www.velomobiel.nl/quest/ and https://velomobil.blog/ueber/velomobil_knowhow/

Many manufacturers of modern velomobiles strive to make their cars very comforttable, protected from wind and atmospheric precipitations, due to new materials for the chassis, hull and cab (fig. 3.49[65], fig. 3.50[66]).

**fig. 3.49.** Hybrid Sport-velomobile *Alleweder 6* by *Akkurad GmbH*

**fig. 3.50.** Classic *Velocar Leiba* by *LEIBA Velomobile GmbH*

And now we recall that in the creation of the movement of the bicycle, and the car, the train, two working bodies participate: WB1 – the wheel and WB2 – the road. And the limitation of the speed of the pedal bike is essentially determined by large losses of energy to overcome the *rolling friction resistance* of the wheel on, in general, *uneven* roads.

We do not discuss here the limitations of power provided by a cyclist in a comfortable mode, not in racing competitions.

But we can bring here one futuristic opportunity to create a *high-speed road* for the classic, hybrid and of course modern e-cars, to show a quite realistic engineering solution[67] in improving the road as WB2. So, it is light and compact vehicle on steel wheels and on string rail (fig. 3.51)! In addition to the built-in power sources, the *UniBike* is equipped with a bicycle generator so that it can be driven by the muscular force of passengers.

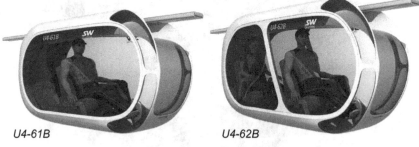

*U4-61B*                                      *U4-62B*

**fig. 3.51.** Single- and double-seat *UniBikes* with bicycle generator by *SkyWay, Minsk, Belarus*

---

[65]  http://www.akkurad.com/velomobile.html
[66]  http://www.leiba.de/ with picture from http://www.velomobile-france.com/pages/infos-vehicules/velomobile-leiba.html
[67]  http://yunitskiy.com/ news/2016/news20160310_en.htm

The *UniBike* here is a portmanteau of the name of string technology inventor *Unitsky* and *bike*.

A string rail[68] (fig. 3.52) combines the features of a flexible thread (at a large span between supports) and a rigid beam (at a small span – under a rail automobile wheel and above the support). A string rail is an ordinary uncut (along its length) steel, reinforced concrete or steel-reinforced concrete beam, or truss, equipped with a rail head and additionally reinforced with pre-stressed (stretched) strings.

*A flat rail head and a cylindrical steel wheel provide minimal energy consumption for movement.*

It make possible to drive on this extremely smooth (even) "rail road" much speedy than on any other "road".

Many other models could be constructed on the base of *UniBike* principle[69].

And to complete this part we have to note that the *UniBike* can also work without mechanical transmission from the pedals to driving wheel but due rotating the pedals to produce electrical energy with alternator built into the pedal hub!

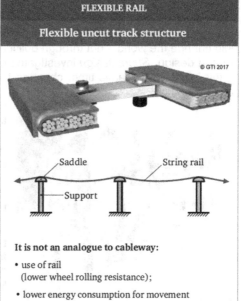

FLEXIBLE RAIL

Flexible uncut track structure

© GTI 2017

Saddle · String rail · Support

It is not an analogue to cableway:

• use of rail
(lower wheel rolling resistance);

• lower energy consumption for movement
(by 3–5 times less);

• possibility of using a gravity engine and gravity brake
(reduced energy consumption by another 3–5 times);

• high durability (by 5–7 times higher).

Motion speed:
from 30 (on support) up to 150 km/h.

Relative structural rigidity:
1/100–1/500.

Track structure curve radius:
$R = 100$ (on support)... 2,000 m.

**fig. 3.52.** Principle of flexible string-rail organization

This idea was implemented e.g. in the solution for the e-bicycle *Mando Footloose* (look earlier at fig. 3.20).

Of course, *UniBike* of this design is intended for movement only on string roads, however, we showed this example to demonstrate that, ***according to the spiral law,*** the "ancient" bicycle principle is applicable for ***the new roads!***

---

[68]  http://yunitskiy.com/news/2017/news20170809_en.htm
[69]  https://www.skyway-uk.com/news_list/3/Unibus

## 3.2.  Engineering and Art of the "Chairless Chair" Inventing

### Story

What[70] started as a rally against the establishment, the *Radical Italian Design* movement (*architettura radicale*) was formed out of the belief that creation could change the world – not through political acts or war, but through architecture and design. *SuperDesign* investigates how and why these designers, during this short yet dynamic time, changed the way we think about the objects that surround us.

**fig. 3.52.** Remembering the ... future: Artifact-prototype *"Wearable Chair"* – art-work by *Gianni Pettena*, 1971

Featuring original perpetrators of the Radical movement – *Gianni Pettena* and *Lapo Binazzi* – the panel is moderated by leading Italian design curator *Maria Cristina Didero*, alongside film Director *Francesca Molteni,* whose documentary will launch in 2017.

And we can see (fig. 3.52) one of the design works[71] by *Gianni Pettena*, artist and architect, as *"Wearable Chairs"* performance in Minneapolis, Minnesota, 1971.

Such wearable chair may be considered as an artefact-prototype – heavy and uncomfortable to wear.

As a design idea and an idea for performance in 1971, such a chair was wonderful, and today such a chair we take it to consider as a prototype of industrial use – uncomfortable to wear. Let it be here as a kind of *"remembering the ... future"* and example of the *spiral development of many technical systems*.

If you work[72] somewhere such as a factory, warehouse, or restaurant kitchen, then you'll know how tiring it can be to stand for several hours at a time. Unfortunately, however, it isn't always practical or safe to carry a stool around with you wherever you go.

That's why *Keith Gunura* started developing the *Chairless Chair* in 2009, when he was a student in the *Bioinspired Robotics Lab* at the *ETH Zurich* research institute. He was inspired to do so by memories of his first job, in which he worked while standing at a packaging line.

Worn as an exoskeleton on the back of the legs, it lets you walk or even run as needed, but can be locked into a supporting structure when you go into a sitting position.

---

[70]   cited from www.expochicago.com/programs/dialogues/dialogues-2017/friday-sept-15
[71]   picture fragment cited from www.youtube.com/watch?v=Ch5ptpbL4g4
[72]   text cited from https://newatlas.com/noonee-chairless-chair-sitting-exoskeleton/33459/

After a long period of development and improvement, the *Chairless Chair 2.0* is being actively developed and marketed by German company[73] with the same name *Noonee*. The device (fig. 3.53) utilizes a powered variable damper to support the wearer's body weight. The user simply bends their knees to get themselves down to the level at which they'd like to sit, and then engages the damper. The *Chairless Chair* then locks into that configuration, directing their weight down to the heels of their shoes, to which it is attached – it also attaches to the thighs via straps, and to the waist using a belt.

**fig. 3.53.** Modern realization of the artifact-result *"Chairless Chair 2.0"* by *Noonee*

A wearable seat[74] especially aimed at workers who are on their feet all day with nary a place to sit down. The *Chairless Chair 2.0* has moved into production and has been put to use at major companies[75] like *Daimler* and *Audi*.

### Extracting and Reinventing

Based on the purpose of application and understanding of the main useful function, we write down the prototype contradictions.

### Informal *Standard Contradiction (SC)*:

Wearable chair ▶ ☺ light and convenient **VS** ☹ durable and reliable

***Radical Contradiction (RC)*:** Wearable chair ▶ light (to wear)
                                       **VS** heavy (for durability)

***Ideal Final Result (IFR)*:** chair shouldn't have too heavy a back, seat and legs
                        **BUT** a function to support the body should be realized

***Radical IFR (RIFR)*:** chair ***should not be*** (we need *"**chairless** system"*)
                     **BUT** function of chair ***should be***

So, according to RIFR we need *"chairless chair"*! What could it be?

The *Noonee* has shown us the answer! Let's see combined and simplified Extracting with Reinventing (fig. 3.54) as an answer from TRIZ in modeling creative content of *Noonee* solution!

---

[73]  https://www.noonee.com/en/
[74]  newatlas.com/wearable-chairless-chair-2018/53521/
[75]  www.youtube.com/watch?v=ZTnoiHd7hFo

**ARTIFACT**                    **Chairless Chair**

**WAS:** Artifact-Prototype                                        **IS:** Artifact-Result

Picture: www.youtube.com/watch?v=Ch5ptpbL4g4

Picture: newatlas.com/wearable-chairless-chair-2018/53521

| **Wearable chair** | **Chairless Chair** |
|---|---|
| Known wearable chairs are relatively heavy and uncomfortable to be durable and reliable. | Invention of *Noonee* has supplied a chair-without-chair as system of levers with seating function and stable chair function to wear and to seat for work. |

**FSC**          Wearable chair ▶ 32. Weight of movable object **VS** 04. Reliability
                 ▶ S03, S12, S13, S28 (from A-matrix)

| D | Navigators from Extracting |
|---|---|
| ++ | 03. Segmentation – it matches the A-matrix |
| ++ | 07. Dynamization – it is obtained only from Extracting |
| ++ | 10. Copying – it is obtained only from Extracting |
| ++ | 12. Local property – it matches the A-matrix |

**Re-INVENTING**

All the models are dominant. **Models 03 and 07** recommend dividing the initial chair in movable parts. **Models 10 and 12** recommend finding the most effective place to support the body with seating function similar to legs functioning. Therefore, parts of the new system copy the structure and lines of the legs and attach to the legs. And on top there is a platform for a seat.

The system has built-in electric drives to facilitate changing the position and fixing the working position while sitting. According to new *Noonee* enthusiasts and producers: *Creating an ergonomic, comfortable, and even more productive workplace without extensive changes at the workplace – thanks to the Chairless Chair, this is now possible. All it takes to work in a healthy position is putting on our device, which only requires a few seconds. The Chairless Chair allows you to effortlessly switch between an active sitting, standing, and walking position.*

**fig. 3.54.** Combined & simplified extracting-reinventing of the
"Chairless Chair"; **FSC** in fig. – Formal Standard Contradiction

**Post-story**

To complete this part about TRIZ-modeling the creative content at this innovation, it should be useful to add some other aspects of inventing and development of this wonderful device. *This is about motivation to invent it and about passion to follow this way.*

Such aspects are especially useful represented in concentrated form to support and motivate the students of Master programs to seek for an opportunity for inventions and innovations.

And here I would like to show a few points from an interview[76] with Keith Gunura to *The CEO Magazine* of 17 December, 2014. We consider all subsequent text as one cited fragment, without dividing it into the original and added fragments, not to make the text intermittent with too many quotation marks.

*"Noonee* is rethinking the way we view furniture with an innovative *"chairless chair"* that is heralding a new era of wearable ergonomic leg devices. It's an innovative and forward-thinking concept: the ability to sit anywhere and everywhere with the aid of such device.

The initial idea was sparked when I got invited by my old professors to one of the lab parties and we started talking about commercializing or trying to get something out of the labs because everything just goes on the shelf. We were just randomly chatting and we both agreed that we wanted to be able to sit anywhere and everywhere.

We were complaining about standing at the train station and having to stand in line in queues. We thought maybe we should make something that people could use in this area.

Of course, in my head I was already thinking of an exoskeleton. You have to think like the person you want to help.

My philosophy is that it might sound crazy now but in the future it's not going to be as crazy. When I think about my idea, I don't even think of it as a stupid or impossible idea. I just think to myself, if people don't accept this then it's the wrong time for such an idea.

My philosophy is to not be afraid of the big things or the stupid questions – just do what you want to do and be happy with that.

After that, there are many other ideas that we believe would contribute to the success of *Noonee.*

> For instance, implementing some more innovative ideas in the future like a full-body exoskeleton.

---

[76] global magazine of information, inspiration and motivation for the world's most successful leaders, executives, investors and entrepreneurs with headquarters in Sydney, Australia
https://www.theceomagazine.com/executive-interviews/engineering/keith-gunura/

The concept itself is that it's supposed to be low-cost and easy to use. We believe that's our strength. I think it's going to push all the other exoskeleton markets to lower their prices to make them competitive.

> I think one day an exoskeleton will be like a bicycle
> that anybody could buy one.

You need the right strategic partners— people who share the same vision as you; people who understand your product and see where it's going. It makes things a lot simpler when you're explaining things to them or negotiating. These could be the people that you merge with in the next 10 years or so, so it's important to have strong relationships with them.

The importance of your partners depends on their reliability. We include them in our development. For instance, our designer, *Sapetti Design*, has been involved and immersed in our development. Our relationship is open and we understand each other's needs fully.

Developing our partnerships is about making sure that we're not just thinking about our success but about our suppliers' and partners' success as well.

We're the middleman when you think about the money flow.

We want three parties to be successful: the consumer, us, and our partners. That's how we think it will grow and to do that we need to handle the needs of each of these parties."

There is also an excerpt[77] from *Sapetti Design,* important for such Master program as *Global Production Engineering* to motivate and prepare the students for team cooperation:

"We believe that design should not only be understood as a 'beautiful shape', but rather as t*he sum of different factors such as engineering, aesthetics, user experience/ergonomics, and communication.* This synergy between design and engineering leads to innovative solutions and well-thought out products.

Key ingredient to success: *Holistic approach to design. Collaboration. Innovation. Positive thinking and hard work. We believe these are the drivers of our work.* In a few words, we strive to become a key partner for innovation and growth, someone the clients can rely on for their next generations of products."

Also the today's *Noonee*-developers[78] say: "One thing is for sure, demographics are changing. Employees are getting older, which puts them at a higher risk of suffering from backpain. Diseases of the muscle tissue and the skeleton are already responsible for one fourth of all sick days.

This creates enormous costs, and therefore, *we need a clever solution – and we don't need this solution someday, we need it today!* "

---

[77] www.sapetti.com
[78] https://www.noonee.com/en/

# 4 ENERGY ENGINEERING[79]

## 4.1. Ocean Current Power Generation System[80]

### Story

There are so many underwater power generators invented for last 50 years! And remembering that the development of water wheels as turbine's prototype has a history of many thousands of years, it is easy to understand that *the spiral development* of hydro-turbines continues successfully today!

Really it is interesting and important because of underwater energy generating could potentially less depend from weather change as at wind power generating or even for hydroelectric stations with dam.

*Marine hydrokinetic energy* producing[81] uses the following ideas:

- *Wave power* is the transport of energy by wind waves, and the capture of that energy to do useful work with a wave energy converter – for example, direct electricity generation or pumping water into reservoirs;
- *Tidal power* turbines are placed in coastal and estuarine areas and daily flows are quite predictable;
- *In-stream* turbines in fast-moving rivers;
- *Ocean current* turbines in areas of strong marine currents;
- *Ocean thermal* energy converters in deep tropical waters.

The intensive development of underwater energy generators is especially promising both because of the depletion of natural oil and gas reserves, and because of possible failures (accidents) at any other types of ground-based power plants.

*IHI Corporation* and *Toshiba Corporation*, together with the *University of Tokyo* and *Mitsui Global Strategic Studies Institute*, have conducted R&D financed by Japan's *New Energy and Industrial Technology Development Organization* (NEDO) in the project "R&D of Ocean Energy Technology – R&D of Next-Generation Ocean Energy Power Generation (Underwater Floating Type Ocean Current Turbine System)" since 2011[82].

*IHI* is the lead company in the co-research project and will manufacture the turbine and floating body. *Toshiba* will supply electric devices, such as the generator and transformer.

Within this framework, the unique "underwater floating type ocean current turbine system" developed by *IHI* and *Toshiba* demonstrates power generation in a real ocean environment.

---

[79] MTRIZ course is provided for international Master of Science Program "Energy Engineering" at El Gouna Center of TU Berlin, Egypt

[80] this work could be considered as co-operative with GPE Master works

[81] https://en.wikipedia.org/wiki/Alternative_energy#Marine_and_hydrokinetic_energy

[82] https://www.toshiba.co.jp/about/press/2014_12/pr2501.htm

© Springer Nature Switzerland AG 2020
M. A. Orloff, *Modern TRIZ Modeling in Master Programs*,
https://doi.org/10.1007/978-3-030-37417-4_4

New approach uses two important ideas. First one is technical and second one is physics-hydraulic and geographical.

Technical idea uses the underwater floating type ocean current turbine system (fig. 4.1) as a power generation device with two counter-rotating turbines. It is anchored to the sea floor and floats like a kite carried and driven by the ocean current.

**fig. 4.1.** Underwater Floating Type Ocean Current Turbine System
of *IHI* and *Toshiba*

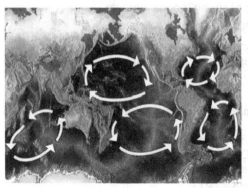

**fig. 4.2.** Kuroshio Ocean Currents

"Geographical" idea uses so called *Kuroshio Ocean Current*[83] as a natural energy resource with little fluctuation in flow regardless of time or season. It is a north-flowing ocean current on the west side of the North Pacific Ocean. It is similar to the Gulf Stream in the North Atlantic and is part of the North Pacific ocean gyre. Like the Gulf stream, it is a strong western boundary current (fig. 4.2).

### Extracting and Reinventing

Combined and simplified Extracting with Reinventing (fig. 4.4) uses a simple cable tethered turbine[84] (fig. 4.3) as a prototype-artifact.

**fig. 4.3.** A cable tethered turbine

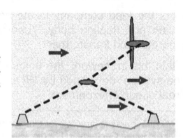

---

83   https://en.wikipedia.org/wiki/Kuroshio_Current
84   https://en.wikipedia.org/wiki/Tidal_stream_generator

| ARTIFACT | Underwater Ocean Turbine System of IHI and Toshiba |
|---|---|

**WAS:** Artifact-Prototype                                                      **IS:** Artifact-Result

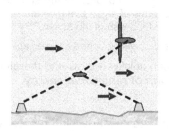

Picture:
en.wikipedia.org/wiki/Tidal_stream_generator

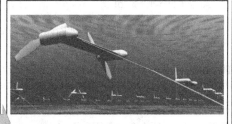

Picture:
https://www.toshiba.co.jp/about/press/2014_12/
pr2501.htm

**Simple tidal stream generator**

This turbine has a single rotor set with fixed pitch blades. During rotation, it can change orientation due to water pressure and deflecting forces.

**Twins generator by IHI and Toshiba**

This generator uses two counter-rotating turbines mutually to compensate the deflecting forces. Next idea is using the kiting principle to raise turbines above the bottom under the ascension of a constant ocean current, here *Kuroshio Ocean Current*.

| FSC | Stream generator ▶ 39. Loss of energy **VS** 31. Tension, pressure |
|---|---|
|  | ▶ S11, S24 (from A-matrix) |

| D | Navigators from Extracting |
|---|---|
| ++ | 10. Copying |
| ++ | 11. Inverse action (from A-matrix) |
| + | 12. Local property |
| + | 14. Use of pneumatic or hydraulic constructions |
| ++ | 18. Mediator |
| ++ | 32. Counter-weight |

## Re-INVENTING

**Model 10** recommends duplicating the initially single generator to supply together with **model 11** stable positioning of the system due to counter-directed rotating the blades of twins-system.

**Model 12** requires fixing the system relatively high above the bottom of the ocean. For this, the system contains a platform with a function of a kite (**model 10** again) according to **model 18** for creating lifting force from the influence of the current according to **model 14** (the effect of water as the effect of wind on a kite) and **model 32**.

**fig. 4.4.** Combined & simplified extracting-reinventing of the "Twins generator by IHI and Toshiba"

## 4.2. Algae-fueled Bioreactor[85]

### Story

> About two-thirds of the oxygen in our atmosphere is produced in the surface waters of the sea by phytoplankton, the minute forms of algae that give the sea its slightly green hue, and which initiate the entire food web of the ocean.[86]
>
> *Jacques-Yves Cousteau, marine biologist & oceanographer*

The biomass used in the production of biofuel can quickly regrow and therefore **the biofuel is generally considered to be a form of renewable energy.**

Biomaterials are divided into generations.

**The raw materials of the first generation** are crops with a high content of fats, starch, sugars. Vegetable fats are processed into biodiesel, and starches and sugars are converted to ethanol. Almost all modern transport biofuels are produced from first-generation raw materials, the use of second-generation raw materials is in the early stages of commercialization or in the research process.

**Non-food remains of cultivated plants, grass and wood are called the second generation** of raw materials. Getting it is much less expensive than that of the first generation crops. Such raw materials contain cellulose and lignin. It can be directly burned (as was traditionally done with wood), gasified (receiving combustible gases), and pyrolyzed.

**The third generation of raw materials is algae.** They do not require land resources, they can have a large concentration of biomass and a high reproduction rate.

**Fourth-generation biofuels are made using non-arable land.** They do not require the destruction of biomass. This class of biofuels includes *electrofuels* and *photobiological solar fuels*. Some of these fuels are carbon-neutral.

It is possible to produce different biofuels from biomass as biogas, syngas, ethanol, biodiesel, green diesel, strait vegetable oil, bioethers, etc. In 2018, worldwide biofuel production[87] reached 152 billion liters (40 billion gallons US), up 7% from 2017, and biofuels provided 3% of the world's fuels for road transport. *The International Energy Agency* wants biofuels to meet more than a quarter of world demand for transportation fuels by 2050, in order to reduce dependency on petroleum.

So, the artifact-invention *Eos Bioreactor* (fig. 4.5) sizing as 3 x 3 x 7 ft (90 x 90 x 210 cm) under studying is a new box-shaped machine[88] of the US company *Hypergiant Industries, Texas, USA,* **that can soak up as much carbon from the atmosphere as an acre of trees.**

---

[85]   this work could be considered as a development of GPE Master work (e.g. in part 8.1) in co-operation with the students of Master program "IT for Energy"

[86]   cited from https://www.hypergiant.com/green/

[87]   https://en.wikipedia.org/wiki/Biofuel#Generations

[88]   cited from https://newatlas.com/environment/algae-fueled-bioreactor-carbon-sequestration/

It comes to organic processes that we can leverage to tackle the runaway problem of climate change, the carbon-absorbing abilities of algae may be one of the most potent tools at our disposal.

For years, scientists have been studying this natural phenomenon in hope of tackling greenhouse gas emissions and producing eco-friendly biofuels (look also in part 8.1).

The reactor uses a specific strain of algae called *chlorella vulgaris*, which is claimed to soak up much more CO2 than any other plant.

**fig. 4.5.** *Eos Bioreactor* by *Hypergiant Industries*

The algae lives inside a tube system and water tank within the device, which is pumped full of air and exposed to artificial light, giving the plant the food it needs to thrive and produce biofuels for harvesting.

*Hypergiant Industries* claims[89]: The *Eos Bioreactor* will help solve the $CO_2$ problem by sequestering carbon more rapidly and more efficiently than trees - as each super-boosted algae bioreactor is 400 times more effective at capturing carbon than trees in the same unit area.

And by using machine intelligence to constantly monitor and manage air flow, amount and type of light, available CO2, temperature, pH, biodensity, and harvest cycles, we maintain perfect conditions for maximum carbon sequestration.

According to *Hypergiant Industries,* the whole process could be described in 6 stages (fig. 4.6):

### 1 AIR INTAKE

Air intake (this can be open air, or could hook up to building exhaust) The air is then bubbled into main tank to diffuse into the water/algae.

### 2 GROWING THE ALGAE

Algae wants $CO_2$ and light. The light can be from the sun, or in this case, artificial light. The algae and water are pumped through a series of tubes to maximize their exposure to light sources lining the inside of the Reactor.

### 3 BIOMASS ACCRETION

As algae consumes $CO_2$, it produces biomass that can be harvested and processed to create fuel, oils, nutrient-rich high-protein food sources, fertilizers, plastics, cosmetics and more.

---

[89]  https://www.hypergiant.com/green/

4 HARVESTING & SEPARATION

Harvesting is a separate system that is completely controlled by the AI in order to have the perfect amount of algae to suck up the most $CO_2$. Also we can swap out harvesters based on the intended use.

5 CLEAN EXHAUST OUTPUT

After the algae consumes 60-90% of the $CO_2$ and other pollutants found in the intake air as vital nutrients, clean oxygen-rich air is released.

6 COMPLETELY A.I. DRIVEN PROCESS

The A.I. optimizes growth of algae and $CO_2$ consumption and harvests the algae without human intervention.

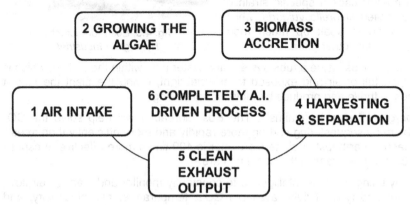

**fig. 4.6.** Integration AI with algae bioproducing is a key innovative move in *Eos Bioreactor* by *Hypergiant Industries*

**Extracting and Reinventing**

Based on the purpose of application and understanding of the main useful function, we write down the prototype contradictions.

**Informal *Standard Contradiction (SC)*:**

Simple bioreactor ▶ ☺ get more biomass **VS** ☹ difficult to operate and control

***Radical Contradiction (RC)*:** Bioreactor ▶ productive
        **VS** not productive (state of traditional manual technology)

***Ideal Final Result (IFR)*:** all essential parameters should be under the
        continuous control

***Radical IFR (RIFR)*:** operator ***should not be*** (for automation)
        **BUT** operator ***should be*** (to "teach" the control system)

Solution has come from *Hypergiant Industries* as self-learning control system through AI with unsupervised training (process 6 in fig. 4.6 and fig. 4.7)!

| ARTIFACT | Eos Bioreactor |
|---|---|

**WAS:** Artifact-Prototype　　　　　　　　　　　　　　　　　**IS:** Artifact-Result

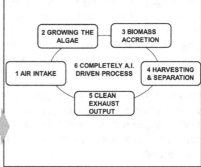

| **Simple laboratory algae cultivation** | **Eos Bioreactor** |
|---|---|
| Growing of algae requires a lot of manual work, while it is difficult to provide the process with the necessary amount of light and heat. | Algae growing in this reactor is supported by a control system with AI, which is leaned to create an adaptive process with the best parameters in different operating conditions of the reactor. |

**FSC**　　Simple bioreactor ▶ +01. Productivity

　　　　　**VS** -34. Temperature **&** -35. Brightness of the lighting

　　　　　▶ S01, S02, S03, S04, S08, S10, S19, S33 (from A-matrix)

| D | S–Navigators from Extracting |
|---|---|
| + | 01. Change in the aggregate state of an object – it matches the A-matrix |
| ++ | 04. Replacement of mechanical matter – it matches the A-matrix |
| + | 08. Periodic action – it matches the A-matrix |
| + | 09. Change in color – it is obtained only from Extracting |
| ++ | 10. Copying – it matches the A-matrix |
| ++ | 18. Mediator – it is obtained only from Extracting |

## Re-INVENTING

According to **models 01, 08** and **09** it is necessary to control the state of the internal environment of the reactor. It is necessary to periodically (**model 08**) change the illumination (**model 09**) and other environmental parameters (**model 01**). To do this, it is necessary to enter "smart control" at the AI level (this can be interpreted as using the **models 10** and **18**) to ensure dynamic (mobile) interaction of the control system with the environment (according to **model 04**). Of course, interpretation and use of the models require professional knowledge.

**fig. 4.7.** Combined & simplified extracting-reinventing of the "Eos Bioreactor"

### Post-story[90]

"*Hypergiant's* R&D team is designing a holistic energy concept that creates a stepping stone to reverse climate change. The key ingredient, algae, converts carbon dioxide into carbon-rich lipids to create biofuel. Every aspect of their algae bioreactor closed system will be automated and controlled by machine intelligence to make it easier to maintain, lowering any manual software burdens. *Hypergiant* cares about "tomorrowing today" and is building towards this bigger vision of collecting data from the interconnected network of algae reactors to globally communicate and continuously optimize devices.

Research shows that algae production is a highly effective means of atmospheric carbon reduction, outperforming trees by a factor of 10. And next generation of algae reactor is smaller and more modular than existing models, using machine intelligence to automate production, optimize output, and reduce maintenance.

Since our ability to affect change is driven by our ability to understand our impact, the future of climate change solutions must be highly connected and data driven. *Hypergiant's* bioreactor comes paired with mobile application that provides status of the bioreactor, detects anomalies, and provides current and historical reporting of $CO_2$ sequestration and biomass production. A cloud based infrastructure connects the bioreactors, allowing them to learn from each other, optimize for new environments, and provide global insights.

So how does the algae activation work? When energy is generated from a renewable source, it allows for consumption of $CO_2$ that was emitted before the gas escapes into the atmosphere. In other words, algae needs three key elements to grow: carbon dioxide, sunlight and water. *Hypergiant* is creating the next generation of algae reactors, smaller and more modular than existing models, using machine intelligence to optimize the benefits of releasing energy back into the system. The end project is a controlled closed system model as a testbed for this vision of the future. The team plans to tap into the *Heating, Ventilation and Air Conditioning (HVAC)* in the office, to use the building exhaust as a point of reference for $CO_2$ density. Best of all, it thrives in the carbon rich dirty air that comes from exhaust pipes.

Using object-based image analysis to analyze satellite imagery, the R&D department is able to seek accurate rooftop space calculations to house algae farms. Machine intelligence can use heat maps from pollution concentration to prioritize algae farm locations as well.

We believe[91] algae are a wildly underutilized natural resource that have untold of potential. The algae have the potential to become the foundation of an infinitely renewable, hyperlocal, and decentralized supply chain. Algae will serve as a source of food, oxygen, and raw materials in future space exploration."

---

[90]   cited about *Eos Bioreactor* from https://www.hypergiant.com/rd/
[91]   https://www.hypergiant.com/green/

## 4.3. Stop Losses – Make Profit

In subsequent cases, graduate students can see the use of the same models in different technical systems. Although, of course, cases with devices for generating electricity are selected here.

First, we will consider these cases, and then we will make a generalization regarding the TRIZ-models that are actually present in the considered technical solutions.

### Case 1. In-pipe hydro-electric power system and turbine by Lucid Energy

*Story*

The history of this case is instructive in itself, as it reflects the searches and inventions of many authors who proposed such ideas during last 50 years at least. The main thing in their direction of thought is the discovery of unexpected sources for generating electrical and other types of energy from existing, but unusual and insufficiently obvious systems for non-specialists.

The invention of company[92] *Lucid Energy Inc., Portland, Oregon, USA,* has become famous due to its nice implementation. It should be said that there is a lot of similar solutions (not shown here) that have not achieved marketing success, but are of interest for both energy engineering and TRIZ-analysis.

Lucid Energy's focus has always been on the development of hydropower technologies. Its initial efforts began in 2007 and focused on the development of small, hydrokinetic turbines for use in open waterways. But open waterways have a high degree of complexity: water flow is unpredictable and weather-dependent, turbines are subject to damage from debris, fish and other inhabitants of water ecosystems and anything placed in a river or stream can harm the environment.

"We knew there must be a better way. So we asked ourselves: Where is there a source of free-flowing water that is clean, controlled and outside of a natural habitat?

*The answer: Gravity-fed drinking water systems.*

For many cities and towns around the world, drinking water comes from mountain reservoirs and high elevation storage tanks. As the water flows downhill through gravity-fed pipeline systems, it builds up considerable pressure – pressure that can be far too high for residential and commercial water delivery. Water agencies use pressure reducing valves (PRVs) to remove some of that pressure, but all that energy potential is lost. There's an opportunity!

By putting turbines inside of those gravity-fed pipelines, we knew we could produce reliable, renewable hydropower that doesn't harm the environment. There was only one catch: how to do it without removing too much pressure or disrupting water delivery.

---

[92] story contains the fragments cited from http://lucidenergy.com/company/

It was time to talk to the water people.

In 2008 Lucid Energy formed a strategic relationship with Northwest Pipe Company, the largest manufacturer of steel water transmission pipe in the United States. This resulted in our introduction to water agency managers across North America. We learned about their needs and concerns. They made it abundantly clear that our product needed to work without impeding their ability to deliver clean drinking water at the proper pressure.

Out of this collaboration came the innovation and development of the spherical, LucidPipe in-pipe turbine design that harvests some of the excess pressure in gravity-fed pipelines and turns it into renewable energy – without disrupting water delivery.

Between 2010 and 2011 we successfully field-piloted three iterations of the 42" turbine design at RPU. We received our first patent in June 2011 and incorporated as Lucid Energy, Inc. in August of that same year. We secured a second patent in January 2013 and are in the process of filing international patents in countries around the world."

The company is now developing the second generation of LucidPipe turbine design. *This design will utilize advanced manufacturing technologies, significantly increase power output and operate across a wider range of pipe sizes and flows, and significantly reduce capital cost.* Dollars-per-kilowatt is its key metric.

The *Gen 2 LucidPipe Power System* will enable Lucid Energy to reduce costs and expand the potential number of installation sites around the world, *helping the company reach its goal to turn the world's gravity-fed water pipelines into generators of clean, renewable energy.*

### Patent and Implementation

We have extracted here only several TRIZ-models from the Lucid Energy patent[93].

First one is itself inventing the idea to realize new functioning as an electrical energy generation from existing water (gas) flow in pipes in accordance with the model **05. Separation:** *separate the "incompatible part" ("incompatible property") from the object or – turned completely around –* **include the only really necessary part (necessary property) into the object.**

Second one is the model **10. Copying:** *d) apply copies, duplicates of the same or another object.* This is a possibility to use turbine connected with generator to produce electrical power from water flow that rotates the turbine.

Next solutions are in connection with the models **12. Local property** and **19. Transition into another dimension:** to use the special device (axis, generator, etc.) mounted on the outer surface of the pipe with the placement of the turbine itself inside the pipe.

---

[93] Patent No.: US 8,360,720 B2, Date of Patent: Jan. 29, 2013

And of course an installation[94] of the turbine inside the pipe (fig. 4.8,a) is clearly the model ***34. Matryoshka (nested doll): a) an object is inside another object that is also inside another, etc.; b) an object runs through a hollow space in another object.*** This is the most important technical idea according to story of the engineers searching.

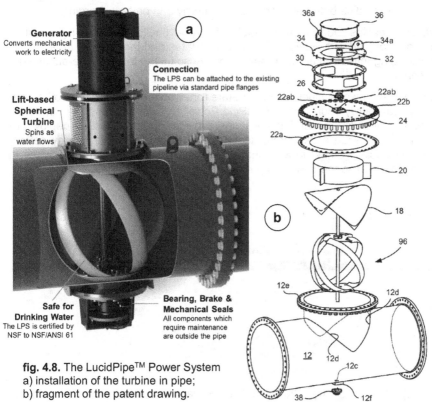

**fig. 4.8.** The LucidPipe™ Power System
a) installation of the turbine in pipe;
b) fragment of the patent drawing.

And now we are extracting the key "philosophical" model for entire system concept – ***21. Transform damage into use: a) use damaging factors, especially damaging influences from the environment to achieve a useful effect,*** which means in our case to transform the water flow kinetic (gravitational) energy, which is always lost, into useful work to rotate the turbine with generator and produce the electrical power!

Really, there are the advantages of LucidPipe:

➢ Hydropower that doesn't harm ecosystems;

➢ Can generate consistent, predictable energy 24/7;

---

[94]  Pictures and some text fragments are cited from http://lucidenergy.com/how-it-works/

> No impact on water delivery;

> *Turns excess pressure into a revenue stream through power purchase agreements;*

> Provides grid-connected or off-grid power;

> Use for distributed electricity, peak-energy and battery charging.

Water agencies today face increasing financial challenges. The high cost of energy, coupled with energy efficiency mandates and the need to repair or replace aging infrastructure all require creative solutions to keep operations sustainable.

By using their water pipelines to generate renewable energy from an otherwise untapped energy source, the LucidPipe Power System can be part of the solution. Pipeline repairs and installations provide opportunities to deploy LucidPipe on a wide scale, producing megawatts of renewable energy nationwide, especially taking in account the possibility of using several turbines in series to increase the power output at one place (fig. 4.9).

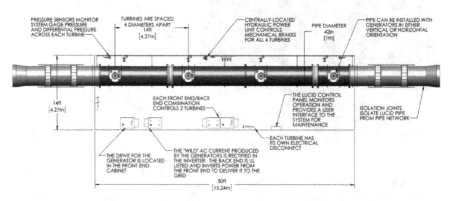

**fig. 4.9.** Turbines can be placed in series.

## Case 2. Electric power generation from sewage water

Micro-turbines, which can be installed in such unusual systems as sewage pipelines in tall buildings and underground, also can have a very large cumulative effect[95]. Calculations show that the generated electricity from wastewater of the standard "nine-story building" could be enough to illuminate, for example, stairwells and porches. Such an approach was also proposed many decades ago by the famous inventor in *Kazakhstan*, professor of the *Karaganda State Technical University, Santay Suleymenovich Zhetesov*, whom I met five years ago during my master classes at KSTU (fig. 4.10).

---

[95] *Wastewater Energy: themes of final qualifying works in the direction of "Services of engineering systems"* – look at https://asb-school-24.ru/pedagogi/tolkachev/zip/stoch-voda.pdf

A particular problem is the possible presence of large contaminants in waste water, such as rags and plastic, which can clog turbine blades. However, a partial solution can also be obtained in the style of TRIZ-thinking (fig. 4.11), presented in one of the Master-student works[96].

The turbine in the figure 4.11 on the right (b) has rotating sub-blades.

**fig. 4.10.** Prof. Santay Zhetesov during MTRIZ-modeling of his inventions with master-students

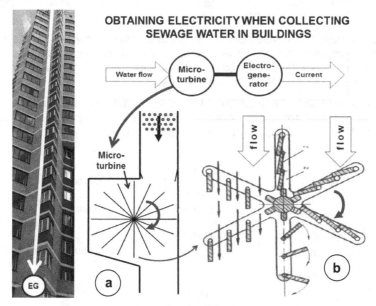

### OBTAINING ELECTRICITY WHEN COLLECTING SEWAGE WATER IN BUILDINGS

**fig. 4.11.** Turbines with conventional (a) and dynamized (b) blades

When the blade moves against the stream, the blade is "opened" and passes the stream. When the blade moves in one direction with the flow, the blade is "closed" and receives the flow pressure to rotate the turbine.

Such a turbine has high efficiency and speed, and at the same time it is cleaned of debris, because it dumps debris during rotation. In addition, it is possible to use different methods of preliminary detention of large garbage.

---

[96] interestingly, when checking this link, I found the educational review work of an 11th grade student: https://docplayer.ru/39345370-Ispolzovanie-stochnyh-vod-dlya-vyrabotki-elektroenergii.html

Extracting the useful TRIZ-models: *03. Segmentation* (to divide the blade into many rotating parts), *07. Dynamization* (rotation of the parts of blade), *12. Local property* ("opened" and "closed" blades), *21. Transform damage into use* (split blade improves overall turbine performance), *24. Asymmetry* (one side of turbine is "opened" and second one "closed" to get an effective stream pressure distribution), *29. Self-servicing* (self-cleaning), *34. Matryoshka* (blade elements are put in its profile).

### Case 3. Utilizing low-potential secondary energy resources[97]

*Existing problem*

Pressure of natural gas on the way from the gas production place to the consumer decreases in several stages: at gas distribution stations  from 2,0 – 7,5 MPa to 0,3/0,6/1,2 MPa, then on gas pressure reducing station and gas pressure reducing units up to 0,005 – 0,6 MPa.

Pressure regulators provide necessary output pressure, however diffuse energy of gas compression to the environment. As a result of it on a global scale, in pressure regulators the energy comparable to energy consumption of Belarus, which is about 36 billion kW/h, is lost annually.

At compressor stations to increase gas pressure (when transporting) from 4 to 8% of its expense are spent. Rated capacity of such compressor stations is from 6 to 12% of rated capacity of a power supply system.

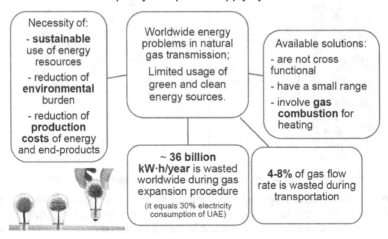

**fig. 4.12.** Losses of energy at the natural gas transportation

*Solution ways*

Limited use of the existing solutions and turbodetanders is connected with a large number of the various consumers what needs an individual approach.

---

[97]  cited from materials at http://ts.energy/en/turbogenerator-unit-turbosphere

The existing competitive elaborations:

- calculated for its use generally on gas pressure regulating station while gas pressure reducing station and gas pressure reducing units have bigger energy potential;

- work in narrow range of pressure and consumption of natural gas;

- problems with gas heating are not solved – gas either isn't warmed up, or warmed up in the portable heat exchanger with gas combustion and, as a result has additional expenses;

- the majority of units are high-speed that has an adverse effect on their working resource and also raises requirements to maintenance and preliminary purification of natural gas;

- need of heating and purification of gas increases investment expenses because it needs acquisition and service of the additional equipment (gas heaters, filters, etc.).

### *Real solution "Stop Losses – Make Profit"*

The **turbogenerator** *units TurboSphere* (**TGU "TurboSphere"**) is the innovative development created by *TurboSphere group* of companies: LLC Scientific and Engineering Center *"EnergoTech"* (the resident of Science and Technology Park of BNTU[98] "Polytechnic", *Minsk, Belarus*) and Skolkovo (*Moscow, Russia*) project participant LLC *"TurboEnergy"*.

This product is intended for a generation of additional electric energy at the expense of pressure difference of natural gas that allows increasing fuel efficiency, to receive an environmentally friendly electric power source and also additional income for the enterprise, equivalent to annual cut on electric energy purchase from external network, or by its sale to a third-party consumer!

TGU "TurboSphere" represents the innovative solution applied at Gas Pressure Regulating Station, Gas Pressure Reducing Station and Gas Pressure Reducing Units in gas transmission systems, the industry, the municipal sector and energetics to productively use energy of natural gas overpressure energy.

The main difference of the TurboSphere units from turbodetanders and detander-generating units is that it is capable to work in the wide range of expenses and gas pressure, keeping at the same time both required parameters and quality of the generated electric power; it is reliable, rather inexpensive unit, the minimum requirements to technical services.

TurboSphere is applicable for both autonomous power supply where the main goal is to ensure needs for the electric power of object's own needs and for parallel work with an external network when the purpose is to generate maximum power using all potential from gas stream while delivering electric power to an internal network of an enterprise and with sale (if needed) of overpressure energy to an external network.

---

[98] Belarusian National Technical University (BNTU), Minsk, Republic of Belarus

The key advantages are reached by:

> The relative simplicity of construction due to minimization of quantity of details and elements and also a turbogenerator low speed (the synchronous rotating speed - 3000 RPM);

> Implementation of the diagram (fig. 4.13) of multistage extension of gas on one driving wheel with a possibility of the intermediate heating of gas in the course of extension by means of the built-in heat exchanger heater, using low-potential heat and thermal waste of the enterprises;

> Unit is intended for operation not only on natural gas, but also with others non-agressive gases.

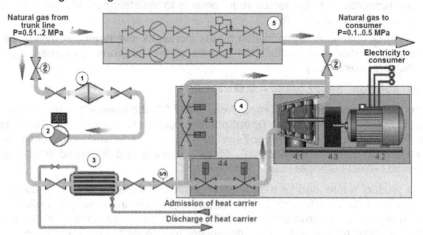

**fig. 4.13.** Work scheme of gas utilization unit in the system *TurboSphere* – flows of energy and energy-supplying materials: 1) filter; 2) gas supply meter; 3) gas heater; 4) gas utilization unit (turboexpander); 5) gas regulation station (GRS)

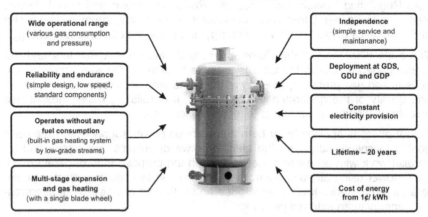

**fig. 4.14.** Main advantages for the system of energy output (5-500 kW)

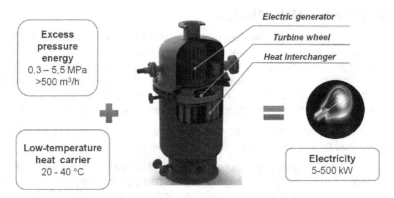

fig. 4.15. 3-in-1: turbine + heat exchanger + electric generator

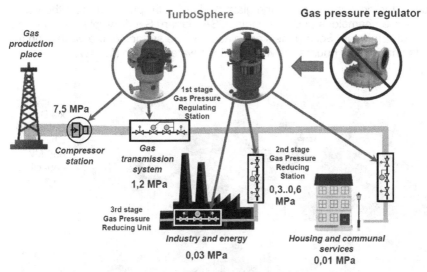

fig. 4.16. Worldwide solution for clean electricity production

So you can make a profit at generating electrical energy
WITHOUT ADDITIONAL FUEL COMBUSTION
but at the expense of
PRESSURE DIFFERENCE of NATURAL GAS !

Is it necessary to say which TRIZ-models are in action here? Or the readers can extract these models ourselves?

Summarizing everything considered here, we note that model **21. Transform damage into use** can be recognized as the most valuable.

Isn't it?!

## 4.4. Stop Sunlit – Make Shadow

About half of the solar radiation in the summer months and about 40% on the hottest days are sufficient for normal ripening of agricultural plants, even at the latitude of Germany. The same applies to all mid-latitudes on all continents and both hemispheres of the Earth. What does it mean? This means that agriculture remains one of the most conservative spheres of human activity, the most costly and wasteful, very harmful ecologically.

So what do plants need? They need a shadow! They will not dry out and be burned by the sun! They will need less water!

They need an adjustable amount of sun and shade! This means that ***agriculture should become a highly organized controlled production***, little dependent on the vagaries of Nature. And this corresponds to the fundamental TRIZ-trend – ***the growth of controllability!***

Today it is no wonder to see the giant fields of solar panels around the world. There are champions in China, India, USA, Japan, Saudi Arabia and the United Arab Emirates. Solar panels are placed on the surface of the earth[99], on the roofs and walls of houses, as well as on the water surface (fig. 4.17).

Longyang-xia Dam Solar Park, China, with 850 MW capacity at 23 sq km

Pavagada Solar Park, India, with 2 GW capacity at 53 sq km (after completing the project)

**fig. 4.17.** Giant solar photovoltaic stations

---

[99]   pictures from https://newsland.com/user/4297805012/content/samye-bolshie-solnechnye-elektrostantsii-na-zemle/6866845

In 2017, China's total solar capacity exceeded that of any other country with 130 GW. And the *"Great Wall of the Sun"* has about 1,5 GW at effective area of 43 square kilometers[100]. The *Longyangxia Dam* in Tibet has **4 million solar panels.** The total number of panels of the "Great Wall of the Sun" is exactly unknown.

It is clear that such stations occupy a lot of the earth's surface, which could potentially be used in the future for agricultural activities. In total, many hundreds of stations, the number of which is growing rapidly in the world, occupy even more land. Another problem is the need for a large amount of water for cooling and cleaning the panels.

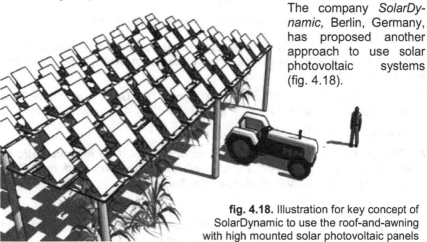

The company *SolarDynamic,* Berlin, Germany, has proposed another approach to use solar photovoltaic systems (fig. 4.18).

**fig. 4.18.** Illustration for key concept of SolarDynamic to use the roof-and-awning with high mounted solar photovoltaic panels

Their approach integrates the following ideas:

➢ raise the panels to a sufficiently high height, for example, 10-20 meters; this will free up the land under the panels for planting and agricultural work using any known machines;

➢ perform panels based on a polymer thick film, which will significantly reduce their weight;

➢ to make the supporting panels inflatable, which will allow them to be folded on non-sunny and rainy days to ensure natural watering of plants under the panels; indeed, usually on sunny days there is no strong wind, and inflatable panels having a sufficiently large size, for example, 5 x 5 $m^2$ or even 10 x 10 $m^2$, can withstand the load of solar components;

➢ perform panels with components that provide for the collection and collection of condensate and rainwater, both for cooling and cleaning panels, and for watering plants.

---

[100] https://www.eurochinabridge.com/en/environment/755-the-great-wall-of-the-sun-china

The temperature on the ground under panels can be 5°C lower than under the sun.

**fig. 4.19.** Pets can feel quite comfortable under tall panels in the shade...

The direction opened by systems similar to those proposed by *SolarDynamic* will make it possible to expand the area for agriculture many times over and also help landscaping desert lands.

The grandiose projects[101,102] with same name *"The Great Green Wall"* are known to fight the deserts in Africa[103] and China[104] (fig. 4.20 and fig. 4.21).

And now we go to MTRIZ-extracting:

*01. Change in the aggregate state of an object, 14. Use of pneumatic or hydraulic constructions* and *25. Use of flexible covers and thin films* (making the panels inflatable from a polymer thick film);

*07. Dynamization* (folding panels);

*10. Copying* and *19. Transition into another dimension* (to copy roofs and awnings and raise panels high above the ground for shading below).

Looking at the pictures in fig. 4.21 it may be obvious that planting requires a lot of manual labor, since it is not easy to apply machine landing in very different terrain conditions.

In fig. 4.21 below the left picture shows the preliminary preparation of conditions for planting seedlings. To do this, ditches of small depth are dug in advance along and across the landing site (such "checkered" fields are obtained) and filled with straw. The straw will decompose and the seedlings will get initial top-dressing. TRIZ-models *02. Preliminary action* and *12. Local property* work here.

Due to the possibility of accumulation of condensed and rainwater in this system, we can say that models *28. Previously installed cushions* and *29. Self-servicing* also work here.

[101] https://en.wikipedia.org/wiki/Three-North_Shelter_Forest_Program
[102] https://en.wikipedia.org/wiki/Great_Green_Wall
[103] https://community.standardbank.co.za/t5/Community-blog/Africa-s-Great-Green-Wall-Symbol-of-hope-and-human-resilience/ba-p/421115
[104] https://theplaidzebra.com/china-is-building-a-great-green-wall-of-trees-to-stop-desertification/

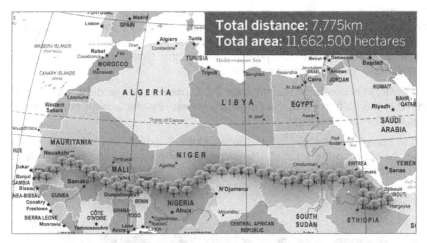

**Total distance:** 7,775km
**Total area:** 11,662,500 hectares

**fig. 4.20.** "The Great Green Wall" in Africa

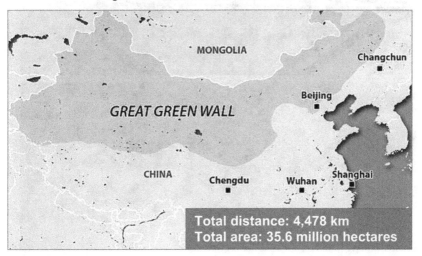

MONGOLIA

Changchun

Beijing

*GREAT GREEN WALL*

CHINA

Chengdu

Wuhan

Shanghai

Total distance: 4,478 km
Total area: 35.6 million hectares

**fig. 4.21.** "The Great Green Wall" in China

Great difficulties are known for the survival of seedlings in unsuitable desert conditions. Therefore the arrangement along the line of planting of new panels according to models **02. Preliminary action** will increase the extrusion of seedlings, as it will not only protect them in the shade, but also ensure the supply of the minimum necessary water.

Moreover, in accordance with models **07. Dynamization** and **12. Local property**, it is possible to move the constructions with inflatable solar panels to new lines for planting seedlings.

And to complete the topic "Stop Sunlit – Make Shadow", let us to consider the cases of applying the inflatable solar panels for protecting the water surfaces from evaporation (fig. 4.22).

**fig. 4.22.** Canal-top solar power plants

Such application is very popular[105] in India. Master-students at Energy Engineering Program have studied and modeled many similar solutions (fig. 4.23) which are **very effective to protect water from evaporation at artificial canals** supplying water to enterprises, fields and settlements.

---

[105] e.g. look at http://visheshnews.in/state-has-a-potential-of-more-than-1500-megawatt-for-installation-of-canal-top-solar-power-projects-power-minister;
http://www.yuvaengineers.com/narmada-canal-solar-power-project-of-gujarat-generates-1mw-power-and-helps-reduce-water-evaporation; http://taiyangnews.info/markets/solar-on-canals-tender-in-punjab/

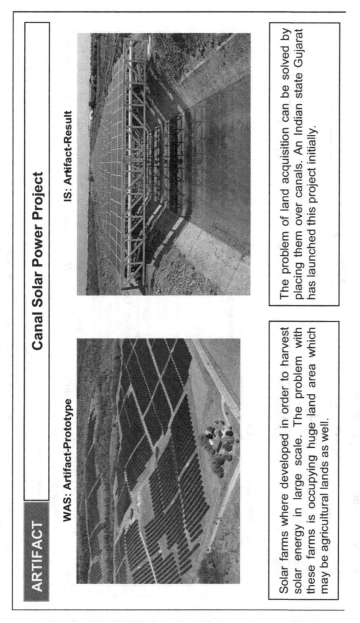

**fig. 4.23 (beg.).** Combined Extracting & Reinventing for the case "Canal-top solar power plant"

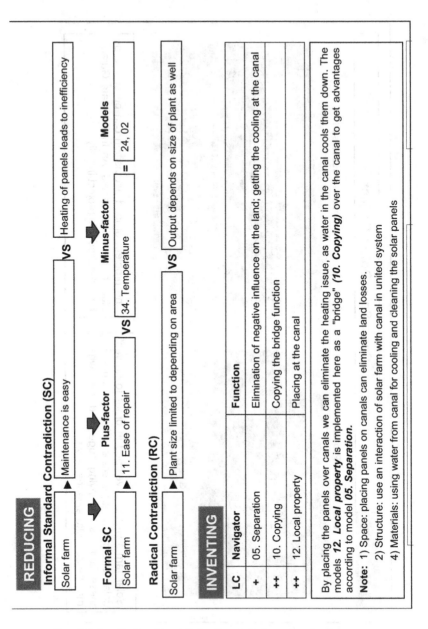

**fig. 4.23 (end).** Combined Extracting & Reinventing for the case
"Canal-top solar power plant"

## 4.5. AirHES by Andrey Kasantsev[106]

*Story*

*Air HydroElectric Station* (HES) is an invention of Russian engineer Andrey Kazantsev (patent WO2013157991A1).

*AirHES* (fig. 4.24; the texts below are from http://airhes.com) provides a downstream (water's outflow) 1, upstream (water's inflow) 2, conduit (pipe, penstock) 3, the turbo generator 4, mesh, fabric or film surfaces 5 to collect water from cloud 8, airship (air-balloon) 6, and the fastening ropes (tethers, lines) 7.

The airship 6 lifts the surfaces 5 at the height of near or above the dew point (condensation level of clouds) for current atmospheric conditions (typically 2-3 km). There super-cooled atmospheric moisture begins to condense from clouds on the surfaces 5. The drainage system on the surfaces 5 assigns the water in a small reservoir (upstream 2), where water under pressure from whole hydraulic head (2-3 km) flows through the penstock or conduit 3 to the downstream 1 on the ground, producing electricity in the turbo generator 4.

The AirHES can be easily mounted in any convenient place for the consumer of electricity and water, simply by lifting and moving it entirely by using the same airship 6.

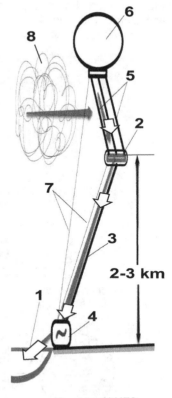

**fig. 4.24.** AirHES
http://airhes.com

If at this point the winds are blowing steady permanent or it is a portable unit (for example, for tourists or military), you can do without the airship 6 and use surfaces 5 like the paragliding wing for self-containment of the assembly in the air (as occurs when you run a kite).

Also, the surfaces 5 can be performed with full or partial metallization (for example, by weaving metal wire). This will increase the structural strength, reduce solar heat, to increase the condensation of water vapor through the filing of an electric field (for example, have experimented with this corona discharge), as well as the need to reduce the ice due to current supply.

The principal difference AirHES is the condensation of moisture from the air that at first glance seems funny and impractical curiosity. Nevertheless, there is

---

[106] this is an improvement and development of an individual and team modeling in class at Master Programs on Energy Engineering and Water Engineering

nothing unusual. In the world there are several great working systems, known as fog collectors. For example, a device for collecting drinking water in Chile[107] was tested in 1987 (fig. 4.25).

**fig. 4.25.** "Atrapanieblas" or fog collection in Alto Patache, Atacama Desert, Chile; Pontificia Universidad Católica de Chile: Rector Ignacio Sánchez visits Alto Patache; foto (up): Nicole Saffie (https://en.wikipedia.org/wiki/Fog_collection#/media/File:Atrapanieblas_en_Alto_Patache.jpg)

It's clear that mountains are not everywhere. And even if they are not so far, how to deliver the water collected by the network?

Is there such a mountain *that will come to Mohammed[108]*?

The answer from AirHES: *yes, it is!* ☺

---

[107] https://en.wikipedia.org/wiki/Fog_collection#Chilean_project
[108] "If the mountain won't come to Muhammad, then Muhammad must go to the mountain," perhaps from a Turkish proverb, retold by Francis Bacon; https://en.wiktionary.org

What AirHES gives:

> ➢ almost eternal and unlimited gratuitous electricity and clean water for drinking and irrigation, and anywhere in the world, where consumers want;

> ➢ minimum space on the ground (as under this HES and under power lines), as well as the ability to use any surface (including the vast areas of deserts, seas, oceans, etc.);

> ➢ modular (you can collect any power plants from standard modules, for example, by 1 MW);

> ➢ mobility (for rapid redeployment, if necessary, or even for use in transport, for example, to supply ocean ships by electricity and drink water); **e.g. could be effective during Great Green Belts development;**

> ➢ cleanness and ecology because of the relatively small local hydro flows (in comparing with conventional HPS) and the complete absence of thermal, chemical or nuclear releases into the environment;

> ➢ increasing the specific hydroelectric power (that is power per liter of water) by using the maximum possible hydraulic head between the upper and lower water level (from the height of the condensation of atmospheric moisture to the ground);

> ➢ significantly lower capital costs per unit of capacity and operational costs by comparing with any other known types of renewable and non-renewable energy;

> ➢ possibility of additional uses for network communication, video surveillance, high-rise advertising, lightning protection, climate protection (for example, against hurricanes and tornadoes in the USA by placement on the seaboard of Gulf of Mexico), regulation of climate (by cutting off rains in St. Petersburg by placement on the dam at the prevailing southwest wind rose), AD (for example, for Israel), shade in hot countries, and much more…

Major trends associated with the transition to renewable energy sources. It then becomes clear that resources are only sufficient for sun and (possibly) wind. Traditional hydropower does not have enough resources.

However, the use of cloud energy changes this assessment. Principally it results from economics.

All three methods of conversion (PV, wind, clouds) are the same order of magnitude of the energy density (~ 100 W/m2) , but only for AirHES all this energy with virtually no loss can be merged into one point (pipe / turbine), making part of the proportional $m^2$ far cheaper than other alternatives.

This implies the 1-2 orders of magnitude smaller payback period that allows quickly rebuild the energetics and successfully withstand the collapse of climate & oil in 2050.

AirHES is an effective and completely green energy source that also gives you clean water that you can drink or use however you want.

It is known that the solar energy reaching to the planet is about 10,000 times greater than the needs of humanity[109] (fig. 4.26). About a quarter of it goes to the evaporation of water and virtually always more or less evenly accumulates in the atmosphere at any point around the world. Since the annual precipitation is about 1 m of rainfall with an average height of 5 km, this gives a potential capacity of about 810 TW, which is more than 60 times greater than all the current needs of humanity (13-16 TW).

Standard hydroelectricity can only use a small fraction of this energy, because all precipitation loses most of their potential energy on the way to the ground to overcome the resistance of the air and hit the ground. In order to use this potential energy more cost-conscious, it is necessary to collect the water at that altitude, where it condenses, and use all possible vertical hydraulic head. This is what constitutes the essence of the decision.

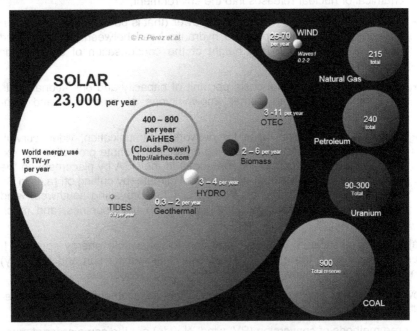

**fig. 4.26.** Comparing finite and renewable planetary energy reserves (TeraWatt-years). Total recoverable reserves are shown for the finite resources. Yearly potential is shown for the renewables.

---

[109] picture here is a collage with a base picture from the work "*A fundamental look at energy reserves for the planet*" by Richard Perez and Mark J.R. Perez, University at Albany, The State University of New York, 2009, at *https://www.researchgate.net/publication/237440187 _A_fundamental_look_at_energy_reserves_for_the_planet*

The inventor *Andrey Kasantsev* says:

"Interestingly, already having gotten this solution, I began to search the Internet for key words such ideas, and unexpectedly found that in 1915 in one of the articles the genius Nikola Tesla was nearly half a step to the realization of this idea.

He was right in principle estimation of the required resource, but had not found a scheme to implement it, although all technics already was ready for it hundred years ago.

It's a pity. If he had pursued the idea, then we would live in a different world— clean, ecological, abundant, without wars for oil, etc...

Unfortunately, humanity has lost a hundred years! "

And we complete this part with MTRIZ modeling and illustrations.

| Extracting-1C | |
|---|---|
| **Navigator** | **Conformity** |
| 01. Change in aggregate state | to use the droplets of condensed water |
| 12. Local property | to use water from cloud with dew point at right height above sea level; gathering the water with fabric and transportation down to generate electricity |
| 19. Transition into another dimension | to use a positioning at high altitude for collecting water from cloud |
| 20. Universality | producing water and electricity |
| 29. Self-servicing | the water cycle is an endless source of energy for mankind |
| 32. Counter-weight | to raise fabric or metal membrane with dirigible or kite |

**fig. 4.27.** Extracting for invention AirHES by Andrey Kasantsev

Two reinventings by two Master-students are shown in fig. 4.29 and fig. 4.30. It is easy to notice the differences in the choice of contradiction factors, in the formulation of the Ideal Final Result, as well as in the results of extraction. But in general, the work is similar in terms of the main conclusions and content of the simulation. It should be borne in mind that the students did these works immediately after the introductory lectures, having absolutely no experience in MTRIZ modeling.

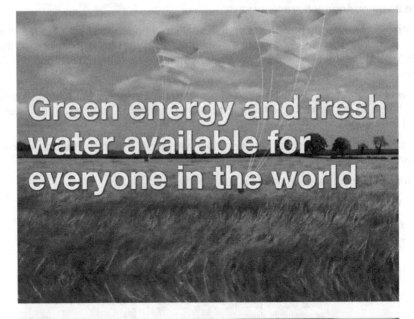

**fig. 4.28.** Ideal foundation of the AirHES by Andrey Kasantsev

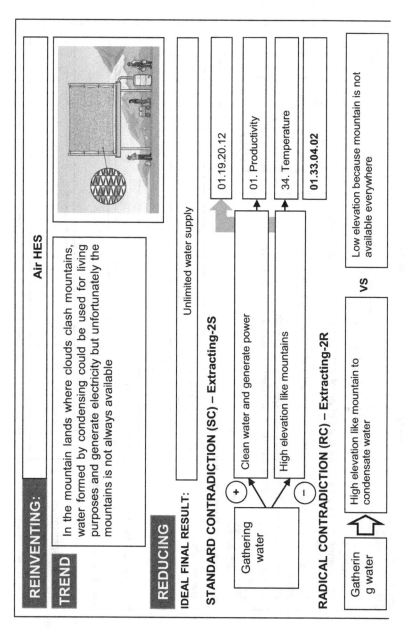

**REINVENTING: Air HES**

**TREND** In the mountain lands where clouds clash mountains, water formed by condensing could be used for living purposes and generate electricity but unfortunately the mountains is not always available

**REDUCING**

**IDEAL FINAL RESULT:** Unlimited water supply

**STANDARD CONTRADICTION (SC) – Extracting-2S**

Gathering water

(+) Clean water and generate power

(−) High elevation like mountains

01.19.20.12

01. Productivity

34. Temperature

01.33.04.02

**RADICAL CONTRADICTION (RC) – Extracting-2R**

Gathering water

High elevation like mountain to condensate water

vs

Low elevation because mountain is not available everywhere

fig. 4.29 (beg.). Reinventing for AirHES (EE-student Omar Muhrat, 19.01.2017)

**INVENTING**

Air HES could be used for high elevation to gathering water

According to:

01. Change in the aggregate state of an object-condensation process from vapor to water

19. Transition into another dimension- doing construction with several parts

20. Universality-gathering water and generate electricity

12. Local property-different parts of an object have different function

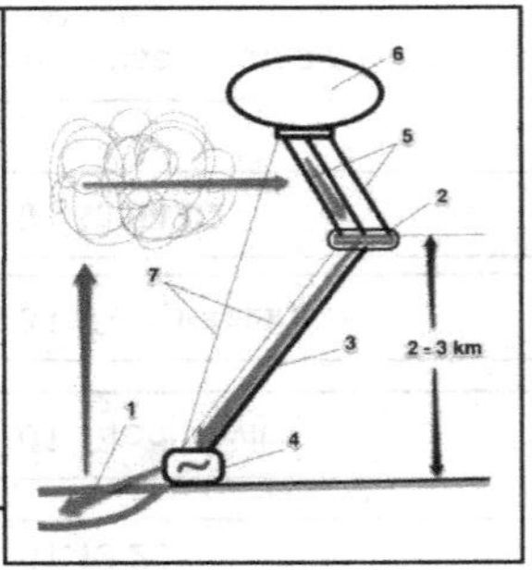

**ZOOMING**

Are the contradictions eliminated? – Yes.

Super-effects: Located in any where

Negative effects: -

Development trends: Simple design

Change of surrounding systems: Used in many locations

Advanced application: To generate electricity

fig. 4.29 (end). Reinventing for AirHES (EE-student Omar Muhrat, 19.01.2017)

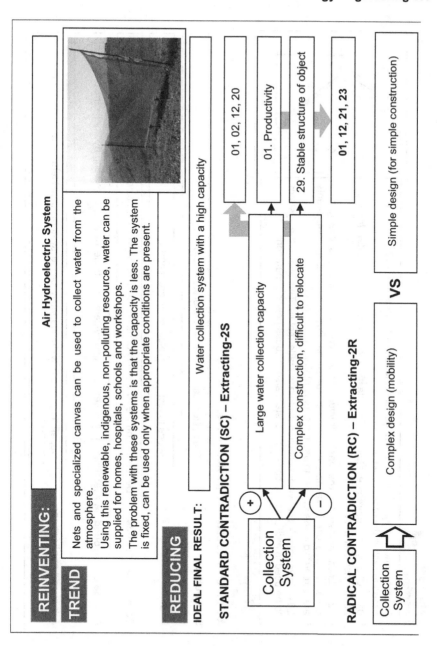

**fig. 4.30 (beg.).** Reinventing for AirHES (EE-student Aniruddh Sridhar, 19.01.2017)

**INVENTING**

In accordance with the selected contradiction models, the models 01, 12 and 20 were assed to realise the required result.

01 - Change in aggregate state – The elastic membranes collect water from the atmospheric air at dew point

12 – Local property – The basic structure of the collection system is changed to accommodate use in variable environment conditions. Power generation components are introduced in the system to use the energy of the water

20 - Universality - The water collection system is made mobile. This allows the systems to be used in multiple conditions. The system also includes power generation.

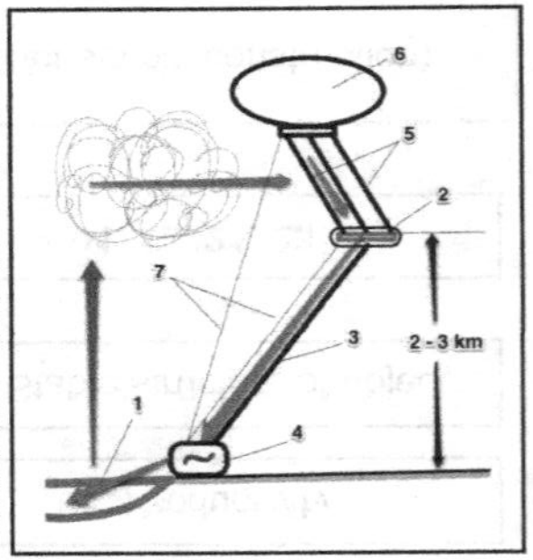

**ZOOMING**

**Are the contradictions eliminated? – Yes.**

**Super-effects:** The system can be used to generate electricity.

**Negative effects:** Needs to test and develop for different applications

**Development trends:** Depending on the availability of moisture in the region, the scale of the system can be varied.

**Change of surrounding systems:** A water collection systems has to be installed in existing hydroelectric plants

**Advanced application:** Depending on the water capacity of the system, considerable amount of electricity can be generated.

**fig. 4.30 (end).** Reinventing for AirHES (EE-student Aniruddh Sridhar, 19.01.2017)

# 5 WATER ENGINEERING[110]

## 5.1. Turn Harm in Use[111]

### Preamble

The following cases inspire me and my Master-students at TU Berlin El Gouna Center in Egypt for more than two years, as we found this information.

At the same time, we understand the technical, economic, organizational and even political difficulties in the implementation of the ideas considered further, but the ideas themselves delight us with their creativity and determination not to surrender to advancing deserts, and also **to abandon the insidious thought to continue living as before just because people here are always like that lived.**

We will not consider here all features and physical-technical details of the projects. But we will definitely find out the creative patterns and development trends contained in them in accordance with TRIZ-approach.

The fundamental aspiration of the author of these projects and his associates and organizations supporting him is to radically improve living conditions in UAE and the surrounding countries. Key climate issues include the onset of deserts and the lack of sufficient drinking water.

The initiator and inventor of the ideas considered further, founder and leader of such companies as *National Advisor Bureau*[112] and *Geowash*[113], engineer *Abdulla Alshehi* says[114]:

"Water, sunlight, and manpower... these are really what we need to realize the vision of a *Full Quarter*[115]. We have all of these things and the many technologies necessary to ensure that we can make the most out of all of that water, sunlight, and manpower available.

Of course, it takes vision, and I hope that some of my outside of the box solutions have inspired you to help see this dream of a blooming desert a modern reality. *It is not of benefit only to the UAE, but to all humanity.*

Anywhere people suffer due to climate change, the Full Desert could begin to help them survive. Anywhere bizarre new weather patterns wreak havoc on the landscape and everything inhabiting it, the Full Quarter could start to undo the underlying causes.

---

[110] using the examination works of the MTRIZ-classes at El Gouna Center of TU Berlin
[111] this is an improvement and development of an individual and team modeling in class at Master Programs on Energy Engineering and Water Engineering
[112] http://www.national-advisor.ae/about_abdulla.html
[113] http://q2wash.com
[114] http://abdulla2alshehi.blogspot.com/2015/08/filling-empty-quarter-initiative.html
[115] according to Alshehi: *the "Empty Quarter" desert, or the Rub' al Khali. A place that stretches more than 650,000 square kilometers and covers areas of Saudi Arabia, United Arab Emirates, Yemen and Oman is almost lifeless.*

© Springer Nature Switzerland AG 2020
M. A. Orloff, *Modern TRIZ Modeling in Master Programs*,
https://doi.org/10.1007/978-3-030-37417-4_5

And terrorism... I already explained that illiteracy and lack of food are two of the main reasons for the ugly head of terrorism. Being able to supply people with food, water, and the means of a good living will help to even out the imbalance leading to so many problems in the world.

The global cooperation needed for this project to come to life will also benefit humanity as well. Working together, we can reclaim the lands of the UAE, stop desertification, bring the landscape to life, and help save ourselves along the way.

With us or without us; *I am confident that one day the Empty Quarter will be Full!"*

There are four main ideas collected below:

❖  proposal to develop the *"Great Green Wall of the UAE"*;
❖  proposal of *"AL Maa"* technology and system for rainwater capturing over areas of water such as the sea and lakes;
❖  proposal of the project *"Endless Rivers"*;
❖  proposal of the project *"ICE Berg"*.

### Case[116] 1. Great Green Wall of the UAE.

Some of the concepts proposed in the book[117] of Alshehi are theoretical at this time, and need far more research or study; you will find that these concepts are new innovative ideas **to solve the water shortage issue.**

Mainly these ideas are unique as product of Alshehi's thinking but some were researched and gather by him via multiple sources.

The United Arab Emirates stands as the most likely pilot run for these projects because of its prime location. If these concepts prove feasible there, their successes can then be duplicated elsewhere in the region.

It's proposed that the United Arab Emirates to take lead on the project; considering the UAE's notable effort towards the environment such as the establishment of clean city of *"Masdar"* and the hosting IRENA (International Renewable Energy Association) headquarter.

It is proposed that a **Great Green Wall of the UAE** shall isolate the country from the Empty Quarter desert (fig. 5.1).

The wall shall stretch from *Al Ain* city up to *Silla* city with the establishment of eco-friendly industries and farms within the western region of *Abu Dhabi*.

And attention: And what is vital for the development of this forest wall? – Yes. **This is water!** Therefore, move on.

---

[116] text fragments here and for next cases cited mainly from *http://abdulla2alshehi.blogspot.com/2015/08/filling-empty-quarter-initiative.html*

[117] Alshehi, A. *Filling the Empty Quarter: Declaring a Green Jihad On the Desert.* – Lulu Publishing Services, 2015. – 164 p.

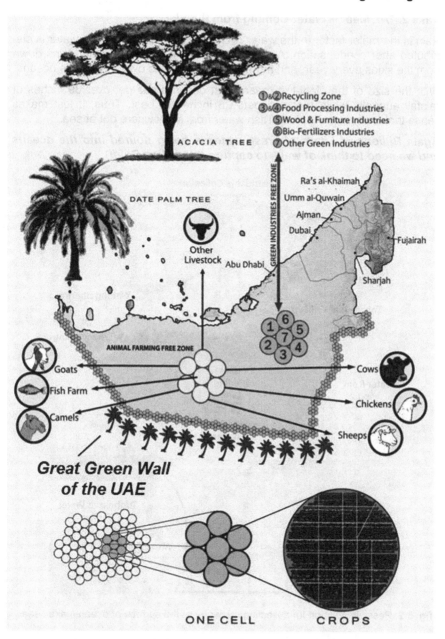

**fig. 5.1.** Infrastructure with food clusters under protection of the Great Green Wall of the UAE

### Case 2. *"AL Maa"* – Water Coming from the Sky.

Rain is the major factor in the water cycle on Earth; it is how fresh water is distributed after condensation. Around 505k cubic km of fresh water rains down from the skies every year, and more than 3/4 of that is over the world's oceans.

With the size of the planet, it means that ocean areas get over 39 inches of rainfall annually while the land gets 28 inches at best. Thus, it just makes sense that we seek to capture fresh water from somewhere out at sea.

*Again Billions of Liters of Fresh water is being poured into the oceans and we need to think of ways to capture this falling Gold!*

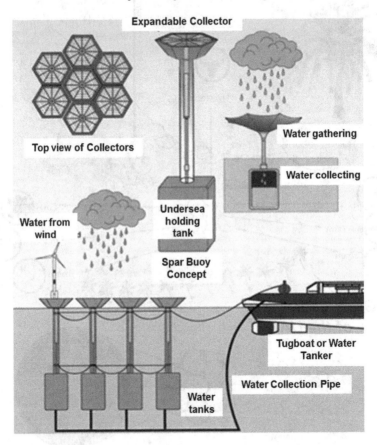

**fig. 5.2.** Possible solution for collecting rainwater at the surface of ocean / lake / sea

Collectors will be collected in mega-clusters over a large area of hundreds of meters or even several kilometers in diameter (fig. 5.3).

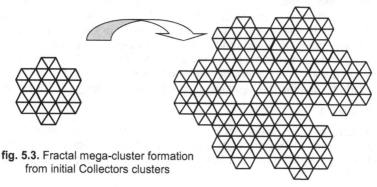

**fig. 5.3.** Fractal mega-cluster formation from initial Collectors clusters

## Case 3. Endless Rivers.

According to the author, the Pakistani river of Dasht is the closest river to the UAE, approximately 500 km from the coast of the emirate of Fujairah, and connecting Dasht water to the UAE via undersea pipelines will restore the flow of rivers in the UAE for the first time after a lapse of thousands of years when rivers were flowing on the Arabian Peninsula.

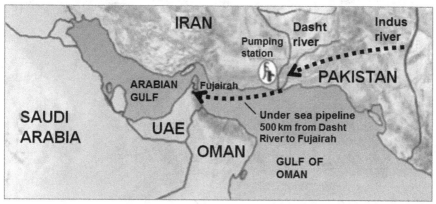

**fig. 5.4.** Main water streams of the Khalifa River

The project will benefit both the UAE and Pakistan, and it is proposed to be called **"Khalifa River",** named after *H.H. Sheikh Khalifa Bin Zayed Al Nahyan,* the president of the UAE. The project will provide the Emirates with a continuous flow of water rich with organic matter and silt, which is great for irrigating crops at the lowest costs, making the UAE a greener place. In addition it would raise the groundwater level. While the Pakistani side of the project will contribute to the reduction of the risk of flooding, excess water shall be pumped into the UAE side in the event of high water levels in the river instead of the water flowing into the Arabian Sea unutilized, as it is currently.

The proposed "Khalifa River" project is part of **the endless river concept**, which is mentioned in the book *"Filling the Empty Quarter"* recently published in the United States, where Abdulla Alshehi calls to reroute the world's river paths to deserts rather than waste them in the oceans (for example, reroute the River Nile to flow into the Sinai desert, and European rivers to pour into the Sahara desert) and thus contribute to the forestation of deserts combating the negative effects of climate change on the world.

### Case 4. ICE Berg.

Billions of liters of fresh water get melted in the oceans every year. Due to global warming many ICE Berg's get disintegrated from the South Pole. This is not only wasting the much needed fresh water but also adding to the sea level rise problem!

For mining water from south-pole icebergs which are already disintegrated due to the global warming, we may crush them and then fill the parts into floating tanks to transport them to the Gulf Region and to use for agricultural purposes.

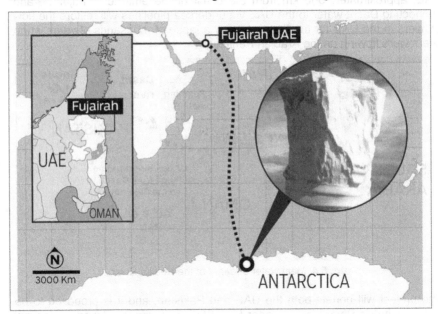

**fig. 5.5.** Iceberg ice delivery route to UAE across the ocean.

Regarding the method of delivering the iceberg fractions or maybe entire ice floating island, the opinions are still divided.

Nevertheless, we will give here two key approaches – delivery of ice over fragments enemy and the entire ice island.

The idea of exploiting icebergs to produce fresh water is not new and is also discussible till nowadays[118].

There were 19th-Century schemes to deliver by steam-boat to India, and to supply breweries in Chile.

In the 1940s, John Isaacs of the Scripps Oceanographic Institute proposed towing an iceberg to San Diego to quench a Californian drought, and also idea goes back to the 1950s with research projects by the US Army.

It gained momentum in the 1970s, notably under the influence of the famous French polar explorer *Paul-Emile Victor*, his friend and *Arts et Métiers* engineer, *Georges Mougin*, and their meeting with the Saudi prince, *Mohamed al-Faisal*. Georges Mougin is convinced that pulling icebergs from Greenland across the oceans to drought-stricken areas is a possible, and practical, solution.

And now Georges Mougin in team with Dr. *Olav Orheim* (director of the Norwegian Polar Institute from 1993 to 2005) and Captain *Nick Sloane*, of Cape Town-based marine engineering company Resolve Marine, are working to arrange new experiment to tow the 85-100 million-ton iceberg to Cape Town at the distance of 2,700 km (1,700 miles) from Gough Island in the South Atlantic.

It should be taken in account that the distance to Fujairah is almost five times longer and the temperature along the path will gradually increase by about two times. This is one of the reasons why Abdulla Alshehi considers the option of shipping in containers as more reliable.

In any case, a huge incentive is that: **an average iceberg holds up to 20 billion gallons (about 91 billion liters) of fresh water, enough for 1,000,000 people for 5 years (!!!)**

Alshehhi says:

> *"The UAE has the ingredients to support creativity and thinking outside the uncommon framework to find creative solutions to the challenges it faces,* including the problem of water scarcity, and the United Arab Emirates would be able to change climate change forever and that would align concerted international efforts with the UAE's efforts."

Supporting such aspirations, we can only recall here that the basis of the philosophy of these projects lies in the magnificent TRIZ model:

*to turn harm in favor!*

This model is exceptionally good because in it:

1) the initial harmful effect ceases and

2) a positive target effect is obtained in another action!

*So, it is always at least double success, double win!*

---

[118] https://www.bbc.com/future/article/20180918-the-outrageous-plan-to-haul-icebergs-to-africa

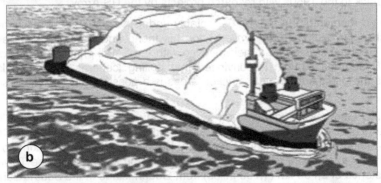

fig. 5.6. Iceberg ice transportation across the ocean:
a,b – defragmented; c – towing the ice island.

P.S. The readers can test yourselves in extracting the dominant and supporting TRIZ-models from these methods of ice delivering and compare with the answers selected from Master-student works below.

| | |
|---|---|
| 34. Matryoshka (nested doll) | constructions |
| 32. Counter-weight | 14. Use of pneumatic or hydraulic |
| 21. Transform damage into use | 12. Local property |
| 16. Partial or excess effect | 07. Dynamization |
| 15. Discard and renewal of parts | 01. Change in aggregate state |

## 5.2. Production of ... Clouds[119]

Deserts[120] already occupy more than one fifth of all terrestrial land (fig. 5.7).

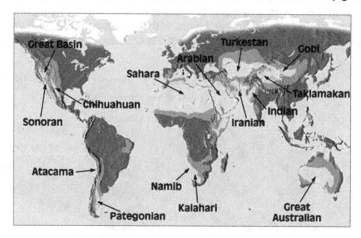

**fig. 5.7.** Largest deserts on the World map

Evaporation from the sea does not reach the desert coast in summer due to the barrier in the form of rising hot air (fig. 5.8). Above the warmed desert surface in the atmosphere, delay layers ("temperature inversions") are formed, which prevent the rise of cloudy moist sea air.

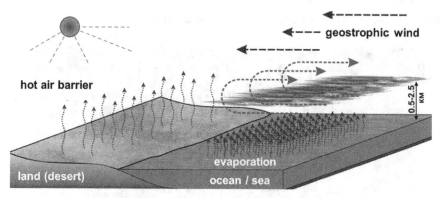

**fig. 5.8.** Hot air barrier prevents wet air from the sea to desert shore

---

[119] this is an improvement and development of an individual and team modeling in class at Master Programs on Energy Engineering and Water Engineering

[120] translated and adapted from the articles by specialists of scientific and engineering enterprise "ATOMENERGOPROM", Moscow, Russia, Vladimir Rogozhkin and Evgeniy Mishin *"Atom vs. Desert (Atomic COOLER)"* at *http://www.atomic-energy.ru/papers/27583* and *"Atomic coolers for Europe"* at *http://www.atomic-energy.ru/papers/29219*

It is possible to "make spring" in summertime in the coastal desert, for example, by lowering (by 20-30%) the level of solar radiation over its territory with creating over the desert a low-water technogenic high-altitude clouds of the required area and density. Figure 5.9 schematically shows this principle of suppressing a barrier from hot air layers with the subsequent penetration of natural low-altitude forms of rain clouds into the coastal territory from the sea.

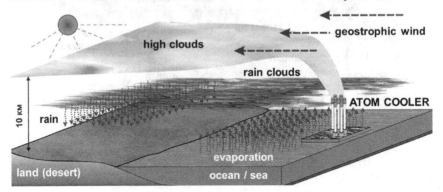

**fig. 5.9.** Producing the high-altitude artificial clouds near desert shore

Artificial high clouds reduces the heat flux from the sun, the surface of the "chilled desert" loses its ability to generate powerful ascending flows. Atmospheric circulation arises, contributing to the unhindered spread of marine air masses on the continent.

Atomic coolers can be used to create artificial clouds (fig. 5.10).

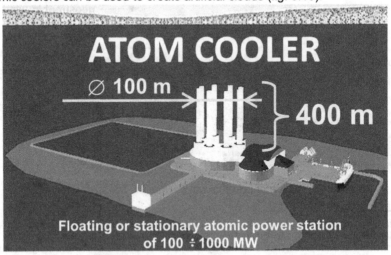

**fig. 5.10.** "ATOM COOLER" near desert shore

"ATOM COOLER" is intended for the use of unlimited reserves of fresh water in the form of steam in the atmosphere over the waters of the seas closest to deserts. The basis of the ATOM COOLER will be injectors (fig. 5.10 and fig. 5.11) – hollow reinforced concrete structures of a cylindrical shape with a height of 400 m and more, a diameter of ~40-100 m, containing electric fan installations and pipe structures for heating and casting humidified air to an estimated height of about 10 km.

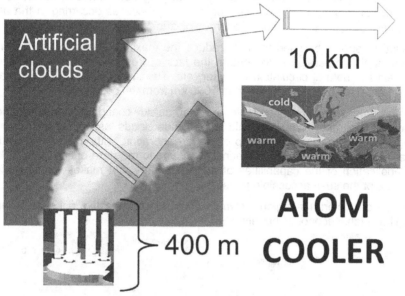

fig. 5.11. Steam injection by "ATOM COOLER" to a height of about 10 km

Nuclear power plants with a total capacity of 200 ÷ 1000 MW can be used as energy sources for COOLERs (depending on the area of shaded areas). Actually, COOLER is an atomic power plant operating on its own cooling towers-injectors.

**ATTENTION – the TRIZ-model *21. Transform damage into use* works with highest efficiency:** *The thermal power spent earlier in the cooling towers is used here to heat the discharged air in order to increase the buoyancy of the air stream.*

At an altitude of about 10 km, humidified rising air will be captured by the geostrophic wind and spread like artificial clouds over a distance of hundreds of kilometers with a width of 100-200 km or more.

Cloud cover, including cloud cover of the upper tier, is an effective natural regulator of the level of direct and diffuse solar radiation. It is known that high-altitude clouds generated, for example, by the activity of volcanoes, can significantly affect the Earth's climate.

Artificial *Cirrostratus clouds (Cirrostratus)* do not have clear outlines and represent a foggy uniform shroud uniformly covering the whole sky through which the disks of the Sun and Moon are visible.

Nuclear coolers will work as ordinary power plants for the use of electricity on the shore, as well as water desalination! *According to TRIZ, this is a direct super effect from the use of "ATOM COOLER".* It is interesting also to note *the explicit use of the TRIZ-model 10. Copying* here, since in the system plan the functioning of the cooler "copies" the processes occurring in the atmosphere during the activation of a large volcano.

COOLERs should be used to softly reduce the maximum summer (and average annual) temperatures, to combat the lack of fresh water, to restore and maintain the natural circulation of water, etc. The influence of COOLER stations can be felt at a distance of up to 1000 km from the coast.

Atomic coolers are fully controllable; they can quickly change the structure and optical transparency of technogenic high-altitude clouds, as well as their operating mode up to a complete stop of cloud generation, without stopping work as a producer and supplier of marine atmospheric moisture condensate. It is the application of the capabilities of nuclear energy that makes atmospheric moisture of the seas accessible to use.

The use of coolers for regional climate control in Europe (fig. 5.12) also can bring invaluable economic benefits.

**fig. 5.12.** Possible placement of the "ATOM COOLER" in Europe

The authors' have finished their articles with manifesto:

It's possible to come up with a lot of good reasons (environmental, economic, political, religious, etc.) to do nothing with climate instability, but only to observe, evaluate, adapt and tolerate. But Man is neither a plant nor a victim.

Therefore:

(1) Humanity does not have the right to passively experience climatic anomalies that have assumed a global supranational character. It is said in the Bible: *"Fill the earth, and possess it, and rule over all the earth."* We are obliged to correct the changing conditions of our lives, otherwise – failure as reason, degradation and death. *Do not adapt to the climate, but regionally adapt the climate!*

(2) Mankind should already today build, use and improve stationary technical means of correction of temperature anomalies of the earth's surface and air.

(3) The goal of climate adaptation is to ensure temperature comfort and stable water circulation in problem regions. *Desalinated seawater from the tap will never replace ordinary rain for people!*

(4) We consider anthropogenic cloudiness of various types in the presence of winds of a stable direction and proximity to the seas as a non-alternative (currently) means for regional climate adaptation.

(5) It is necessary to adopt international agreements on the criteria for non-interference in natural affairs, on the level of permissible natural discomfort.

(6) It is necessary to centrally finance the design and construction of "ATOM COOLER" stations as a means of compensating for climatic discomfort in the territory of the peoples of Europe and around the World.

And I really want to complete this section with graphic quotes from work by *Abdulla Alshehi:*

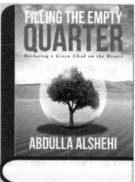

# 6  URBAN DEVELOPMENT[121]

## 6.1. Inventive Archibiotecture

*There so many architectural and functional ideas in the world to study esthetically and to model creatively with MTRIZ-approach!*

*The master students have unlimited opportunity to find, collect and model hundreds of wonderful cases!*

Let's look at just a few cases to discover a way to model their achievement.

So, the **"archibiotect"** is the new trans-disciplinary approach invented[122] by *Vincent Callebaut* in 2008. Archibiotect is a word combining the prefixes of the words **ARCHI**tecture + **BIO**technologies + **TEC**hnologies.

Whereas the primary reason of architecture has been to protect man against nature, **the contemporary city strives to reconcile human beings and their natural ecosystems.**

*The garden is no more placed side!*

### Case 1. Tao Zhu Yin Yuan Tower[123]

The archibiotect Vincent Callebaut discussed his work with CNN Style: "[He] wants his buildings to be more than your average tower block. His vision is ambitious: create an energy-saving, carbon-absorbing civilization to fight global warming.

*I want to give hope for a better tomorrow,* he says.

In 2050, we will be 9 billion human beings on our blue planet, and 80% of the world population will live in megacities. It's time to take action against climate change, to invent new eco-responsible lifestyles and to incorporate nature into our cities.

It's not a trend. It's a necessity!

To think outside of the box, to shake up the old structures.

To make society evolve is the most difficult thing to do in this world... but it is possible step by step."

Known for his *ecovisions* that surpasses the human imagination and designs that champions sustainability, his cutting-edge ideas give hope for a better future.

Vincent Callebaut designs and builds with the future in mind. Biomimetic and plus-energy buildings that produce their own power, vertical forests, pollution-removing towers and boats, floating cities and ocean scrapers, vertical food farms that meet the major urban and ecological challenges of the 21st century.

---

[121] using the examination works of the MTRIZ-classes at El Gouna Center of TU Berlin

[122] cited in several points from http://vincent.callebaut.org

[123] http://vincent.callebaut.org/zoom/projects/190320_taozhuyinyuansite/taozhuyinyuansite_pl011.jpg

© Springer Nature Switzerland AG 2020
M. A. Orloff, *Modern TRIZ Modeling in Master Programs*,
https://doi.org/10.1007/978-3-030-37417-4_6

Each of his green project showcases its green credentials in combining bioclimatic rules such as solar cycle and prevailing wind directions on the one hand, with renewable energy technologies in the form of wind turbines, thermal and photovoltaic solar energy, rainwater recycling, geothermal energy, biomass, upcycling of biomaterial on the other hand.

Biomimicry specialist Vincent Callebaut is definitely an active proponent of environmentally-friendly and people-oriented architecture that ushers in new eco-lifestyles and new circular economy.

Among his most celebrated projects are the self-sufficient amphibious city called Lilypad, a Floating Ecopolis for Climate Refugees – a long-term solution to rising water levels and the four challenges of climate, biodiversity, water, and health that was laid out by OECD - and Dragonfly, a Metabolic Eco-concept Farm for Urban Agriculture in New York City.

Like an unstoppable force, he then went on to create the award-winning project Tao Zhu Yin Yuan Tower under construction in Taipei that received many international awards.

Collage made with the pictures from https://www.german-design-award.com/en/the-winners/gallery/detail/16227-tao-zhu-yin-yuan.html and http://vincent.callebaut.org/zoom/projects/190320_taozhuyinyuansite/taozhuyinyuansite_pl011.jpg

**fig. 6.1.** Tao Zhu Yin Yuan Residential Tower's design by Vincent Callebaut

Tao Zhu Yin Yuan Residential Tower's design[124] is comprised of a large cylindrical central core, flanked on opposite ends by two nearly rectangular living spaces (each of these will make up 40 apartments). Each floor rotates 4.5 degrees relative to the prior floor for a total rotation of 90 degrees through the 21 floors. The tower identifies as **forest architecture** – showing great combination of science and technology as well as harnessing nature to help the surrounding environment.

The Tower, also known as *"Agora Garden"*, presents a pioneering concept of sustainable residential eco-construction that aims at limiting the ecologic footprint of its inhabitants by researching the right symbiosis between the human being and the nature.

The wide planted balconies of suspended orchards, organic vegetable gardens, aromatic gardens and other medicinal gardens makes tower look like a twisting vertical forest.

The main target of this build has been devoted to promoting carbon-absorbing, in order to help tackle global warming. The buildings balconies will be covered in 23,000 trees and shrubs – and it is estimated that because of this the building will absorb around 130 tons of carbon dioxide a year.

There are 165 meters of garden space for each unit, and over ten trees besides shrubbery planted on each floor. Even on ground level, the surrounding area has meters of foliage. As well as reducing the levels of $CO_2$, the plants will provide a partition between the noise of the city and the peaceful apartments.

It's not just the plants that will help reduce the buildings carbon footprint, the Tao Zhu Yin Yuan Tower has several innovative sustainability enhancing features aimed at reducing energy and water consumption.

There's a solar and wind power system on its roof which will supply electricity to the building, and reduce carbon dioxide intake by 35 tons. The core cylinder will be used to create natural ventilation that can reduce indoor temperature and decrease air conditioning consumption. There's a rainwater recycling system installed to feed the ground floor sprinklers.

According to the *"Cradle-to-Cradle"* concept where nothing is lost and everything transforms itself (a philosophy Callebaut follows); all the construction and furnishing materials for the building are selected through recycled and / or recyclable labels. By imitating the processes of natural ecosystems, it deals thus with reinventing in Taiwan the industrial and architectural processes in such a way that it produces clean solutions, and creates an  industrial cycle where everything is reused, either back to the ground as non-toxic organic nutrients, or back to the industry as technical nutrients able to be indefinitely recycled.

The first of its kind, the project reveals the symbiosis of human actions and their positive impact on the nature.

---

[124] here it's cited from https://premierconstructionnews.com/2019/03/08/tao-zhu-yin-yuan-tower/

Main comment from TRIZ: this concept is very strongly and clearly based on the TRIZ-law *"Integration of Alternative Systems"*.

All the following examples illustrate this law in the demonstrated projects.

We can extract also many other TRIZ-models from such *"archibiotecture"* that will be demonstrated in the next cases.

And it should be noted, however, that architecture has always been an example and area of *trans-disciplinary integration*. Really, we must accept as an imperative that future urban systems should only develop *in integration with natural systems, observing the best environmental standards with comfort and aesthetics for human habitation.*

### Case 2. "Forest Cities" in China[125]

The eastern Chinese city of Nanjing, like many of the country's urban areas, suffers from intense smog.

The Italian design firm *Stefano Boeri Architetti* believes that building towers covered in plants could help the city reduce its pollution.

The company recently announced that it will build two skyscrapers that will hold a total of 1,100 trees and 2,500 cascading shrubs on their rooftops and balconies.

**fig. 6.2.** Bird's-eye view of the Nanjing Green Towers

[125] cited mainly from https://www.sciencealert.com/china-may-build-a-smog-eating-forest-city-filled-with-tree-covered-skyscrapers

**fig. 6.3.** Really it's a forest on the tower

On flat land, the plants from each tower would cover over 75,000 square feet (7000 square meters). Collectively, the plants on Nanjing's towers will eat 25 tons of carbon dioxide each year and produce about 60 kilograms of oxygen daily, according to the firm.

Though only two forest-like towers are currently underway, *Boeri's* ultimate goal is to create an entire ***"forest city"*** in Nanjing and other Chinese cities. *Boeri* envisions that this futuristic development could contain up to 200 vegetation-covered towers, a train system, and lots of green space.

Some TRIZ-models especially interesting here:

***05. Separation:*** include the only really necessary part (necessary property) into the object – *making steel, concrete and glass covering by "forest".*

***09. Change in color:***  a) change the color of an object or its environment – *the greening.*

***10. Copying:*** d) apply copies, duplicates of the same or another object – *copying the forest.*

***11. Inverse action:*** a) instead of an action prescribed by the conditions of an assignment, complete a reverse action – *if the building cannot go to the forest, let the forest come to the building!* ☺

***12. Local property:*** b) different parts of an object have different functions; c) every object should exist under conditions that correspond best to its functions – *the "forest" should get an appropriate condition to grow and function.*

*19. Transition into another dimension:* a) … It is also possible to improve the transition from a surface to a three-dimensional space; b) … use the back of the space in question – *use of all sides of facade.*

It is not all but I hope enough to understand the approach to modeling in this case. Of course it is possible also to use more complex models like *mimicry* or *BioTRIZ effects*, etc.

### Case 3. The Domes

This is very important trend to supply to habitants the sustainable conditions to communicate with landscape at the urbanization and to protect the green plants from negative natural influences especially in condition of climate change (fig. 6.4).

**fig. 6.4.** Glass dome with climate control for the garden in "Zarjadje", Moscow
(https://www.mos.ru/news/item/13013073)

Very impressive functionality is supplied in the project of "gardening" in China[126] (fig. 6.5). This is a multifunctional space for living in the green atmosphere with flowers, trees, pools, bushes, grass, vegetables, etc. like village habitants! Psychological and social functions of such "garden" inside glass dome are extremely valuable! Stability and safety, aesthetics and comfort, healthy conditions for relax and fitness, for meetings and work (creative work!) – it becomes possible in new architectural creative solutions.

**fig. 6.5.** US-firm "Sasaki" Unveils Design for Sunqiao,
a 100-Hectare Urban Farming District in Shanghai
(https://www.mos.ru/news/item/13013073)

---

[126] cited mainly from http://www.archdaily.com/868129/sasaki-unveils-design-for-sunqiao-a-100-hectare-urban-farming-district-in-shanghai

**fig. 6.6.** The Edem Project established in 2000 in Cornwall, England
(https://en.wikipedia.org/wiki/Eden_Project)

The biomes (bio + domes) consist of hundreds of hexagonal and pentagonal, inflated, plastic cells supported by steel frames. This is a project by James Tennant Baldwin (born 1933) – an American industrial designer and writer. He was a student of Buckminster Fuller (1895-1083) – a famous American architect, systems theorist, designer, inventor and futurist.

**fig. 6.7.** Manhattan Dome, by Buckminster Fuller and Shoji Sandao, 1960
(look e.g. at https://medium.com/designscience/1960-750843cd705a)

It's easy to extract from similar solutions e.g. such creative models:

**34. Matryoshka (nested doll):** a) an object is inside another object that is also inside another, etc.; b) an object runs through a hollow space in another object;

**35. Unite:** a) unite similar objects or objects for neighboring operations; b) temporarily unite similar objects or objects for neighboring operations.

### Case 4. Floating overwater and underwater settlements

**First example** for this case is selected again from the works of "archibiotect" Vincent Callebaut (look case 1).

**fig. 6.8.** *Lilypad* – the floating city by *Vincent Callebaut;*
bottom left picture – view from under water.
(look e.g. at https://www.iloboyou.com/lilypad-the-floating-city)

"There are very few urban design solutions[127] that address housing the inevitable tide of displaced people that could arise as oceans swell under global warming. Certainly none are as spectacular as this one. The *Lilypad*, by *Vincent Callebaut*, is a concept for a completely self-sufficient floating city intended to provide shelter for future climate change refugees. The intent of the concept itself is laudable, but it is Callebaut's phenomenal design that has captured our imagination.

*Biomimicry* was clearly the inspiration behind the design. The Lilypad, which was designed to look like a *waterlily*, is intended to be a *zero emission city* afloat in the ocean. Through a number of technologies (solar, wind, tidal, biomass), it is envisioned that the project would be able to not only produce it's own energy, but be able to process $CO_2$ in the atmosphere and absorb it into its titanium dioxide skin.

Each of these floating cities (main and also bottom right picture in fig. 6.8) are designed to hold approximately around 50,000 people. A mixed terrain man-made landscape, provided by an artificial lagoon and three ridges, create a diverse environment for the inhabitants.

Each Lilypad is intended to be either near a coast, or floating around in the ocean, traveling from the equator to the northern seas, according to where the gulf stream takes it.

They inspire creative solutions, which at some point, may actually provide a real solution to the climate change problem."

**Second example** is selected from the works[128] of architect E. Kevin Schopfer and Tangram 3DS well-known also with a project NOAH (New Orleans Arcology Habitat; *https://www.designboom.com/architecture/e-kevin-schopfer-noah-new-orleans-arcology-habitat*), based on the principle of **Arcology** (a mix of **ARC**hitecture and ec**OLOGY**).

"It's been almost a year now since Haiti was ravaged by a horrific earthquake[129], and while its citizens are still picking up the pieces, the good news is that there is no shortage of creative ideas about how to rebuild an even better, more sustainable infrastructure for the country.

One of these ideas comes from architect *E. Kevin Schopfer* and *Tangram 3DS*, who envision the new Haiti to have a floating city on which people could produce food and promote industry. Called *Harvest City*, the collection of islands would be a fully functioning community of 30,000 residents and could be a key player in Haiti's recovery.

---

[127] this fragment is cited from https://inhabitat.com/lilypad-floating-cities-in-the-age-of-global-warming/

[128] pictures and text are cited from https://inhabitat.com/harvest-city-floating-islands-to-rebuild-haiti/

[129] The *Haiti earthquake* was a catastrophic magnitude 7.0 earthquake, occurred at 16:53 local time (21:53 UTC) on Tuesday, 12 January 2010, with an epicenter approximately 25 kilometers (16 mi) west of Port-au-Prince, Haiti's capital. By 24 January, at least 52 aftershocks measuring 4.5 or greater had been recorded. An estimated three million people were affected by the quake. Death toll estimates range from 220,000 to 316,000, to Haitian government figures. Cited from https://en.wikipedia.org/wiki/2010_Haiti_earthquake

Harvest City would be a place for Haitians to live and start their lives again, but it would also be a place for agriculture and jobs to thrive.

Two thirds of the city would be dedicated to farming and one third to light industry. The city would be composed of a collection of tethered, floating modules that span a diameter of 2 miles.

**fig. 6.9.** *Harvest City Haiti* – the floating islands by *E. Kevin Schopfer;*
(look e.g. at https://inhabitat.com/harvest-city-floating-islands-to-rebuild-haiti/)

Divided into four zones interconnected by a linear canal system, neighborhoods would be made up of four story housing complexes. The outer perimeter of the city would be composed of crop circles with secondary feeder canals, while the city center with schools, offices, and public space would be located in the inner harbor area.

The floating islands of Harvest City will be secured to the sea bed by a cable and were designed to weather hurricanes and typhoons. A low profile, low draft dead weight capacity and perimeter wave attenuators are some factors that Schopfer incorporated into the city to ensure it would be safe from storms. A breakwater using the concrete rubble debris from the earthquake would also be constructed to add to the city's stability."

There are many examples to be collected today in this direction.

But next ***third example*** is the last for this case collection. This is[130] *The Environmental Island, GREEN FLOAT, by Shimizu Corporation*, Japan, contributing into urban development, energy saving and $CO_2$ reducing among the main company's values. *Shimizu* is changing the future and changing the nature of abundance by taking on the challenge in two areas of innovation – *"Plant-like City"* and *"Floating City"*.

You can see that futuristic floating cities seek to provide a solution to many of our environmental problems, like rising sea levels, increasing temperatures and dwindling resources.

*Shimizu Corporation* has been hard at work coming up with some pretty crazy concepts lately, and *Green Float*, the Environmental Island is one of them (fig. 6.10). Designed for the equatorial pacific, presumably near Japan, *Green Float* is a concept for a series of floating islands with eco skyscraper cities, where people live, work and can easily get to gardens, open space, the beach and even "forests". Islands are connected together to form modules and a number of modules grouped together form a "country" of roughly 1 million people.

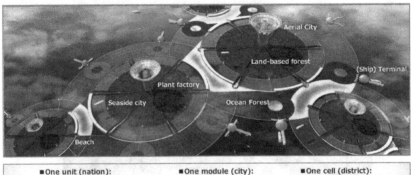

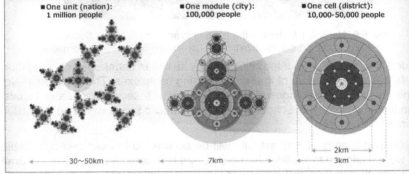

**fig. 6.10.** *The Environmental Island, GREEN FLOAT, by Shimizu Corporation*

---

[130] pictures and text fragments sited from https://www.shimz.co.jp/en/topics/dream/content03 and https://inhabitat.com/futuristic-floating-city-is-an-ecotopia-at-sea/floatingisland-ed01

A 1,000 m tower (fig. 6.11) in the center of the island acts as both a vertical farm as well as a skyscraper with residential, commercial and office space. The green space, the beach, and the water terminal on the flat plane of the island are all within walking distance.

Energy for the islands would be generated from renewable sources like solar, wind, and ocean thermal, and they also propose to collect solar energy from space, presumably from their own fantastic idea to install a solar belt on the moon.

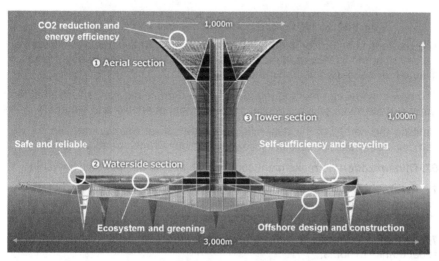

**fig. 6.11.** Main areas at *GREEN FLOAT*

So, it should be a completely self-sufficient floating *"ecotopia"* (**ECO**logy and u**TOPIA**) that is covered in vegetation, generates its own power, grows food, manages waste, and provides clean water.

There are so many opportunities for master-students of dozens of specialties to model creative engineering solutions in this project. Let's consider one such opportunity using the example of assembling this island on the water.

This is *Offshore Ultra-High-Rise Construction ("Smart" System Float-Over Dock)* – a special method used for building ultra-high-rise towers offshore.

The building is not built upward off the ocean. The frame of the structure built at ground level, and the assembled structure is temporarily submerged in the ocean. When the frame assembly has been completed, the buoyancy of the ocean is used to lift it up in one motion.

This method enables construction work to always proceed at ground level without people or materials being raised, so it enables safe and efficient construction (fig. 6.12).

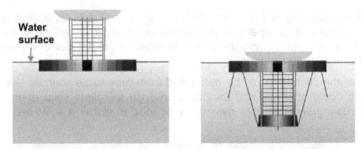

**fig. 6.12.** Method of *GREEN FLOAT* assembly "on / in water"

The previous construction method (fig. 6.12 on the left) is applied at a height of several hundred to thousand meters; this method requires the transportation of a large number of people and materials.

The new construction method (fig. 6.12 on the right) is always performed at the water surface level, so it minimizes the movement of people and materials.

Omitting the details of the assembly process (look at *Shimizu*-website[131]), we can confidently extract here the following dominant transformation models:

**07. Dynamization:** to make giant construction and its components movable;

**11. Inverse action:** not to build the components higher and higher, but mount on top and to lower the entire structure down under water; and entire giant tower will be raised after the completing assembly;

**14. Use of pneumatic or hydraulic constructions:** the whole structure stays afloat thanks to floats;

**18. Mediator:** the floats and floating mega-platform; the cables;

**32. Counter-weight:** the floats and cables;

**34. Matryoshka (nested doll):** the entire structure is gradually mounted inside the central space of the mega-platform below the surface of the water;

**37. Equipotentiality:** assembly at ground and water levels eliminates the rise of materials and people to an increasing height, as it would be in the case of traditional assembly.

To all this, we can add that the actual ***placement of such a city on the water*** means the exact implementation of model **05. Separation:** *separate the "incompatible part" ("incompatible property") from the object or – turned completely around – include the only really necessary part (necessary property) into the object*; and model **12. Local property:** *c) every object should exist under conditions that correspond best to its functions.*

> ***Master-students can and should by such modeling
> increase their creative skills!***

---

[131] https://www.shimz.co.jp/en/topics/dream/content03/

## 6.2. Urbanization with String SkyWay Systems by Unitsky[132]

### 6.2.1. Story

The ancient states grew along roads, and I think that modern and future states will build *String SkyWays* to better lands.

Such or nearly such a concept became the guiding principle of an engineer and inventor Anatoly Unitsky more than 40 years ago.

And now we can see a great progress in the practical implementation of Unitsky's engineering concepts and inventions.

Strategic goals are:

1) effective and ecology friendly transportation systems – urban, regional, interregional or even intercontinental in different categories: passenger and cargo, moderate in speed and high-speed, individually-oriented and massive in capacity;

2) "ideal" transport communication net between metropolises and the surrounding areas within a radius of 100-200 km.

These goals can be achieved both locally and globally only on the basis of Unitsky's string transport systems.

### 6.2.2. String rail

Key concept of new transportation systems is a *string rail*.

The essence of this invention is the completely new rail construction (fig. 6.13).

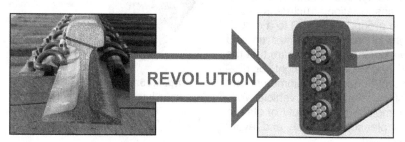

**fig. 6.13.** String rail construction

Pursuant to the TRIZ-model that we know as ***"Matryoshka"*** (named after the famous Russian nested doll), the new rail is made of dozens and hundreds of wires. Each such wire is tight as a string, and all of them together make one powerful "string rail" which is completely straight!

---

[132] this part is based on the materials of *SkyWay Group of Companies* established by *Anatoly Unitsky*; cited mainly from https://rsw-systems.com, www.yunitskiy.com, https://www.skyway-uk.com/technologies/linear_cities.html and monograph *"String Transport Systems: on Earth and in Space"*, edition of 2019, by Anatoly Unitsky

In fig. 6.14 you can see the relative "sag" radii of motorway surface (Ram = 1, used as the base unit), high-speed railroad rail (Rrr = 3), and Unitsky string rail (Rsr = 10).

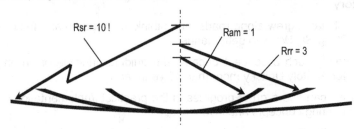

**fig. 6.14.** Rigidity and straightness of Unitsky's road is 10 times better than those of motorway road surface, and 3 times better than those of a rail of high-speed railway

### 6.2.3. SkyWay

Second key concept is providing passenger and cargo transportation in a separate space – on the "second level" above the ground – *SkyWay!*

The second dominant TRIZ-model which is de facto realized here is *"Transition into another dimension"* – in all senses of the word!

Technically, this means that the entire transportation structure is raised above ground (fig. 6.15) to a certain height in order to minimize up and down sloping, and to assure that the road is completely straight for dozens and hundreds of kilometers and that flight in a SkyWay vehicle while contemplating the beauty of Nature gives pleasure to each and every passenger!

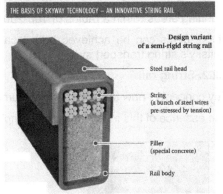

**fig. 6.15.** Explanation of smart structure of modern string rail (doc. 2017_59.pdf from http://www.yunitskiy.com/news/2017/news 20170809_en.htm)

The fact that SkyWay vehicles move above the ground on a specially designed rail-string overpass ensures a number of advantages:

- optimized aerodynamics,
- increased speed,
- unprecedented safety,
- rational use of land and resources,
- minimized environmental damage caused by transport.

In addition, the cost of construction and operation is significantly lower compared to the existing transport solutions.

**fig. 6.16.** SkyWay – string roads ... "in the sky"

### 6.2.4. Transportation modularity

Third key concept is modularity and standardization of classes and type of transportation modules (fig. 6.17-6.19).

The SkyWay systems can meet a wide range of transportation demands offering a possibility of highly efficient passenger and cargo transportation for any distances under any natural and climatic conditions.

All types of SkyWay systems are distinguished by energy efficiency, minimal adverse environmental impact and a high safety level of passenger and cargo transportation.

The SkyWay modular vehicles on steel wheels characterized by highly aerodynamic design, equipped with an anti-derailment system and a smart system of safety, control, power supply and communications.

## HIGH-SPEED UNIBUS
U4-362

Passenger capacity .........................................…..6 people
Maximum speed ........................................... 500 km/h
Type of track structure ........................... truss overpass
Power usage ...............................0.11 kg / 100 km x pass

**fig. 6.17.** SkyWay high-speed vehicle example

## UNIWIND
U4-651

Passenger capacity ......................................…..2 people
Maximum speed ........................................…..... 150 km/h
Type of track structure ..................... super-light rail-string
Power usage ..................….............0.3 kg / 100 km x pass

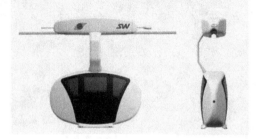

**fig. 6.18.** SkyWay urban vehicle example

## UNITRUCK
U4-131

Load-carrying capacity ...........................…..... 1700 kg
Maximum speed ................................... 150 km/h
Type of track structure ...... semi-rigid, flexible, truss
Power usage ....................….............1.05 kg / 100 km x t

**fig. 6.19.** SkyWay cargo vehicle example

### 6.2.5. Transportation infrastructures

Fourth key concept is infrastructures organization.

There are many projects developed by company SkyWay to propose newest urban, natural and industrial infrastructures with highest transportation efficiency.

## URBAN TRANSPORT

**Designed for short distance passenger transportation (up to 200 km)**

It fits harmoniously into the existing infrastructure of any megalopolis.

It solves transport problems of large cities by developing a network of high-rise building connected with each other by elevated (air) transportation

**SkyWay for DUBAI**

**fig. 6.20.** One of the SkyWay demonstrations for UAE

**fig. 6.21.** Plan for SkyWay Innovation Center in Sharjah, United Arab Emirates (https://www.actu-transport-logistique.fr/ferroviaire/systeme-de-transport-innovant-skyway-marque-des-points-dans-les-emirats-arabes-unis-511679.php)

**fig. 6.22.** No traffic jams in the urban SkyWay (demo)

### 6.2.6. MTRIZ MODELING

To reveal the creative engineering thinking in key string concepts is possible with MTRIZ-modeling which cases are shown in fig. 6.23 and fig. 6.24.

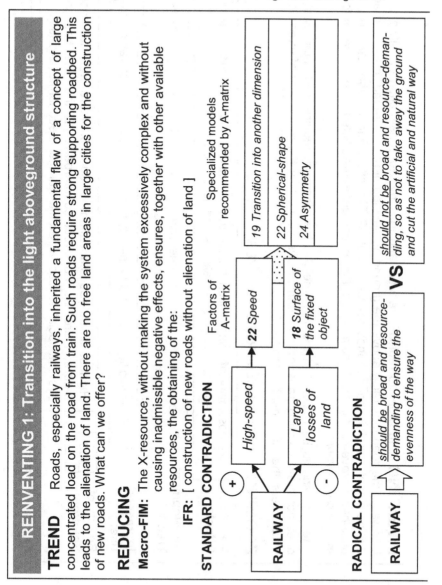

**REINVENTING 1: Transition into the light aboveground structure**

**TREND** Roads, especially railways, inherited a fundamental flaw of a concept of large concentrated load on the road from train. Such roads require strong supporting roadbed. This leads to the alienation of land. There are no free land areas in large cities for the construction of new roads. What can we offer?

**REDUCING**

**Macro-FIM:** The X-resource, without making the system excessively complex and without causing inadmissible negative effects, ensures, together with other available resources, the obtaining of the:

**IFR:** [ construction of new roads without alienation of land ]

**STANDARD CONTRADICTION**

RAILWAY
(+) High-speed → **22** Speed
(−) Large losses of land → **18** Surface of the fixed object

Factors of A-matrix

Specialized models recommended by A-matrix

**19** Transition into another dimension
**22** Spherical-shape
**24** Asymmetry

**RADICAL CONTRADICTION**

RAILWAY

*should be broad and resource-demanding to ensure the evenness of the way*

**VS**

*should not be broad and resource-demanding, so as not to take away the ground and cut the artificial and natural way*

**fig. 6.23 (beg).** Reinventing of the light aboveground structure of SkyWay

**INVENTING**   In order to radically reduce land loss by the construction of railways and highways, the road may raise above the ground as development of "air-land" class of roads – the dominant principle corresponds to the model *19 Transition into another dimension: a) possible improvement in the transition from the movement in the plane to the space; b) to use multi-storey layout.*

The road should be realized at the level (height) that does not require land withdrawal and does not break existing transport communications, as well as the movement of streams and small rivers, the movement of animals.

**ZOOMING   Contradictions Removed:   YES**

| | |
|---|---|
| **Super-Effects:** | The possibility of construction high-speed railway because of straightness of the road realized without steep slopes and abrupt turns. |
| **Development Trend:** | To relieve traffic flows in large cities. To unload the passenger roads by construction of separate cargo roads. |
| **Increased use:** | The possibility of use within the city to relieve the problem of traffic jams. |
| **Beauty of Solution:** 100 | Highest grade. Substantiation: 1) such structure was not known before, there are no direct counterparts; 2) the idea has great system, social, functional, architectural and economic potential. |

**fig. 6.23 (end).** Reinventing of the light aboveground structure of SkyWay

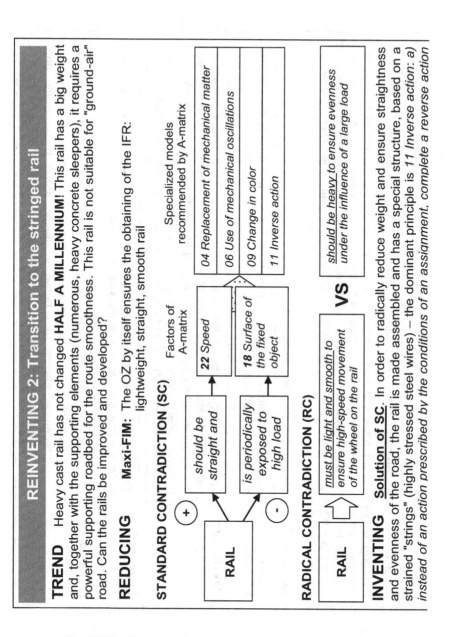

## REINVENTING 2: Transition to the stringed rail

**TREND** Heavy cast rail has not changed **HALF A MILLENNIUM!** This rail has a big weight and, together with the supporting elements (numerous, heavy concrete sleepers), it requires a powerful supporting roadbed for the route smoothness. This rail is not suitable for "ground-air" road. Can the rails be improved and developed?

**REDUCING** **Maxi-FIM:** The OZ by itself ensures the obtaining of the IFR: lightweight, straight, smooth rail

**STANDARD CONTRADICTION (SC)**

RAIL

(+) *should be straight and*

(-) *is periodically exposed to high load*

Factors of A-matrix

**22** *Speed*

**18** *Surface of the fixed object*

Specialized models recommended by A-matrix

*04 Replacement of mechanical matter*

*06 Use of mechanical oscillations*

*09 Change in color*

*11 Inverse action*

**RADICAL CONTRADICTION (RC)**

RAIL

*must be light and smooth to ensure high-speed movement of the wheel on the rail*

**VS**

*should be heavy to ensure evenness under the influence of a large load*

**INVENTING** **Solution of SC.** In order to radically reduce weight and ensure straightness and evenness of the road, the rail is made assembled and has a special structure, based on a strained "strings" (highly stressed steel wires) – the dominant principle is *11 Inverse action: a)* *instead of an action prescribed by the conditions of an assignment, complete a reverse action*

**fig. 6.24 (beg).** Reinventing of the string rail for SkyWay road

(do not make rail heavier, and radically lighten it); *04 Replacement of mechanical matter: c) to move from unstructured fields to fields with a specific structure* (rail should be a composite structure with controlled parameters, in particular, evenness; www.yunitskiy.com).

**Solution of RC. 01 In space:** load-bearing part of the rail (string structure) made high-strained, and the entire structure – with normal tension of materials; and rail as a whole system is becoming highly-even in string tension process by assembly directly on the masts of future road; **03 In structure:** rail is radically structured; **04 In material:** cheap wire instead of a expensive cast rail.

**ZOOMING    Contradictions Removed:    YES**

| | |
|---|---|
| **Super-Effects:** | The possibility to achieve extremely high speeds for a pair of "wheel - rail" (for example, up to 1000 km / h). |
| **Development Trend:** | **The possibility to build the roads through forests, rivers, swamps, permafrost, etc.** The possibility of construction high-speed freight roads (for example, multi-ton units – oil, ore, products, fruits and vegetables, etc., drinking water! – that is particularly important – at a speed of 500 km / h) – **this goal is set for the first time in the history of land transport.** |
| **Increased use:** | The possibility of building a highly smooth and durable airfields, highways, buildings, the ports remote in the sea, and others. |
| **Beauty of Solution:** 100 | The idea of outstanding beauty and efficiency, meeting the TRIZ laws of systems development: inventive use of special physical-technical effects (the behavior of the string under static and dynamic loads), separation, structuring, controllability. |

**fig. 6.24 (end).** Reinventing of the string rail for SkyWay road

### 6.2.8. Linear Cities

*A **linear city*** is an urban settlement of cluster type in which the surface of the earth is intended for pedestrians and green plants, and transport, energy and information communications are located above the ground on the "second level" (on special supports).

***The basic principle*** of the construction of each infrastructure cluster is a pedestrian quarter, in which comfortable low-rise buildings with the widespread greening of urban areas and the use of renewable energy sources are located between multifunctional high-rise buildings, all connected by horizontal lifts.

***The key element*** of the system are horizontal lifts (transport arteries), connecting the neighboring high-rise buildings, settlements, residential, shopping, entertainment and other clusters, allowing people to comfortably move between them in a matter of minutes. An important advantage is that the cost of public transport can be included, as with conventional elevators in the buildings, into the cost per square meter of a quarter of the linear city while maintaining the average comparative cost of new housing.

**fig. 6.25.** SkyWay linear city in the most comfortable for living green area

**fig. 6.26.** Next idea as residential cluster in SkyWay Linear City (due to its size the SkyWay tracks are virtually almost invisible at these pictures)

### Advantages of the SkyWay transport and infrastructure complex:

### Low net cost of construction and transport service

Construction of the SkyWay transport complex is less costly by 2–3 times compared with railroad (tram) complexes; by 3–5 times – than automobile complexes, by 10–15 times – than a monorail road and by 15–20 times – than a magnetic levitation train, with simultaneous reduction of transport net cost by 3–5 and more times.

### Low power consumption

Power consumption is by 5–7 times lower than in the existing transportation systems that use steel wheels or magnetic levitation; by 15–20 times lower in comparison with vehicles on pneumatic tires (motor transport) or using air cushion (aircraft, ground effect vehicles, helicopters).

### Minimum land acquisition

The acquisition of land for the transport and infrastructure complex is less by approximately 100 times in comparison with motor-road and rail (tram) systems. Moreover, in case of using a suspended system for communication between urban high-rise buildings there is no costly land acquisition for transport at all.

### Complete automation

Automation of the transport and infrastructure complex and optimal transport logistics; in the framework of the linear city the average time spent by a passenger in transit from home to work is not more than 15–20 minutes at distances up to 15 kilometers.

### Highest safety level

High transport, environmental and anti-terrorism safety, due to the exclusion of intersections, pedestrian crossings and oncoming lanes, where frontal collisions of rolling stock are possible. In addition, the high-rise string-rail track structure has a tenfold safety factor and is inaccessible to vandals, and the rolling stock is equipped with an anti-derailment system.

### High speed performance

Urban transport – up to 120–150 km/h; high-speed intercity transport – up to 450–500 km/h.

### Minimum operating costs

Operating costs are reduced by 5–7 times compared with automobile and by 2–3 times in comparison with railway (tram) transport.

### Use of renewable power sources

The SkyWay transport complex with electric motors for efficient and cost effective movement on a 'steel wheel – steel rail' design (including the application of a unique motor-wheel) can cover 100% of its power needs by renewable energy sources – solar and wind.

### Recovery of live fertile soil

Live fertile soil and natural ecosystems (flora and fauna) can be recovered in any natural environment, including a desert.

### Lack of harmful effects on humans and the natural environment

The harmful effects of exhaust gases, noise, vibration, electromagnetic and other radiations are reduced by 15–20 times compared with motor transport and by 2–3 times – with rail (tram) and monorail ways.

### Saving time and financial expenses

Savings for passengers and cargo shippers: time travel (cargo delivery) within city limits is reduced by 1–1.5 hours per day or by USD 20–30 per day calculated per one passenger or a ton of cargo.

### Short payback period

The SkyWay transport and infrastructure complex has an unprecedented payback period: from 2 to 3 years.

### Presentation of SkyWay Linear City

Specialists of company SkyWay have prepared a presentation of SkyWay Linear City, that was presented to the top-management of Ministry of Transport in Abu Dhabi. Hard-to-reach island Al Hudayriat to the south of Abu Dhabi in the United Arab Emirates was chosen as an example for the construction.

**fig. 6.27.** Bird's-eye view of the SkyWay Linear City built in the desert with the restored 30 cm thick (!) layer of fertile soil

Here it is very important to underlie the **environmental benefit** due to zero carbon emissions when using SkyWay technology.

Pedestrian city will eliminate the problem of gas pollution of cities, eliminating the need for residents to use vehicles with internal combustion engines for urban travel.

Linear city logic suggests walking distance to the nearest SkyWay transport station of no more than 250—400 meters (there are 4 transport stations servicing different lines within walking distance).

In hot climate and under the bright sun all the streets of each Linear City block (cluster) are connected to the nearby transport stations with pedestrian green "tunnels" made of climbing and flowering plants, vines and shrubs offering comfortable walking to the urban transport station.

Any resident of the cluster block can use individual environmentally friendly vehicles (electric car/Segway/bicycle) to move across the Linear City (parking spaces for private vehicles provided next to the station). At the same time in each cluster 210 tons of fuel will be saved within one year in comparison with conventional transportation, with all the ensuing environmental benefits (see also above).

Furthermore, SkyWay transport system unibuses are a variation of the high-efficiency electric cars, just running on steel wheels. They have a 70-90 times higher energy efficiency, than for instance, Tesla cars — 4-5 times higher efficiency due to replacement of pneumatic tires with steel wheels and 18 times higher efficiency due to slacking track between the adjacent city transport system towers.

And next extremely important point: **biospheric advantage** die using the innovative agricultural technologies for landscaping large desert areas.

Restoring fertile soil on the territory of the Linear City, capable of providing the residents with natural products grown in the gardens within residential areas and on the roofs of private houses (berries, fruits, vegetables, salads, etc.) on the high-quality natural soil without the use of GMOs, chemicals, herbicides and pesticides.

Creating natural fertile soil (e.g. 5% humus delivered from Belarus, +95% local soil) in the territory of Linear City for growing gardens and creating parks, including on the roofs of buildings, putting up hedges and green pedestrian tunnels.

When mixed with local soil in certain proportions the humus soil reproduces and gives the soil black earth fertility and stimulates rapid plant establishment, growth and development of all planted trees, shrubs and plants, substantially reducing the need for irrigation.

Consumption of water when using the proposed technology is reduced 5-7-fold on average. This is due to the specific and unique property of accumulation and retention of moisture by the powdered soil. And by adding small amounts of a liquid activator to irrigation water the need for daily watering of plants and trees is eliminated, however, despite this, the roots get sufficient daily amounts of water and substances required for successful growth and development.

The cost of creating a 30-cm fertile soil layer on the area of 1 hectare (10 000 m2) in the desert of Abu Dhabi:

1) Soil (humus) – 200 tons.  Cost – 250 thousand  USD (including the delivery from Belarus).

2) Activator – 500 liters. Cost – 5,000 USD (incl. the delivery from Belarus).

3) Local soil (at 20/1 ratio with the restoring agent) – 4,000 tons. Cost (including local transportation and mixing) – 40,000 USD.

Total: the cost of creating a 30-cm fertile soil layer on the area of 1 hectare – 295,000 USD.

*FYI:* The construction costs per 1 m2 of asphalt city road average to 1,500 USD. The costs of creation of 1 m2 of 30-cm thick natural fertile soil is 29.5 USD. This a is 50-fold difference! It iis extremely inexpedient to "bury" the city soil under the asphalt.

This soil will grow a garden that will bear fruit and serve to the benefit of the people for centuries. Its green plants will produce oxygen, capture dust, absorb the noise of the city, create micro-climate and provide the biologically active substances beneficial for health and needed by every city dweller – phytoncides that kill and inhibit the growth and development of bacteria, microscopic fungi, and protozoa.

The proposed technology can significantly reduce the need for irrigation water (tenfold), and, in some cases, completely abandon or significantly reduce the use of various kinds of chemical, mineral and organic fertilizers.

**fig. 6.28.** Linear City with fertile soil in the ground and off the shore

**fig. 6.29.** Linear City in place of the former desert, with residential clusters on the shore, on islands and off the shore

**fig. 6.30.** Project demo-case: National Park on the Al Hudayriat Island of Abu Dhabi with SkyWay Linear Cities around the perimeter of the island and off the shore (2019)

### 6.2.9. Looking to the Future

To make resume for previous materials it would be useful to recall some facts about role of TRIZ for the works of engineer and inventor Anatoly Unitsky who received in 1985 a diploma of higher education in the second specialty – "Patent Law and Inventing" and was in 1987-1988 Head of the patent and licensing service of the Institute of Mechanics of Metal-Polymer Systems of the Academy of Sciences of the BSSR in Gomel.

Anatoly mastered[133] the TRIZ and made a rich treasure trove of practical patenting experience, just like Einstein who spent several years working in a Swiss patent office and later described that period with these words: *I learnt to separate the wheat from the chaff.*

In all probability, without that background the great physicist would not have developed his relativity theory, nor would Unitsky have designed his string transportation systems.

PROSPECTS FOR SKYWAY TECHNOLOGY APPLICATION:

➢ All innovative SkyWay components can be manufactured at the places of project implementation using the existing technological base

➢ Exploration and development of underdeveloped and hard-to-reach territories, creation of a single network of cargo, urban and high-speed intercity tracks.

➢ Maximal reduction of capital and operating expenses for transport and infrastructure construction.

➢ Qualitative change in the economic structure of countries and increased GDP.

➢ Integration of countries into international transport corridors, creation of a fundamentally new logistics of the 21st century.

➢ Development of related branches for track structure and rolling stock manufacture (metallurgy, chemical, petrochemical and radio-electronic industries, machine building, construction, etc.).

And to complete this case studying we would like to address to Master program students the words[134] of *Thomas Welser*, the Head of the company *Welser Profile Austria GmbH*:

*History is made by people with vision. Thank you for your book – a document with a clear vision comes true and a strong impact on history in transportation. Thank you for being part of it.*

---

[133] invention motivation and development by Anatoly Unitsky is written in TRIZ-chapter *"TIPS-based (Theory of Inventive Problem Solving) history of Unitsky's String Technologies"* of Unitsky's monograph *"String Transport Systems: on Earth and in Space"*, edition of 2019

[134] cited from and about the Unitsky's monograph *"String Transport Systems: on Earth and in Space"*, edition of 2019; https://rsw-systems.com/news/welser-profile

# 7 INDUSTRIAL CASES[135]

The cases discussed below have been selected from dozens of projects, as they do not contain confidential information sensitive to customers. Next idea was do not use too particular, professional and difficult examples. Nevertheless, these cases illustrate the purposes, character and sometimes the solutions of including the TRIZ-modeling in certain projects.

Almost all projects were accompanied by introductory trainings of customer's specialists on the basis of MTRIZ. Also, some cases show seminars and trainings at which students received MTRIZ-solutions of problems directly in the classroom or lecture hall.

## 7.1. Does SIEMENS like simple solutions?
## Case 1. Sorting Drum

This case is demonstrated at the Master lessons as an educational example in simplified format because of customer's know-how restrictions. Nevertheless I think that it is not out of place to say that this case is a small example from a many-stage project completed 22 years ago.

In fig. 7.1,a you can see a rotating drum with a diameter of about 2 meters and a length of 5 meters. The cylindrical surface of the drum has longitudinal slots of a given width. Envelopes are fed inside the drum for sorting by thickness.

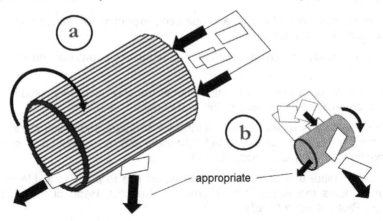

**fig. 7.1.** Simple and effective sorting machine on the TRIZ-model "Inverse action"

If the envelope is thinner than the slot width, then it falls down onto the conveyor for transmission to automatic processing and dispatch. If the envelope is thicker than the slot width, then such an envelope passes along the drum and goes to another conveyor for manual processing.

---

[135] MTRIZ courses were provided for last 20 years in Master programs at dozens of universities and companies in different countries

© Springer Nature Switzerland AG 2020
M. A. Orloff, *Modern TRIZ Modeling in Master Programs*,
https://doi.org/10.1007/978-3-030-37417-4_7

There are the following problems: 1) envelopes sometimes get stuck in the slots inside the drum, these envelopes are difficult to get, these envelopes obstruct the movement of other envelopes, so that the productivity of the machine decreases; 2) the high energy consumption of the machine and 3) the machines take up a lot of space in the envelope processing workshop.

Preliminary training of a small group of specialists allowed for a long time to solve such problems for many real situations. In this case, a new drum (fig. 7.1,b) was proposed in which the envelopes are not fed in, but onto the surface of the drum. It is implemented, as the dominant, TRIZ-principle **"Inverse action"**.

As a result, the new drum has approximately the same total length of the sorting slots, but its diameter is three times and its length is 5 times smaller, and in general, its multi-parameter efficiency coefficient is 55 times better!

Here, envelopes get stuck very rarely, because if the envelope is stuck in the slot, then after the drum is rotated, this envelope itself falls down from the slot under the influence of its weight and the inertia of movement from the rotation of the drum.

If the envelope is firmly stuck in the slots, then a special brush removes it, and this is much easier than removing the jammed envelope in the old drum.

Generally saying this is also an implementation of key TRIZ-law of **F-expansion & P-compression** (look in part 1.7.2).

## 7.2. Does SIEMENS like simple solutions?
### Case 2. Labeling Machine

This device (fig. 7.2) attached the label to envelop with a compressed air impulse.

When the specialists tried to increase the speed of the envelope by 2 times, it turned out that the label often misses the desired place or attaches obliquely. This happens both because the impact of the air impulse on the surface of the label was not always uniform, and because the thin label was easily deformed.

After training we came up with a solution using the Meta-Algorithm of Invention T-R-I-Z as follows:

*Step 1: Trend (diagnostic). Things will get harder!*

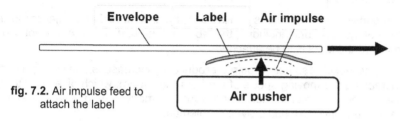

**Envelope**   **Label**   **Air impulse**

**fig. 7.2.** Air impulse feed to attach the label

**Air pusher**

The main causes of the problem are as follows:

First, the label becomes deformed, and its gluey side touches the surface of the envelope too late (or, sometimes, too early).

Second, there emerge, in the air adjacent to the surface of the moving envelope, aerodynamic effects preventing normal attaching the label to an envelope.

Third, sometimes the air impulse bends the label in the middle, and when such label, attached only at the center, starts moving together with the envelope, its leading edge gets hit by the oncoming rush of air. As a result, the label curves up, shifts sideways or even comes unstuck, and the entire operation ends in a failure.

### *Step 2: Reducing (reforming). What cannot be changed?!*

Two models give a relatively good idea of the roots of the problem.

One of the possible Standard Contradictions:

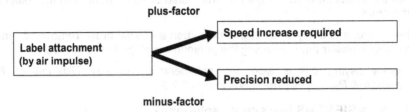

**fig. 7.3.** Possible wording of the Standard Contradiction for the labeling problem

A Radical Contradiction variant:

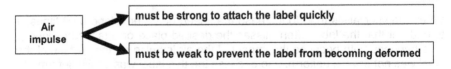

**fig. 7.4.** Possible wording of the Radical Contradiction for the labeling problem

Required "ideal result": the label is attached **"quickly-AND-precisely"** at any speed, however high!

Let us consider the resources available in the *Operative Zone* adjacent to the surface of the envelope, including the label and the portion of the envelope surface where the label is to be attached.

*Object (systemic) resource*: required productivity increase is achieved primarily by increasing the moving speed of the item (envelope), which gives rise to the problem (poor precision). There is no "room for maneuvering" here, as even a 2-times speed increase has proven problematic.

*Informational resource*: it may be possible to improve precision of measurements to optimize the air impulse timing.

*Functional resource*: air impulse parameters can be fine-tuned, but this looks too complicated. It is difficult to generate an air burst of the "right shape" and release it exactly at the "right moment".

*Structural resource*: the directions in which the envelope and the label are moving are mutually orthogonal, i.e. they move at right angles with respect to each other (see bold arrows in fig. 7.2). When the envelope begins to move faster, it becomes increasingly difficult for the label to "hit the target". Changes in this area are quite possible.

*Spatial resource*: it is possible to manage the "shape" of the air burst, but this solution does not seem viable (see *Functional resource*).

*Temporal resource*: there are virtually no "additional" possibilities (see *Informational resource*).

*Material resource*: the customer has forbidden to change the material of which the label is made.

*Energy resource*: increasing the power of the air burst will only exacerbate the problem. However, it is possible to align the directions of energy flows, which is close to the results of structural resource analysis.

Bottom line: within the framework of the original operating principle of the machine, only one resource – *structure of movement* – shows any promise at all in terms of finding a suitable solution.

To formulate an express solution, we can apply the MITO method and to build two Formal Standard Contradictions (below at Step 3).

**Step 3: Inventing (transformation). Creating conditions for the "ideal result"!**

**FSC1:** Labeling ▶+22. Speed **VS** -05. Precision of manufacture ▶02, 04, 09, 29

**FSC2:** Labeling ▶+22. Speed **VS** -14. Internal damaging factors ▶01, 05, 18, 33

Combining of two "results" we have a tuple as following: 01, 02, 04, 05, 09, 18, 29, 33.

Pursuant to model *02,b. Preliminary action*, the following recommendation is quite promising: *prepare objects in advance so that they can be put to work from the best position and are available without loss of time.*

Our attention is drawn to navigator *04,c. Replacement of mechanical matter* which can be interpreted as follows: *replace... unstructured fields with fields with a specific structure.* In our interpretation: *replace "badly" structured and "uncoordinated" movements with "well" structured and "coordinated" movements.*

And in any case, model *18. Mediator* looks very promising.

There is a nascent trend to modify and coordinate the trajectories of movement of the envelope and the label. The trajectory of movement of the envelope cannot be changed. All we can do is change the trajectory of movement of the label.

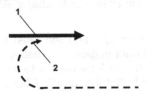

fig. 7.5. New matching of trajectories for efficient passage of energy (material)

Here is an idea for matching the trajectories – the trajectory of movement of the envelope is straight (1), and the trajectory of movement of the label could be made "rotational" (see fig. 7.5).

During such coordinated movement, the label is "rolled onto" the envelope (getting glued to it in the process), and then the envelope and the newly-attached label move together to the site of the address printing operation.

However, implementation of these trajectories in the "old" machine proves impossible. Moreover, subject to the overall trend (further increase of the speed at which the envelope moves along the process line), we come to the conclusion that the air impulse label attachment operating principle has exhausted itself.

We need a new assembly. A possible new technical solution is presented in fig. 7.6.

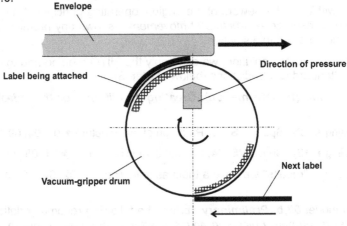

**fig. 7.6.** Label attachment mechanism in the form of a vacuum-gripper drum

It is proposed that the label attachment device be made in the form of a rotating drum with one, two or more "windows" in its surface (cross-hatched areas in the drawing). The labels are pressed to the surface of the drum by atmospheric pressure, as air from the inside of the drum is sucked out by an inbuilt pumping. The drum with sucking function realizes the model *18. Mediator!*

And really we have got a model **11. Inverse action** – not to push the label but to suck it and transport to labeling place at envelop.

And last but not least: there is a clear realization of TRIZ-trend **"effective closed paths and through-routes of energy"** (see at 1.7.2).

**Step 4: Zooming (verification). Is this a good solution?**

Analysis reveals the following: First, the original contradictions have been completely eliminated.

Second, the new principle is very promising in terms of further increase of the speed of operation! The customer conceded that the idea underlying this solution was efficient, even though it required a radical modification of the operating principle used by the label attachment assembly.

Third, the customer greeted the emergence of a "powerful" positive "super-effect": the letter processing shop would now become much quieter, because there would be no more incessant "rub-a-dub-dub" noise produced by the air burst generator of the old machine!

Incidentally, the consultant did not know about the noise problem produced by the "old" machine, and the "quiet" nature of the new machine was truly an unexpected and pleasant "surprise" side effect. This became the very useful super-effect of the new machine!

## 7.3. Does SIEMENS like simple solutions?
## Case 3. Current Collector of Trolley Boom

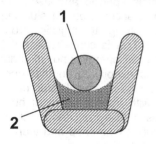

fig. 7.7. Damaging the contact strip under the abrasive influence of contact line

Typically made of graphite, contact strips (bushes) 2 are abraded during operation by the contact line 1 and are therefore subject to wear which requires replacement after certain period (fig. 7.7).

The extending along the contact line grinding pieces (graphite strips) of rod current collectors have a relatively short working area, which is ground continuously over its entire length and is therefore more worn. Such contact strips are replaced about every 1000 km, which means high maintenance.

During the studying and projecting many variants of new contact devices for current collecting, one of the most important goals was to increase the durability of contact strip.

Elementary TRIZ-analysis shows that increasing the durability could be connected with volume or mass of contact strips. It means change in space or/and material. Change in space means transforming shape, size and location.

A very simple solution[136] can be proposed as a graphite rod located so that one end of it could be in contact with the current wire, and the other end could be spring-loaded so that the sliding contact between the wire and the end of the graphite rod could be reliable and constant (fig. 7.8).

Despite the simplicity of this proposal, it has certain logic of invention and development, namely: applying the models such as *02. Preliminary action, 07. Dynamization, 12. Local property, 19. Transition into another dimension, 34. Matryoshka (nested doll)* and *40. Uninterrupted useful function*, as well as such TRIZ-lines of system resources development as "occupation of void" and "increase of the working tool dynamization".

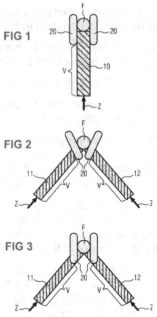

fig. 7.8. Useful change of contact shoe with long contact rod(s)

## 7.4. Training, Discussion and Solution of a Logistic Problem for an Automobile Concern at the Ajou University[137], Korea

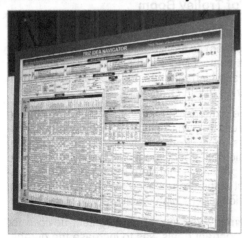

What is specific and useful to take in account during studying and solving to logistic problems? This is consideration the paths of energy and material from one point to another one. Here it was a complicated problem with transportation of hundreds details from feedstock to manufacturing shop and then to assembly shop.

fig. 7.9. Basic instrumental TRIZ-models from "bird's-eye view" on a wall poster in the lobby of Structural Mechanics Laboratory

---

[136] Application DE102011083894A filed by *Siemens AG* of 2011-09-30 "Current collector i.e. trolley boom, for track-bound trolley bus, has contact shoe provided with wear reserve part, and support comprising advancing mechanism for implementing advancing movement of contact shoe"; inventor – *Michael Orloff*

[137] unfortunately, Prof. S.H. Yoo, one of the TRIZ-pioneers in Korea, ex-president of Korea TRIZ association, suddenly passed away four years ago

And after joint MTRIZ-training, discussing and problem analysis one very important thing was revealed – *ineffective trajectories of objects (in space and time) at energy criterion.* In this case a wall poster "TRIZ Idea Navigator" has been used for problem solving by multi-professional team (fig. 7.9 and fig. 7.10).

After using such models as *02. Preliminary action, 05. Separation, 11. Inverse action, 12. Local property the trajectories, 19. Transition into another dimension 32. Counter-weight* and *37. Equipotentiality*, main workflows have been well optimized and made more energy effective and without many point for movement of the details up and down according to TRIZ-trend *"effective closed paths and through-routes of energy"* (see at 1.7.2).

Since workshops and warehouses previously developed gradually, without a single plan, now, on the one hand, the problem of optimizing space and workflows has become more complicated, but on the other hand, the availability of a new "ideal" plan has allowed such optimization to be carried out with an understanding of all the improvements and reasonable concessions.

**fig. 7.10.** Master' and Doctor' students at Ajou University under the supervising of Professor Seung-Hyun Yoo during MTRIZ-training and solving to logistic problem for automobile concern

## 7.5. Machine for Assembly of the Liquid Crystal Screens[138], Korea

The customer had not found a solution for assembly of the liquid crystal screens (LCS) during about three years at engineering department of 30 specialists. After discussing the problem during meeting with engineers, the TRIZ-project was realized for ONE MONTH and together with working up and handling the report to customer it required about two months.

In total, 27 solutions were developed on various principles, of which the first 7 solutions with the some simplest principle were incorrect and even inoperative.

Recall now the diagram in fig. 1.4 in part 1.3: the first wrong solutions were created because the customer did not provide us with complete and accurate information about the operation of this assembly machine. Moreover, I could not see the real size and operation of the machine and the assembly shop, and that was strange and unusual for my practice as a consultant.

Thus, at once two top aspects in a definition of a problem according to fig. 1.4 were actualized – **insufficient information** and **unreliable (false) information.**

As soon as the customer expressed his opinion about the first solutions and explained the disadvantages and courses of disadvantages, it immediately became possible to change the strategy of developing the solutions, and the next 20 solutions were developed.

**Problem story**

The company produced the machines of LCS-manufacturing for displays. Special technology of manufacturing uses two very thin (with thickness of less than 1 millimeter), flexible, being easily broken, glass sheets with size of every side to 2 meter.

Many screens of different size could be resulted from such pair of glass sheets. The schemes at the fig. 7.3 and fig. 7.4 explain the initial situation.

The liquid crystal fields are regularly drawn on the surface of the lower sheet according to the sizes of screens to be (fig. 7.3).

The epoxy glue lines are drawn on the lower surface of the upper glass according the contours of the screens to be (fig. 7.3).

Both sheets are pressed against each other in vacuum chamber to get a workpiece before cutting this workpiece into several screens (fig. 7.4).

So, the sheets should be introduced into the vacuum chamber and placed one above the other.

A robot grabs the first (lower) glass sheet (fig. 7.5,a) with dozens of fingers with vacuum suction cups, carries the sheet to the lower platform and presses the sheet to the platform.

---

[138] this case is demonstrated to the Master students as a sample of solving to "unsolvable" problem

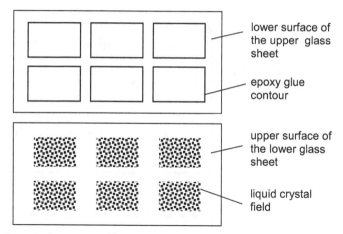

lower surface of the upper glass sheet

epoxy glue contour

upper surface of the lower glass sheet

liquid crystal field

**fig. 7.3.** Glass sheets for manufacturing the liquid crystal screens

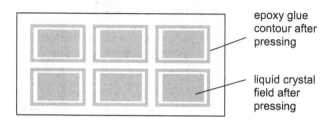

epoxy glue contour after pressing

liquid crystal field after pressing

**fig. 7.4.** Location of epoxy glue contours and liquid crystal layers after the alignment and pressing the glass sheets

After that, the platform attracts and holds the sheet by vacuum suction, namely by pumping air through the outlet tubes from the chamber and creating reduced air pressure in the openings of the platform adjacent to the glass.

After that, the robot grabs and transfers the second (upper) glass sheet to the upper platform, and the upper glass is sucked by vacuum suction cups (similar holes in the upper platform) and, thus, is held by the upper platform with a suction.

Air is pumped out of the chamber.

Then the upper sheet is gradually lowered to a very small distance of several tens of microns from the lower sheet (fig. 7.5,b), aligned in plan and then pressed against each other.

The problem lies in the fact that the suction must be realized *in vacuum chamber* to increase the defect-free screens. When creating a deep vacuum in the chamber, it becomes impossible to hold the upper glass by pumping air through tubes in the upper platform (fig. 7.5,c).

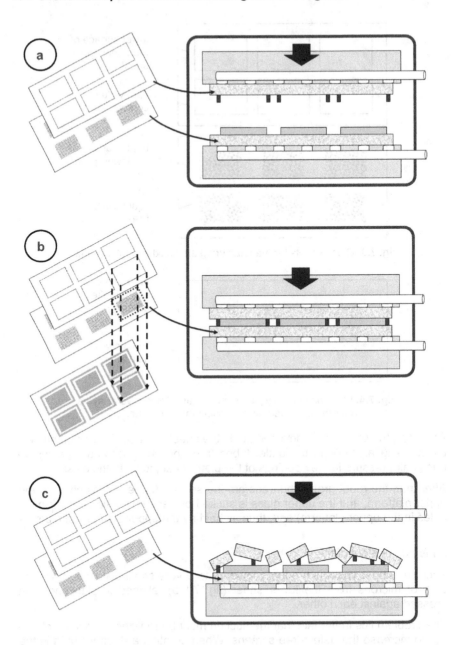

**fig. 7.5.** Transfer of the glass sheets to holding platforms in vacuum chamber

The customer has attempted to use the electrostatic forces to hold the glass sheets. This required much energy consumption.

According to TRIZ-trends, idea with electrostatic forces was very progressive and could be improved to make less energy consumption. But unfortunately, the customer prohibited using electrostatic field in this machine without any explanation.

I have thought that maybe a very strong electrostatic field could damage some electronic components placed at the glass sheets.

After that, it was necessary to start a new solution to this problem.

## TREND

### Problem extraction from initial situation

Lower sheet lies on the platform, and it only must be kept from the horizontal displacement.

Upper sheet must be held completely by unknown forces in deep vacuum, and this is a problem.

The force of gravity pulls upper sheet downward (fig. 7.5,c), but this sheet must be tightly, accurately and evenly held by upper platform.

### *How could it be done?*

Strategy of the solution depends on the determination of OZ, operative time and other resources.

The following versions are possible:

1) working only with the upper sheet at the operative time;

2) working with two sheets at the operative time;

3) joint studying pre-operative, operative and post-operative times;

4) combined studying of possibilities.

## REDUCING

### Operative Zone and the resources

Upper surface of the upper sheet (to hold!); space between this sheet and the platform; the surface of the platform above the sheet; gravity (weight of sheet); low vacuum.

The material of sheet (glass) and the material of platform (aluminum) must be taken in the attention.

### Inductor and receptor

Inductor – platform.

Receptor – glass sheet.

**Standard Contradiction(s)**

Platform S2 must hold a sheet S1 (fig. 7.6), but there are no effective interaction between platform and upper sheet. In other words: platform must hold the glass sheet, but gravitational field counteracts this goal.

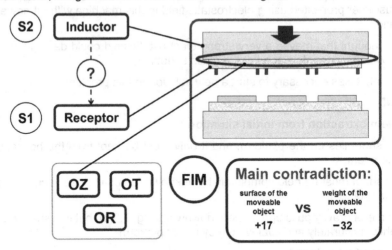

**fig. 7.6.** Main modeling components to describe the structure of problem "holding the glass sheet in deep vacuum under the technological platform" taking in account also selecting the 3-rd SC from the table below

| Nr. SC | Effect, Condition, Object | | | | | |
|---|---|---|---|---|---|---|
| | holding the upper glass sheet under upper platform in deep vacuum | | | | | |
| | ( + ) - factor | | | ( - ) - factor | | |
| 1 | to hold in va-cuum | 09 | ease of manu-facture | harm action of gravitational field | 13 | external damaging factors | 05, 18 |
| 2 | to hold in va-cuum | 09 | ease of manu-facture | expensive energy con-sumption for electrostatic holding | 37 | energy use by the mo-ving object | 03, 04, 10, 13 |
| 3 | large surface of the sheet | 17 | surface of the movable ob-ject | harm action of gravitational field | 32 | weight of the mo-vable object | 04, 05, 14, 19 |

**fig. 7.7.** Selected Standard Contradictions for problem modeling

## Radical Contradiction

Platform S2 must hold a sheet S1 to perform the main positive function, **BUT** platform S2 must not hold a sheet S1 because of there are no visible resources (fields, forces).

## Ideal Functional Result

Macro-FIM: X-resource, without producing the inadmissible effects and without complicating a system, ensures together with other existing resources

[ *the reliable holding the upper sheet under upper platform in deep vacuum* ].

## Detailed structure modeling (fig. 7.8)

The constructed model is incomplete, since there are only two "substances": S1 – receptor and S2 – inductor, and the field of interaction **Fm** between them is absent (!). There is also the harmful gravitational field **Fg**, which moves S1 away from S2.

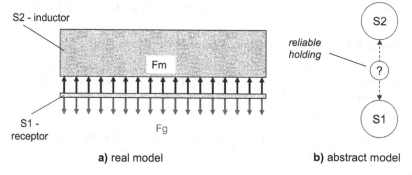

a) real model          b) abstract model

**fig. 7.8.** Substance and fields (forces) in Operative Zone

## INVENTING

## Solution to Standard Contradictions

From the table (fig. 7.7) we could extract ranged set of the navigators: 03, $04^2$, $05^2$, 10, 13, 14, 18 and 19.

The composing "portrait" of a solution to be:

*03. Segmentation* – disassemble an object into individual parts;

*04. Replacement of mechanical matter* – c) replacement of static fields with dynamic ones, from temporally fixed to flexible fields, from unstructured fields to fields with a specific structure; d) use of fields in connection with ferric-magnetic particles;

*05. Separation* – separate the "incompatible part" ("incompatible property") from the object or – turned completely around – separate the only really necessary part (necessary property);

**10. Copying** – a) use a simplified and inexpensive copy instead of an inaccessible, complicated, expensive, inappropriate, or fragile object;

**13. Inexpensive short-life object as a replacement for expensive long-life one** – replace an expensive object with a group of inexpensive objects without certain properties, for example, long life;

**14. Use of pneumatic or hydraulic constructions** – use gaseous or fluid parts instead of fixed parts in an object: parts that can be blown up or filled with hydraulic fluid, air-cushions, hydrostatic or hydro-reactive parts;

**18. Mediator** – a) use another object to transfer or transmit an action; b) temporarily connect an object with another (easily separable) object;

**19. Transition into another dimension** – b) do construction on several floors; tip or turn the object on its side; use the back of the space in question.

The possible solutions could be formed on the base of navigators 03, 04, 05, 10, 13, 18 and 19: *to use some copy of the sheet like mediator with a feature to hold the sheet on the whole upper surface ("a back") and to be held to a lower surface of the upper platform.*

### Solution to Radical Contradiction in general form

In this case it is necessary to find and use *new physical-technical effects* for obtaining the field Fm (fig. 7.8,b), with enough force for holding the sheet by the upper platform under the conditions of deep vacuum.

### Ideas of technical solutions

According to above mentioned models, many different solutions could be generated. We show one of them based on electromagnetic field.

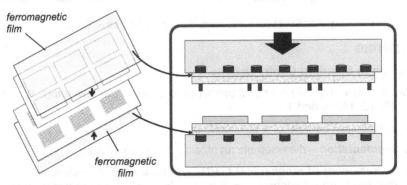

**fig. 7.9.** The use of thin ferromagnetic film cladded on the free side ("a back") of a glass sheet and "field" of small film electromagnets embedded in platforms

The ferromagnetic film can be applied to the free surface (on the "back") of a glass sheet using one of the known methods, and after the operations in vacuum the film should be removed.

## ZOOMING

The economic effect of replacing an electrostatic solution with an electromagnetic one is estimated at approximately 250-400 thousand euros per platform, and up to 0.8 million euros per machine.

At the same time, the economic effect of reduced electricity consumption by the buyer of this machine is estimated at about 1 million euros over 5 years, which is beneficial.

## 7.6. How Searches Become Reality at SAMSUNG, Korea

As a positive case, the use of TRIZ on SAMSUNG should be pointed out. I remember classrooms for TRIZ during two-year period my working for Samsung Advanced Institute of Technology (SAIT) in Suwon, the walls of which were hung with huge, many-meter-long posters showing TRIZ-trees-of-evolution of various electronic products and technologies for potential use at SAMSUNG enterprises.

It would be amazingly important for the TRIZ researcher and teacher to use these cases for research and teaching. But, alas, all this was and remains confidential information. Passing into the territory of SAIT, sadly, but with understanding, of course, I parted with my mobile phone and camera.

It remains only to recall these walls and posters, as well as pleasing meetings with engineers who showed me my Springer-book[139] on their table.

Several projects were developed on order of SAIT. It is interesting to see how, gradually, year after year, some ideas appear in the company's products. I do not have the opportunity to explain and even comment on these projects, but it is possible today to show the names of some of them (fig. 7.10) and at some simple examples of the application of TRIZ-laws or models (fig. 7.13-7.29).

This does not mean that it was I, who proposed these ideas, I did not always know this, but they seemed new to me, and my expert-listeners never noted which ideas were new for them and which weren't. But I was ordered new and new projects next months for research and proposals for the development of entire technical and technological areas.

And the main thing that I consider it necessary to tell the students of the Master programs here, is that TRIZ was, remains and will be the strongest tool for developing products and technologies at any enterprise. Therefore, it is possible and necessary to study TRIZ, MTRIZ. And for managers of Master programs at universities and in the industry I can recommend including MTRIZ in their programs.

And now, for more than 10 and even 15 years, every SAMSUNG enterprise has its own TRIZ-school.

---

[139] Orloff, M. *Inventive Thinking through TRIZ: A Practical Guide* // 2nd issue. – SPRINGER-Verlag Inc., New York. – 352 pp., 2006 (1st issue in 2003)

**fig. 7.10.** Examples of the titles of MTRIZ-projects for prediction of system development and evolution

**fig. 7.11.** Training at Korea University of Technology and Education, Cheonan, with participation of SAMSUNG specialists

**fig. 7.12.** First meeting with TRIZ-specialists from Russia and Belarus, who worked at SAMSUNG at that time

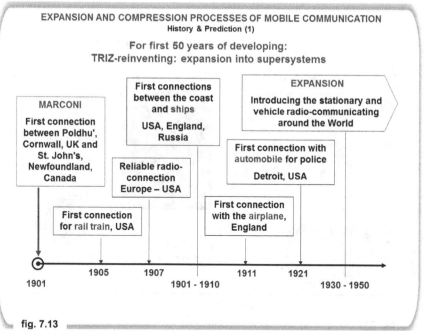

fig. 7.13

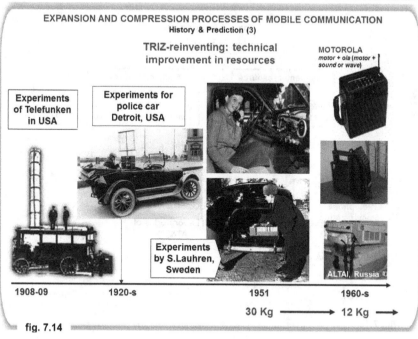

fig. 7.14

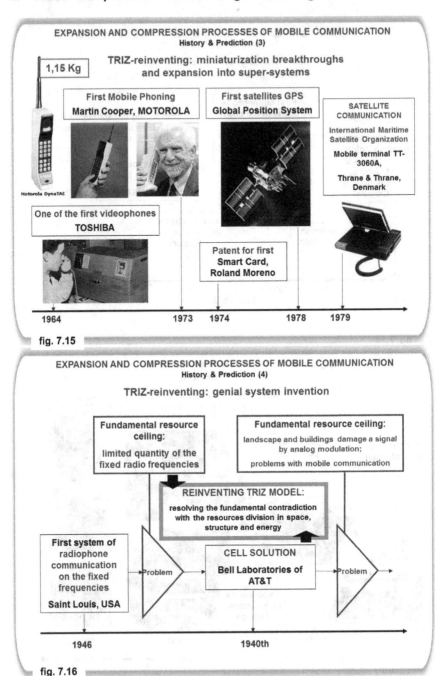

fig. 7.15

fig. 7.16

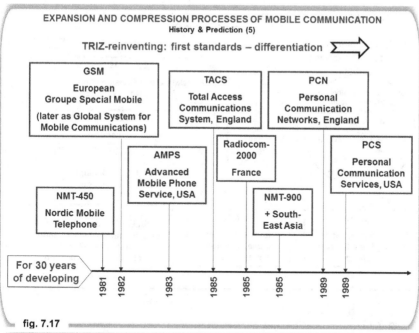

fig. 7.17

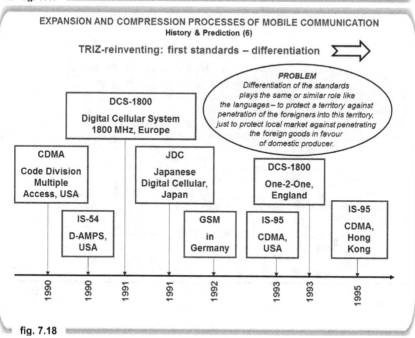

fig. 7.18

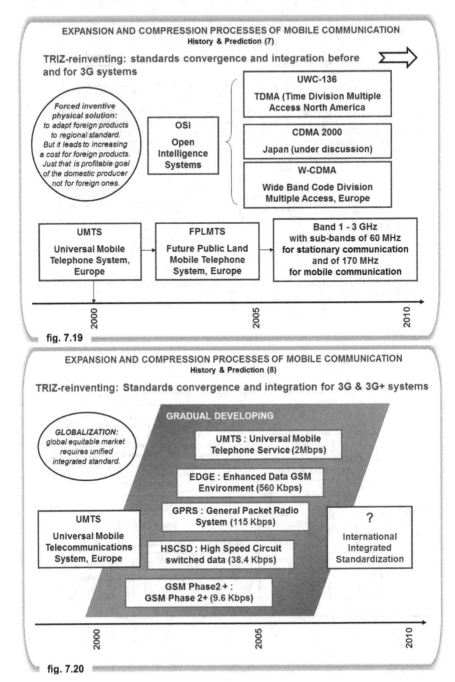

**EXPANSION AND COMPRESSION PROCESSES OF MOBILE COMMUNICATION**
History & Prediction (7)

TRIZ-reinventing: standards convergence and integration before and for 3G systems

*Forced inventive physical solution: to adapt foreign products to regional standard. But it leads to increasing a cost for foreign products. Just that is profitable goal of the domestic producer not for foreign ones.*

OSI
Open Intelligence Systems

UWC-136
TDMA (Time Division Multiple Access North America

CDMA 2000
Japan (under discussion)

W-CDMA
Wide Band Code Division Multiple Access, Europe

UMTS
Universal Mobile Telephone System, Europe

FPLMTS
Future Public Land Mobile Telephone System, Europe

Band 1 - 3 GHz
with sub-bands of 60 MHz for stationary communication and of 170 MHz for mobile communication

2000        2005        2010

fig. 7.19

**EXPANSION AND COMPRESSION PROCESSES OF MOBILE COMMUNICATION**
History & Prediction (8)

TRIZ-reinventing:  Standards convergence and integration for 3G & 3G+ systems

GRADUAL DEVELOPING

*GLOBALIZATION: global equitable market requires unified integrated standard.*

UMTS : Universal Mobile Telephone Service (2Mbps)

EDGE : Enhanced Data GSM Environment (560 Kbps)

GPRS : General Packet Radio System (115 Kbps)

UMTS
Universal Mobile Telecommunications System, Europe

HSCSD : High Speed Circuit switched data (38.4 Kbps)

GSM Phase2 + :
GSM Phase 2+ (9.6 Kbps)

?
International Integrated Standardization

2000        2005        2010

fig. 7.20

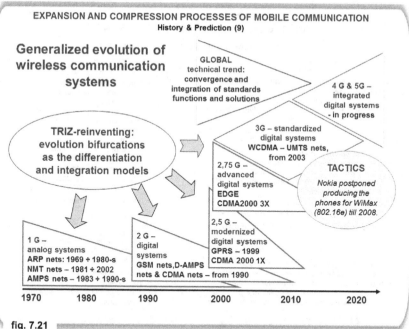

fig. 7.21

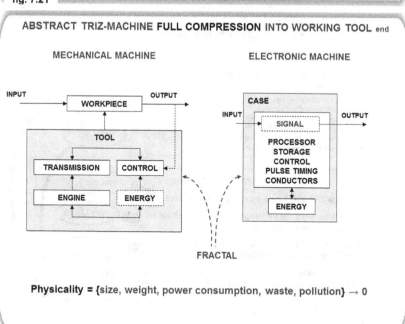

fig. 7.22

EXPANSION AND COMPRESSION PROCESSES

**TRIZ Express-Prediction for MOBILE PHONE**

REDUCTION (resource-oriented analysis - continuation 3)

### PURPOSE & FUNCTIONALITY

- *main functionality: wireless communication with voice, pictures, video and texts,*
- *specialization in additional functions, e.g. in photography, watching and controlling trough the Internet, etc.*

### DESIGN EVOLUTION

- *shaping, coloring, materials etc.,*
- *segmentation to sliding and turning, etc.*
- *dynamization of details and surfaces,*
- *adaptation in usability and ergonomically to different clusters of users, etc.*

### CONTROLABILITY IMPROVEMENT

- *change between waiting and working modes, etc.*

### MONO – BI – POLY – TRANSITIONS

- *two screens (with a shifted parameters),*
- *two kameras (with a shifted parameters),*
- *two, four, six and more functional surfaces,*
- *call signal – acoustical monophonic & polyphonic, mechanical (vibration), electrical (weak discharge), etc.*

### SYSTEMS INTEGRATING

- *voice communication PLUS:*
- *live video communication,*
- *photo-making,*
- *video-recording,*
- *music and video players,*
- *radio- and TV-receiving,*
- *watch,*
- *3D-recording and demonstration,*
- *texts messages (SMS),*
- *health-sensors, etc.*

### TRANSIT TO MICRO-LEVEL

- *storages & microprocessors, etc.*

### DYNAMIZATION IMPROVEMENT

- *sliding,*
- *turning, rotation, deploying, etc.*

### TRANSIT TO SUPER-SYSTEM

- *Internet-connections,*
- *positioning,*
- *remote house watching & control,*
- *gaming, etc.*

**fig. 7.23**

EXPANSION AND COMPRESSION PROCESSES

**TRIZ Express-Prediction for MOBILE PHONE**

REDUCTION (resource-oriented analysis - continuation 4)

### REPLACEMENT OF HUMAN

- *voice recognition,*
- *gestures recognition, etc.*
- *help in timing and scheduling,*
- *supporting with function of intelligence searching the information in storage and Internet, etc.*
- *increasing the level of personal and group security,*
- *recognition and help for health and mood, etc.*

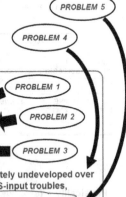

### CRITICAL RESOURCES   PROBLEM 1

- **too weak and short-time working power source,**

- **too little screens for veiwing the pictures and video,** PROBLEM 2
  **the documents in size of A4, etc.**

- **the loud speakers are too close placed apart – that** PROBLEM 3
  **unsufficient for high quality stereosound, etc.**

- very little and non-ergonomic functional keys, absolutely undeveloped over the minimum size limit, that leads in particular to SMS-input troubles,

**fig. 7.24**

EXPANSION AND COMPRESSION PROCESSES

**TRIZ Express-Prediction for MOBILE PHONE**

TRIZ-PREDICTING THE DIRECTIONS OF PROBLEM RESOLVING

TRANSFORMATION FOR PROBLEM 1 „POWER SOURCES"

To predict the directions for this problem solving the
following TRIZ-laws could be used:

➤ LAW OF F-EXPANSION & P-COMPRESSION

➤ LAW OF TRANSITION TO SYSTEM JOINING UP
  AND INTEGRATION

➤ LAW OF TRANSITION TO SUPER-SYSTEM

➤ LAW OF COMPRESSION INTO WORKING TOOL

➤ LAW OF TRANSITION TO MICROLEVEL

And additionally:

➤ ENERGY THROUGH PASSAGE AND CLOSED
  PATHES (in particular: recuperation of energy)

➤ Functional Ideal Model:

  ALL, THE SYSTEM NEEDS, IS IN OPERATIVE ZONE

PREDICTING A TREND TOWARDS
CONVERGENCE AND INTEGRATION

IDEALITY CONCEPT
„PERPETUAL POWER SOURCES"
FROM THE BODY AND
NEAR ENVIRONMENT:

➤ THOUSANDS THERMOCOUPLES,

➤ PIEZOELECTRIC CONVERTERS OF
  MECHANICAL MOVEMENT &
  OSCILLATIONS (incl. acoustic),

➤ THE RADIO-WAVES ENERGY
  TRANSFORMERS,

➤ SOLAR ENERGY BATTERIES, etc.

Integration into
the WORKING TOOLS:
covering surfaces and screens.

**fig. 7.25**

EXPANSION AND COMPRESSION PROCESSES

**TRIZ Express-Prediction for MOBILE PHONE**

TRANSFORMATION

TRIZ-PREDICTING THE DIRECTIONS OF RESOLVING TO
PROBLEM 2 „VIEWING THE BIG SIZE PICTURE " AND
PROBLEM 3 „HIGH-QUALITY SOUND"

To predict the directions for this problem solving the
following TRIZ-laws could be used:

➤ LAW OF TRANSITION TO SYSTEM JOINING UP
  AND INTEGRATION

➤ LAW OF TRANSITION TO SUPER-SYSTEM

➤ LAW OF COMPRESSION INTO WORKING TOOL

And additionally:

➤ ENERGY THROUGH PASSAGE AND CLOSED
  PATHES (in particular: recuperation of energy)

➤ Functional Ideal Model:
  ALL, THE SYSTEM NEEDS, IS IN OPERATIVE ZONE

Integration into
the WORKING TOOLS:
covering surfaces
and screens.

PREDICTING A TREND TOWARDS
CONVERGENCE AND INTEGRATION

IDEALITY CONCEPT
„BEST PLACE FOR BEST
FUNCTIONALITY"
BETWEEN-ON-IN
THE BODY AND MEDIA-PHONE:

➤ WIRELESS STEREO-SOUND
  TRANSMISSION INTO THE EARS,

➤ PICTURING OF ANY SIZE WIRELESS
  TRANSMISSION INTO THE EYES WITH THE
  CYBER-SPECTACLES or onto the any
  suitable surface – paper, wall, air, etc.

➤ WIRELESS VOICE TRANSMISSION – FROM
  THE AREAS NEAR EARS, OR FROM
  ANOTHER PARTS OF THE FACE.

**fig. 7.26**

EXPANSION AND COMPRESSION PROCESSES

**TRIZ Express-Prediction for MOBILE PHONE**

TRANSFORMATION

TRIZ-PREDICTING THE DIRECTIONS OF RESOLVING TO
PROBLEM 2 „VIEWING THE BIG SIZE PICTURE " AND
PROBLEM 3 „HIGH-QUALITY SOUND"

Natural "communicating tools"

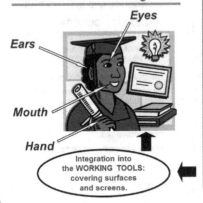

Eyes

Ears

Mouth

Hand

Integration into
the WORKING TOOLS:
covering surfaces
and screens.

PREDICTING A TREND TOWARDS
CONVERGENCE AND INTEGRATION

IDEALITY CONCEPT
„BEST PLACE FOR BEST
FUNCTIONALITY"
BETWEEN-ON-IN
THE BODY AND MEDIA-PHONE:

➢ WIRELESS STEREO-SOUND
TRANSMISSION INTO THE EARS,

➢ PICTURING OF ANY SIZE WIRELESS
TRANSMISSION INTO THE EYES WITH THE
CYBER-SPECTACLES or onto the any
suitable surface – paper, wall, air, etc.

➢ WIRELESS VOICE TRANSMISSION – FROM
THE AREAS NEAR EARS, OR FROM
ANOTHER PARTS OF THE FACE.

fig. 7.27

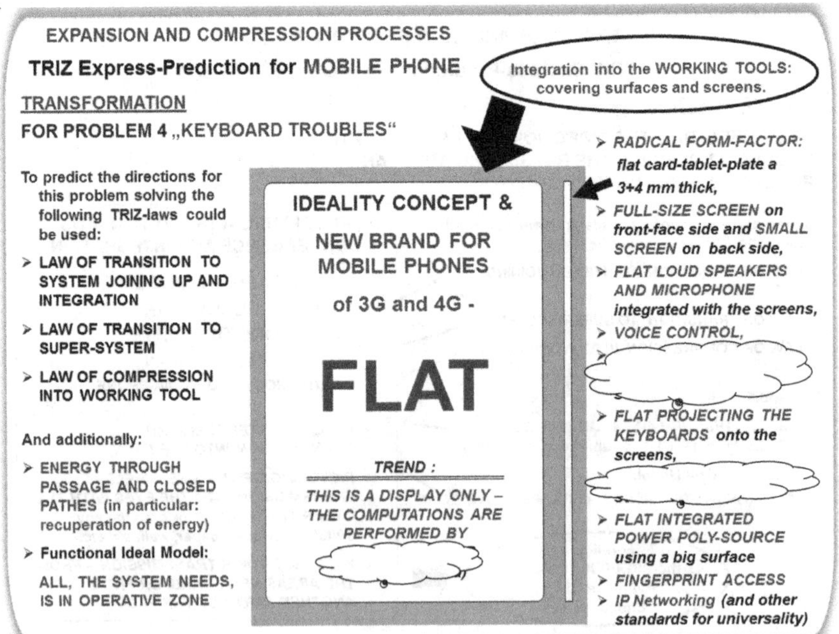

---

EXPANSION AND COMPRESSION PROCESSES

**TRIZ Express-Prediction for MOBILE PHONE**

TRANSFORMATION

FOR PROBLEM 4 „KEYBOARD TROUBLES"

Integration into the WORKING TOOLS:
covering surfaces and screens.

To predict the directions for
this problem solving the
following TRIZ-laws could
be used:

➢ LAW OF TRANSITION TO
SYSTEM JOINING UP AND
INTEGRATION

➢ LAW OF TRANSITION TO
SUPER-SYSTEM

➢ LAW OF COMPRESSION
INTO WORKING TOOL

And additionally:

➢ ENERGY THROUGH
PASSAGE AND CLOSED
PATHES (in particular:
recuperation of energy)

➢ Functional Ideal Model:

ALL, THE SYSTEM NEEDS,
IS IN OPERATIVE ZONE

IDEALITY CONCEPT &

NEW BRAND FOR
MOBILE PHONES

of 3G and 4G -

# FLAT

TREND :

THIS IS A DISPLAY ONLY –
THE COMPUTATIONS ARE
PERFORMED BY

➢ RADICAL FORM-FACTOR:
flat card-tablet-plate a
3+4 mm thick,

➢ FULL-SIZE SCREEN on
front-face side and SMALL
SCREEN on back side,

➢ FLAT LOUD SPEAKERS
AND MICROPHONE
integrated with the screens,

➢ VOICE CONTROL,

➢

➢ FLAT PROJECTING THE
KEYBOARDS onto the
screens,

➢ FLAT INTEGRATED
POWER POLY-SOURCE
using a big surface

➢ FINGERPRINT ACCESS

➢ IP Networking (and other
standards for universality)

fig. 7.28

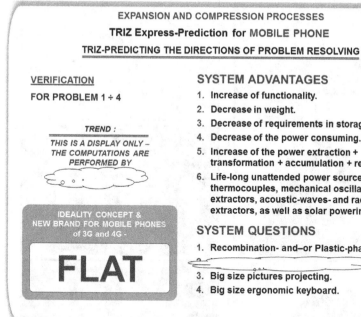

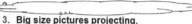

EXPANSION AND COMPRESSION PROCESSES

**TRIZ Express-Prediction for MOBILE PHONE**

TRIZ-PREDICTING THE DIRECTIONS OF PROBLEM RESOLVING

VERIFICATION

FOR PROBLEM 1 ÷ 4

TREND :

*THIS IS A DISPLAY ONLY –
THE COMPUTATIONS ARE
PERFORMED BY*

IDEALITY CONCEPT &
NEW BRAND FOR MOBILE PHONES
of 3G and 4G -

# FLAT

## SYSTEM ADVANTAGES

1. Increase of functionality.
2. Decrease in weight.
3. Decrease of requirements in storages capacity.
4. Decrease of the power consuming.
5. Increase of the power extraction +
   transformation + accumulation + recuperation.
6. Life-long unattended power source with the
   thermocouples, mechanical oscillations
   extractors, acoustic-waves- and radio-waves-
   extractors, as well as solar powering.

## SYSTEM QUESTIONS

1. Recombination- and–or Plastic-phases.

3. Big size pictures projecting.
4. Big size ergonomic keyboard.

fig. 7.29

## 7.7. Some Dreams Come True Elsewhere[140], Korea

For one of the companies, the author developed an idea of a "flying component" as an additional device to a mobile phone (word "drone" wasn't then).
The prototype was "selfie stick", which also only recently became popular.

The MTRIZ-idea from laws of system evolution was to design and push to
market very small "flying device" like a beetle – better "May beetle" ☺, made as
a detachable device for a mobile phone. I made MTRIZ-presentation for leading managers (about 20 persons) of the company with slide-by-slide "animation" (fig. 7.29), where a guy sees his girlfriend walking on the seashore. This
could only be done if she had such a "beetle" flying nearby and transmitting the
image to her mobile phone.

And some of them said: *This is an empty dream! No practical use!*

Today this idea is very close realized by other companies. *But! The beetle is
still too big! It's not my complete idea yet!*

Ten years later, one of the managers called me and said: *Michael, you were
right!* I answered: *Thank you, mister L.! But it's too late, that dream came true
elsewhere.*

---

[140] this case is demonstrated to the Master students as a sample of an inventive TRIZ-idea considered once by managers as "dream", "idle fancy", "useless", which became effective today

fig. 7.30

## 7.8. How Can Hundreds of Specialists be Quickly Introduced into MTRIZ? Cases of Intensive Mass Training[141], China

Work in China has shown that it is possible to organize sufficiently long multi-day and short 1-2 days, but intensive training and certification programs to introduce large contingents of specialists, undergraduates and professors into the basis of MTRIZ.

All courses can be divided into three types: 1) multi-day trainings at the university or at the continuing education center in one of the technology parks; 2) training, supplemented by express consultations at enterprises and 3) solving problems on the order of enterprises (with or without the participation of customer's specialists).

In longer courses, the work of the students was organized in a special way, namely: every day after the lecture, the students gathered in small groups to study the material for the day; usually a student with a better understanding of the material would explain examples and answer questions. Such nightly work was carried out always and everywhere, at all enterprises and educational centers. By the way, I forgot to note that all my Korean groups at universities and enterprises worked in the same way.

**fig. 7.31.** Dozens of trainings at the *Harbin Polytechnic University industrial park*

Trainings in the halls for 200-300-400 students for several days always left a great impression: excellent discipline, working silence, thoughtful questions and comments.

I studied with these students and always openly and sincerely thanked the entire audience for their lessons for me, creative cooperation and respectful communication.

Dozens of students made reports based on the results of MTRIZ modeling, drawing contradictions and the main selected models on large plastic boards or electronic screens – girls and guys, undergraduates and specialists of different professions! That was awesome!

---

[141] This section is addressed to managers of companies' and university' education, showing examples of organization of trainings and certification programs for many hundreds of specialists, undergraduates and professors

Work at many large enterprises throughout Heilongjiang province was distinguished by the fact that immediately certain tasks were solved in the interests of enterprises. A particularly striking case occurred at one of the plants of company *Huawei*. After my lecture for top managers (two pictures below in fig. 7.32), I was asked a question about a very difficult challenge, the solution of which was vital for the company to receive new orders. I explained the effective way to solve this problem. And then the Chairperson of the Directors' Board of this plant stood up and showed on my notebook the principle of the solution, which she immediately found now after many months of unsuccessful searching! And this solution alone costs tens of millions of dollars. It was really great!

**fig. 7.32.** Dozens of trainings with express-consulting at the industrial companies

A highly effective mass training was organized for more than 50 (!) enterprises, as well as for universities, whose employees arrived in the morning and left in the evening by buses. The huge theater hall was rented at the Northeast Forestry University. 1000 students worked in the hall for 2 days. In fig. 7.33 with registration team, there are in my hands the certification works – almost 500 (the final test was voluntary). It was an amazing picture when test forms began "to fly" (passed from hand to hand) from the upper rows to the stage, like hundreds of white birds! *The result: not a single error in MTRIZ-methods! This is possible by the training with standard MTRIZ-formats only!*

**fig. 7.33.** Two-day training for 1000 attendees at the theater of
Northeast Forestry University. Harbin

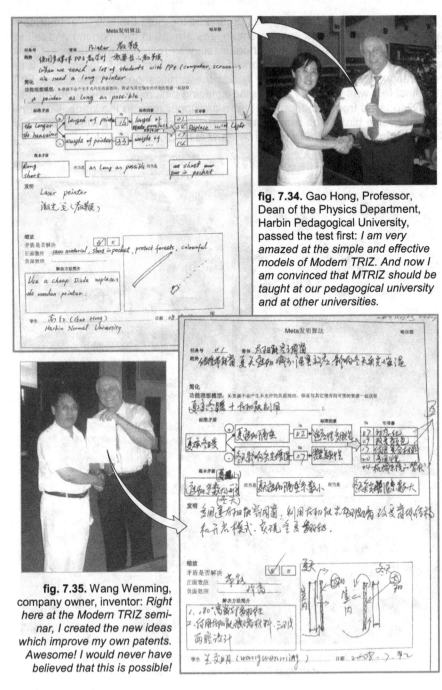

**fig. 7.34.** Gao Hong, Professor, Dean of the Physics Department, Harbin Pedagogical University, passed the test first: *I am very amazed at the simple and effective models of Modern TRIZ. And now I am convinced that MTRIZ should be taught at our pedagogical university and at other universities.*

**fig. 7.35.** Wang Wenming, company owner, inventor: *Right here at the Modern TRIZ seminar, I created the new ideas which improve my own patents. Awesome! I would never have believed that this is possible!*

## 7.9. New Waves and Generations in Training and Software, Belarus

In the history of TRIZ, Belarus is one of the centers of the first acceptance of TRIZ ideas, the application and teaching of TRIZ, almost without interruption, starting in the 1960s.

In Minsk, the phenomenal software *Invention Machine* was developed as a global pioneer initiative and a project by Dr. Valery Tsourikov, the founder of company *Invention Machine Laboratory* in Minsk in the mid-1980s, founded later the famous company *Invention Machine Co.* in Boston, USA, in 1991.

The subsequent development of software *Invention Machine* became several software systems *TechOptimizer, CoBrain* and *GoldFire*.

Before and today, continuous improvement of all these software systems was and is carried out in Minsk[142].

Software *Invention Machine, TechOptimizer, CoBrain* and *GoldFire,* based on TRIZ-concepts and enhanced by artificial intelligence algorithms, have become a powerful tool for supporting engineering creativity for dozens of universities and many of the largest companies in the world.

The author of this book supported *Invention Machine Laboratory* from the very beginning and later and was one of the first distributors of software *Invention Machine* and *TechOptimizer* in Europe in 1994-1997 together with supplying lectures and consulting on the TRIZ grounds.

It should be noted that many well-known TRIZ-specialists came out of *Invention Machine Laboratory*. *Igor Devoyno* has been a scientific leader in development of software *Invention Machine, TechOptimizer, CoBrain* and *GoldFire* for more than 30 years. *Valery Sushkov* founded in the mid-1990s his first TRIZ-company in Netherland, and subsequently such well-known companies as *xTRIZ* and *ICG T&C. Dmitry Kucharavy* worked for LG. *Nikolai Shpakovsky* made a significant contribution to the development of TRIZ at SAMSUNG and other South Korean companies, and is currently actively working with China and Russia. *Georgy Severynets* and *Alexander Skuratovich* were also repeatedly invited to work with Korean companies.

Dr. Valery Tsourikov (left, fig. 7.36) is developing in Minsk a new generation of intelligent software called *TrueMachina* for automatically creating inventive concepts.

**fig. 7.36.** At the presentation of *TrueMachina* in Minsk, January 29, 2019

---

[142] to date (September 2019), the development and marketing promotion of *GoldFire* continues under the auspices of the company *IHS Markit*

At the beginning of 2000-s software *EASyTRIZ Junior* and *EASyTRIZ Practitioner* of Modern TRIZ Academy were developed in Berlin with participation of several groups of Belarus programmers from Minsk. Thousands of copies of this software have been delivered to students and customers of the MTRIZ Academy over the past 15 years.

The mission of the MTRIZ Academy is to develop and promote the foundations of Modern TRIZ. *The fact is that the basic TRIZ models have been and will remain an indispensable tool to support the creative thinking of engineers for an immeasurable time in the future.*

However, these models need to be improved, made more reasonable and more accessible in use. For this purpose, almost 20 years, standardized formats for presenting information on the process of creating an invention have been used. These formats structure the Extracting and Reinventing procedures, which form the basis of training at MTRIZ, as well as the basis for the discipline of creative engineering thinking in solving real practical problems.

To continue new software development, the company Global Enterprise for Mastery in TRIZ (GEMTRIZ) has been established by author in Minsk in 2019.

**fig. 7.37.** Master' and Doctor' students training in MTRIZ at Belarusian State University of Informatics and Radio-electronics (BSUIR), Minsk, 2017

**fig. 7.38.** Training at Design Bureau of Precision Electronics Machinery (KBTEM), Minsk, 2018

**fig. 7.39.** Professors and researchers training in MTRIZ at BSUIR, Minsk, January, 2019

## 7.10. International Master Trainings, Germany

For more than 12 years, the MTRIZ Academy has been conducting trainings in Berlin for master's programs from dozens of universities. More than 1,500 Master' students from different countries underwent (in class and distantly) training at the MTRIZ Academy.

At present, the MTRIZ Academy is conducting a pilot project for the distant e-education of MTRIZ Masters in short and complete programs both in a universal format and in specializations for a number of engineering areas.

**fig. 7.40.** ECM Space Technologies, Berlin: training for specialists of China aero-space enterprises and institutions (2012-2018)

**fig. 7.41.** Regular seminars organized by company *s2m GmbH* for the specialists from Chinese aero-space industry and institutes at *Charlottenburg Innovations-Centrum (CHIC)*, Berlin.

**fig. 7.42.** TU Berlin, Aerospace Institute (ILR): trainings for specialists and Master' students of aerospace institutions from Uzbekistan, Kazakhstan, Belarus and Russia

**fig. 7.43.** Master' students of Caspian University, Almaty, Kazakhstan, in Berlin

**fig. 7.44.** Master' students at European programs in cooperation with ECM, Berlin

**fig. 7.45.** Lectures for top-managers at ESMT
European School for Management and Technology, Berlin

## 7.11.  Mass Training at the Industrial Holding, Kazakhstan

In March of 2017 the group of specialists of metallurgy holding *CASTING*, Kazakhstan, invited by ECM (Berlin) and Caspian University (Almaty), visited AiMTRIZ for training in MTRIZ (fig. 7.46).

**fig. 7.46.** The group of specialists from *CASTING*, Kazakhstan,
at two-day MTRIZ-training, in Berlin

Already at first training two effective modernization ideas were invented and discussed with author of this book to get first small consulting in use of TRIZ for solving to industrial problems for this company.

During last 2 years several trainings were provided for this company and after that special MTRIZ-education programs were planned and realized as pilot projects.

The first project with an *EASyTRIZ Junior course* for 150+ students was designed to achieve the following goals:

1) the mass mastering of the basic models of MTRIZ for the development of proposals for the modernization of technological operations in workshops, warehouses, transport routes, in marketing, etc.;

2) separation of proposals for immediate and deferred implementation;

3) certification of "students" and selection of promising employees for additional training and inclusion in creative groups of further modernization;

4) selection of initiative employees capable of managerial work, leaders in team work.

Employees recommended for training listened to author lectures in rented university halls (fig. 7.47). The same lectures were held for other factories of the company.

Each employee was given a set of books[143] and handouts by the author including the samples of previously performed similar work. Further, the training was carried out remotely with the accumulation of a library of certification works, preparation of certificates and recommendations for employees promotion.

**fig. 7.47.** Introductory lecture for the steel factory of company *CASTING* at *S. Toraighyrov Pavlodar State University*

The main results of such training are as follows:

- increased interest of employees in participating in the modernization of the enterprise;

- some employees made proposals with an annual economic effect of tens of thousands of euros;

- several employees were promoted;

- a number of employees were recommended for further studies at the university;

- compiled orientation groups for future creative solutions to possible new challenges.

***The cost of training the entire contingent paid off at least 10 times!***

Particular attention was paid to the training of senior staff in the Master MTRIZ program. Each highly qualified specialist invited to study under the Master program had to not only perform certification tests, but to solve one or more urgent challengers with the help of MTRIZ.

---

[143] the Junior course kit includes two books: "ABC-TRIZ" and special issue of AiMTRIZ under the title " Flash-TRIZ" of 48 pages only (!)

To understand the processes in factories, the author of this book visited all the most important workshops (fig. 7.48).

**fig. 7.48.** In the workshop for rolling billets

The students of the Master program were equipped with a full set[144] of author's books and a special training program, in which all the training steps, tests and expected results were tabularly described.

Special handout also included a poster in A1 format (fig. 7.48) to get a look at key models of (M)TRIZ from a "bird's eye" view.

Such a poster hangs in the office of the General Director – Chief Designer of the company (fig. 7.49), as well as in the offices of other directors and all the main specialists (fig. 7.50).

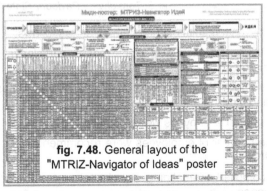

**fig. 7.48.** General layout of the "MTRIZ-Navigator of Ideas" poster

For the work of students of the Master program, one full working day per week was allocated and increased (!) wages were introduced for the period of study.

Also, for Master' students, a separate room (fig. 7.51) was rented in the coworking center with all the amenities so that these employees are completely disconnected from everyday worries and anxieties. The co-operation was arranged distantly with my periodic visits the company.

---

[144] three Springer-books: "Inventive Thinking through TRIZ", "Modern TRIZ", "ABC-TRIZ" and AiMTRIZ's digest "TRIZ-Flash"

**fig. 7.51.** Separate working room in the coworking center for the students of MTRIZ Master Program. And with a wonderful slogan!

After initial 10 months, *the annual efficiency of first developed ideas were estimated at a total of several tens of millions of euro!* The project goes on.

## 7.12. Space Industrialization with String SpaceWay Systems by Unitsky, Belarus

This small section of the book for Master program students is devoted to the industrialization of space based on the **GPV (General Planetary Vehicle)** concept. This great invention of Anatoly Unitsky was described in my books previously.

But importance of his invention, engineering and experimental works is so valuable that it is extremely useful to reproduce key ideas here.

After more than a century of space systems evolution, there emerged a rather intricate problem: rockets proved to be a powerful air, land and water pollutant. "The Second conference on non-rocket space exploration" was devoted to discussing this problem[145].

**fig. 7.52.** Among the participants of the conference, June 21, 2019

Creation of a circumterrestrial civilization needs:

- transferring of certain industrial processes into the near space;
- development of new communication systems;
- accumulation of solar energy for the Earth;
- creation of weather management systems, etc.

In the foreseeable future it will require thousands of rocket launches per year, and those future rockets will be carrying much heavier payloads than today. This option appears to lead into a dead end because, as a minimum, it threatens to destroy the Earth's ozone layer.

Initial request and an answer for it were and are to exclude the rockets as a main transportation mean for development of future industrial civilization in near space.

---

[145] The second conference on non-rocket space exploration, June 21, 2019, Minsk – Maryina Gorka, Belarus; https://rsw-systems.com/news/spaceway-conference

The basic principles of the GPV are quite simple. Anatoly Unitsky formulated them back in 1982 in his article *"Interchange, Space, Ring"*, published in the journal *"Inventor and Innovator"*.

The "ideal transport" should be the one that uses *"only internal forces"*, says the inventor. To do this, it must have the shape of a ring!

So, the GPV is a geocosmic aircraft covering a planet in the plane of the equator (fig. 7.53). The peculiarity of its functioning is that entering space is carried out by increasing the diameter of its ring and achieving the circumferential velocity of the body, equal to the first space, at the calculated height (with passengers and cargo).

In this case, the position of the center of mass of the GPV does not change during space entering: it always coincides with the center of mass of the planet. The optimal driving internal force for the GPV is the excess centrifugal force from the "belt flywheel", accelerated around the planet in a vacuum channel using a linear motor and a magnetic cushion to the speeds, exceeding the first space speed – up to 10-12 km/s, depending on the ratio of the linear masses of the body and flywheel.

To transfer the pulse and the moment of the pulse to the body of the GPV when entering orbit in order to obtain an orbital speed equal to the first space, the second "belt flywheel" is required, also covering the planet.

So, imagine a gossamer trestle with a diameter of slightly more than 40,000 kilometers circling the Earth along the equator. The trestle supports a vacuumized pipe housing a series of linear electric motors and a rotor suspended in a magnetic field. Inside the rotor less than half a meter "thick", there are millions of tons of payload. The rotor is covered with heatproof coating.

The linear motors gradually – over the course of several days or weeks – boost the rotor to circular velocity – 10 km per second. The pipe opens up, and the rotor, now free of its magnetic field fetters, shoots up and several dozens of minutes later reaches the near space and attains the height of 300 km!

In the process, the rotor expands (fig. 7.53,a) – first due to the elastic properties of its material (phase 1), and then by operation of special telescopic systems (phase 2).

And if we define the rotor boosting system as an "antigravity" engine, the rotor itself will surprisingly turn into a centrifugal propulsion unit. The dynamized (TRIZ!) rotor launches itself into the Earth orbit together with its cargo in the same way as *Baron von Münchhausen* pulls himself out of the bog together with his horse!

The same mechanism can be used to launch into near space the capsule which houses the rotor – now rotating around the planet. Cabins holding thousands of passengers and/or cargo bays carrying millions of tons of payload can be placed inside the capsule or suspended on its outer surface!

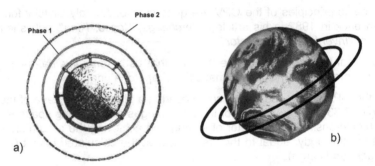

**fig. 7.53.** Unitsky planetary space transportation system: a) rotor orbit launch phases – view from one of the poles; b) rings of circumterrestrial space civilization.

Clearly, the trestle, the rotor and the space-based industrial and residential infrastructure can only be implemented on the basis of string transportation systems that we discussed in the previous section! Incidentally, due to unique string ways properties, the trestle (floating) can span even three oceans! Look illustration in fig. 7.54.

All we have to do to make it happen is to… disarm the Earth! And to invest into the future of the GPV those massive amounts of cash now spent on murder and wars. Anatoly Unitsky: *"The world, constituting the essence of our technocratic civilization, is created by engineers. However, this world is often ruled by others – those for whom personal enrichment is at the forefront; those who naively believe that in a situation where the planet will be on the brink of disaster, their money can save them. They are confident that, together with their families, they will be able to shelter on personal islands, in underground bunkers, on submarines and "Boeing" with antimissile defense. But they are wrong. The planet is one big room without walls. Beforetime, primitive people, along with their leaders, burned fires in their caves and died of lung cancer at the age of 20. They were able to survive only due to the fact that they had guessed to move their primitive technologies – usual fire – out of their home. So now we, the terrestrial civilization, must move the technosphere out of our common home – the biosphere.*

*There is only one cardinal way out of the current situation: it is necessary to provide the technosphere with an ecological niche outside the biosphere. There is no such ecological niche on the Earth. But it is in near space, at a distance of 300-500 km from the surface of the planet, where there are ideal conditions for most technological processes: weightlessness, vacuum, unlimited raw materials, energy and spatial areas. Thus, we come to the conclusion that there is a need to industrialize space."*

Extracting the creative TRIZ-models is shown in fig. 7.55. There are many possibilities for Master students of dozens of specialties to model space industrialization with GPV "Unitsky Ring".

**fig. 7.55.** Illustrations of invention "Unitsky Ring".
At the stage of approach of GPV with expanding string load-carrying rotor
towards the Unitsky space industrial ring

| ARTIFACT | UNITSKY SPACE TRANSPORT – *Unitsky Ring* |
|---|---|

**Description** In the late 1970s A.E.Unitsky  proposed the idea of *a string space vehicle* – "pan-planet vehicle". The special *platform-ring – a vehicle!* – on the basis of the string structure by A.E.Unitsky, is set around the Earth following the Equator (there are other options). Ring is accelerated for a few weeks to orbital (first cosmic) velocity, is released from the holding fields and takes off into space by centrifugal force. When taking off, the ring expands. The landing and maneuvering the ring(s) are possible. Potentially, the ring can transport millions of tons of cargo in both directions.

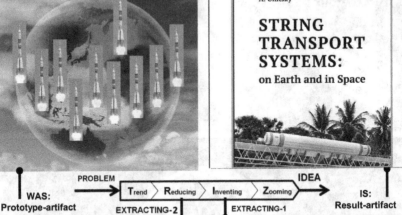

A. Unitsky

# STRING TRANSPORT SYSTEMS:
## on Earth and in Space

PROBLEM → Trend ⟩ Reducing ⟩ Inventing ⟩ Zooming → IDEA

WAS:
Prototype-artifact          EXTRACTING-2          EXTRACTING-1          IS:
Result-artifact

| D | No | | Navigator | Function / Substantiation |
|---|---|---|---|---|
| + | 07 | | Dynamiza-tion | During the launch and the landing, the rotor changes in size by up to 5-10% |
| + | 11 | | Inverse act. | Increase – not decrease! – the payload by millions of times |
| + | 19 | | Transition into another dimension | The ring moves horizontally, rotating in a circle, and vertically, in radial direction (vertical movement raises it above the Earth as an expanding/collapsing rotor) |
| + | 29 | | Self-servic. | Yunitsky ring is the only self-transporting vehicle! |
| + | 37 | | Equipoten-tiality | The ring is imparted equipotential acceleration above the equator (unlike a rocket which is launched vertically) |
| + | F | 01 In Space | | Each component of the ring is rigid, while the ring as a whole is dynamic in terms of its dimensions |
| + | F | 03 In Structure | | The ring is segmented. The "Matryoshka" principle is used to create a telescopic structure |
| + | F | 04 In Material | | The system uses the string concept to form a rotor which self-organizes into a regular ring when in motion |

| | Contradiction / Description |
|---|---|
| SC | The general planetary conveyance system must assure *gigantic lifting capacity and attainment of the circular velocity to enter the Earth's orbit*, but none of the existing rockets *have sufficient energy* to support the required cargo traffic. |
| RC | The general planetary conveyance system *must* transport to space and back hundreds of thousands and millions of tons of cargoes, and it *must not* do it, as the rocket method of transportation threatens to inflict irreparable damage on the planet Earth. |

**fig. 7.55.** Extracting for invention "Unitsky Ring"

# Part III

# Samples of MSc Theses with MTRIZ Modeling

A creative person must be able to solve complex problems. Until recently, there was nowhere to learn this.

Today, the science of strong thinking has been created. True, it was created on the basis of technology. But ... TRIZ principles are applicable in all areas of activity.

Everyone can learn to solve creative problems today.[146]

*Genrikh Altshuller*

---

[146] Compiled by M.Orloff from the books of Genrikh Altshuller, founder of TRIZ.

# 8 Modeling the Green Automotive Innovations [147]

Extracting and Reinventing based on the Meta-Algorithm of Inventing is a modern "TRIZ-Tomography" of inventive thinking. It works like a "Time Machine": reinventing allows us to explore creative thinking of any inventor from any time and era!

*Ruwim Kisselman,*

*PhD in engineering, inventor, head of the Inventors Club[148] "Schöpfer" ("The Creators" – German) from Bonn (Germany), organizer of the yearly seminars of the Modern TRIZ Academy in the German Museum in Bonn[149]*

---

### Technische Universität Berlin

*"Wir haben die ideen für die zukunft"*

Global Production Engineering
International Masters Program

## Modeling the Green Automotive Innovations with Modern TRIZ

by

**KAMARUDIN, KHAIRUL MANAMI**

A thesis submitted in partial fulfillment of the requirements for the degree of 'Master of Science in Global Production Engineering'

Under the guidance of
**Professor, Dr. Dr. Sc. techn. Michael Orloff**

**Fakultät V – Verkehrs- und Maschinensysteme**
Produktionstechnisches Zentrum

**Technische Universität Berlin**

**Global Production Engineering**

**Dean: Prof. Dr.-Ing. Günther Seliger**

**Department of Assembly Technology and Factory Management**

**Berlin, Germany**

---

[147] Part 8 is based on the edited and supplemented fragments of the master thesis by Khairul Kamarudin (Malaysia), a certified graduate of my course in MTRIZ at GPE Program, TUB
[148] Association of Inventor Clubs created in Germany earlier within the framework of the former INSTI Program (*Innovationsstimulierung* – state program for the support of innovative movement in Germany
[149] www.deutsches-museum.de

## MODERN TRIZ INVESTIGATION AND MODELING IN CASE STUDIES OF AUTOMOTIVE INNOVATIONS.

To strengthen the understanding of TRIZ method in the field of green technology in automotive sectors, here are several case studies of "Re-inventing" elaborated clearly. Throughout many automotive new green technology, regardless of any fields be it chemical, electrical, mechanical or others, TRIZ implementation can guide engineers and designers to focus on the problems, contradictions and later decide on the best solution which is far better, faster and economical to human. The new invention or innovation certainly will heal and preserve the environment.

### 8.1. NEW FUEL ALTERNATIVE

In this analysis with TRIZ, the case studies are focused on new source of fluid or new fuel alternative for renewable energy.

### The Prototype – Fossil Fuel

Fossil fuel is from natural sources, mainly from coal and dead organisms (oil and gas) aged 650 million years. Coal is crushed to a fine dust and burnt while oil and gas are burnt directly. This fuel contains high percentage of Carbon and Hydrocarbon and it is non-renewable. The problem with fossil fuel is that the sources are decreasing while the demands doubled every 20 years since 1900. It's widely usage, for running transportation, electricity even producing plastics and many other products contributes to environmental effect to the earth, air pollutants such as Nitrogen Oxides, Sulfur Dioxide, volatile organic compounds and heavy metals. The fossil fuel main drawback is it always emits Carbon Dioxide ($CO_2$) that is very harmful for the earth ozone layers, causes depletion and acid rain. $CO_2$ are mostly in air and hard to absorb entirely by all forest & trees naturally.

**Example.** The fossil fuel comes in the major area - coal, fuel oil and natural gas and fossil fuels provides around 66% of world's electrical power and 95% of the world's total energy demands. According to Environmental Canada, mostly the usage of fossil fuel that contributes to air pollutants are mainly comes from electricity activities.

**fig. 8.1 (left).** Coal mining – even the earlier process of coal energy is harmful to the environment.

**fig. 8.2 (right).** Coal is one of fossil fuel sources.

*"The electricity sector is unique among industrial sectors in its very large con-
tribution to emissions associated with nearly all air issues. Electricity generation produces a large share of Canadian nitrogen oxides and sulfur dioxide
emissions, which contribute to smog and acid rain and the formation of fine
particulate matter. It is the largest uncontrolled industrial source of mercury
emissions in Canada. Fossil fuel-fired electric power plants also emit carbon
dioxide, which may contribute to climate change. In addition, the sector has
significant impacts on water and habitat and species. In particular, hydro dams
and transmission lines have significant effects on water and biodiversity "* [150]

## The Result of Contradiction Analysis

The "plus-state" of the fossil fuel can be determined as in A-Matrix as number **37; energy used of the movable object**, because the fossil fuel is the major role as energy
provider to transportation, electricity and others. The "minus-state" of the fossil fuel can
be determined with A-Matrix number **13; external damaging factors**, understood as
usage of energy threatening the environment, the external factors.

### Inventing Idea

Combining the plus and minus state factors above, the results will show several points in As-navigators. The target is to have alternative fuels that comes
correspondingly from natural sources but can be a great replacement of world's
energy. The new alternatives will mass produce but does not harm the atmosphere, renewable and recyclable. After combining the A-Matrix 37 and 13, the
idea navigator resulted:

- **Navigator 13. Inexpensive short-life object as a replacement for expensive long-life one** - according to this navigator, uses of inexpensive oil
  such as vegetable oil can replace the expensive and hard to produce fossil
  fuel.
- **Navigator 20. Universality** - according to this navigator, universality, the
  vegetable oil can be found anywhere even in supermarket.
- **Navigator 01. Change in the aggregate state of an object** - and according to this navigator, the alternative fuel of vegetable oil is harmless than the
  fossil fuel. The oil can be touched, even can be eaten, opposites from fossil
  fuel, it is highly flammable, dangerous and poisonous.
- **Navigator 10. Copying** - additional navigator - the form of the new fuel is
  as well as in liquid form (oil).

## The Artefact (solution) – Biodiesel

Biodiesel is typically made by chemically reacting lipids e.g., vegetable oil (soy,
sunflower, rapeseed, corn, palm, etc.) and animal fat with mixture alcohol. Biodiesel can be used alone in pure form - B100 or blended with petroleum diesel
at any concentration in most injection pump diesel engines. In 2005, Daim-

---

[150]  "Electricity Generation" from website Environment Canada (http://www.ec.gc.ca/cleanair-
airpur/Electricity-WSDC4D330A-1_En.htm)

lerChrysler had launched a Jeep with engine that uses Biodiesel and the usage of Biodiesel extended to other car manufacturers, railway transportation "Virgin Trains" a biodiesel train from British Train Operating Company, aircraft usage and also as heating oil. Biodiesel has better lubricating properties and much higher "cetane" ratings than today's lower sulfur diesel fuels. Biodiesel addition reduces fuel system wear, and in low levels in high pressure systems increases the life of the fuel injection equipment that relies on the fuel for its lubrication. In February 2010, Environmental Protection Agency (EPA) reported that biodiesel from soy oil results, on average, in a 57% reduction in greenhouse gases compared to fossil diesel. The result of re-inventing in standard form is represented at the figure 8.6.

**fig. 8.3.** The chemical chain animation of Biodiesel.

**fig. 8.4.** Sunflower is one of bioethanol source with its seed contains 36 to 42% oil and 38% protein meal.

National Renewable Energy Laboratory (NREL) have done a lifecycle analysis found that B100 biodiesel emits 78.5% less than fossil fuel and reduces petroleum use by 95% throughout lifecycle. Biodiesel also improves fuel lubricity of engine resulted in reduces moving parts wears and raise "cetane" number of the fuel. Biodiesel is nontoxic and will cause far less damages as fossil fuel when spilled or natural gas released into the environment. It is safer because it is less combustible, safe to handle, store and transport.[151]

Car manufacturers that produce biodiesel powered cars: BUICK, Cadillac, Chevrolet, Chrysler, Dodge, Ford, Hammer, GMC, Isuzu, Jeep, Lincoln, Mazda, Mercedes-Benz, Mercury, Nissan, Pontiac, Saturn, Toyota and some others.

Then, some questions should be asked to confirm that the solution is right.

---

[151] Resources from U.S Department of Energy (www.afdc.energy.gov) and National Renewable Energy Laboratory (www.nrel.gov).

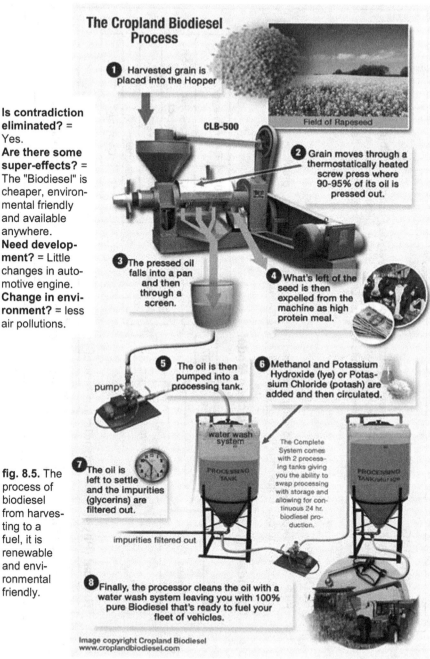

**The Cropland Biodiesel Process**

**1** Harvested grain is placed into the Hopper

CLB-500

Field of Rapeseed

**2** Grain moves through a thermostatically heated screw press where 90-95% of its oil is pressed out.

**3** The pressed oil falls into a pan and then through a screen.

**4** What's left of the seed is then expelled from the machine as high protein meal.

**5** The oil is then pumped into a processing tank.

pump

**6** Methanol and Potassium Hydroxide (lye) or Potassium Chloride (potash) are added and then circulated.

water wash system

The Complete System comes with 2 processing tanks giving you the ability to swap processing with storage and allowing for continuous 24 hr. biodiesel production.

PROCESSING TANK

PROCESSING TANK/storage

**7** The oil is left to settle and the impurities (glycerins) are filtered out.

impurities filtered out

**8** Finally, the processor cleans the oil with a water wash system leaving you with 100% pure Biodiesel that's ready to fuel your fleet of vehicles.

Image copyright Cropland Biodiesel
www.croplandbiodiesel.com

**Is contradiction eliminated? =** Yes.
**Are there some super-effects? =** The "Biodiesel" is cheaper, environmental friendly and available anywhere.
**Need development? =** Little changes in automotive engine.
**Change in environment? =** less air pollutions.

fig. 8.5. The process of biodiesel from harvesting to a fuel, it is renewable and environmental friendly.

## TARGETING

Fossil fuel sources are decreasing and its wide and everyday usage contributes to environmental effect to the earth, air pollutants, such as Nitrogen Oxides, Sulfur Dioxide, volatile organic compounds and heavy metals. The fossil fuel always emits Carbon Dioxide ($CO_2$) that is very harmful for the earth's ozone layers.

## REDUCING

FIM: X-resource, without producing the inadmissible negative effects, provides together with other existing resources obtaining [to have alternative of environmental friendly fuel].

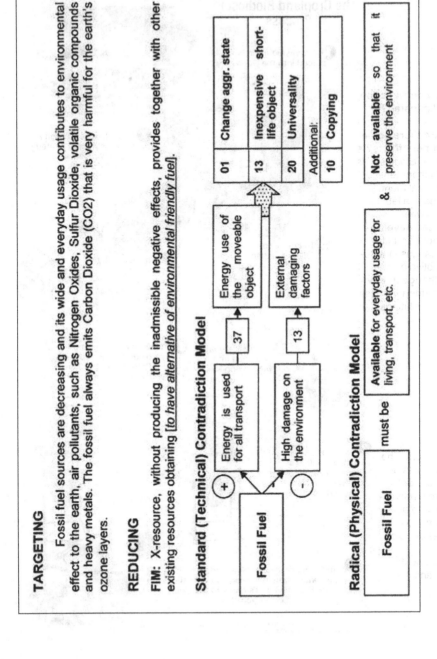

## Standard (Technical) Contradiction Model

| | |
|---|---|
| 01 | Change aggr. state |
| 13 | Inexpensive short-life object |
| 20 | Universality |

Additional:

| 10 | Copying |
|---|---|

Not available so that it preserve the environment

Energy use of the moveable object

External damaging factors

Energy is used for all transport — 37

High damage on the environment — 13

(+) (-)

**Fossil Fuel**

## Radical (Physical) Contradiction Model

**Fossil Fuel** must be | Available for everyday usage for living, transport, etc. | & | Not available so that it preserve the environment

## INVENTING

According to navigators: **13. Inexpensive short-life object as a replacement for expensive long-life one**: uses of inexpensive oil such as vegetables oil, **20. Universality**: the vegetable oil can be found anywhere, **01. Change in the aggregate state of an object**: vegetable oil harmless than the fossil fuel and **10. Copying**: (additional navigator): the state or density of the new fuel is also in liquid form (oil)

## ZOOMING

Are contradictions removed?: (Yes.) – No.

Super-effects: The biodiesel is cheaper, available anywhere and harmless even to touch.

Negative effects: just little changes on vehicle's engine.

## Picture

## Brief description

The effective innovative idea is to produce fuel type called biodiesel from quite any vegetables, to replace natural resources like petrol according to navigators 13, 20, 01 and 10.

**fig. 8.6.** Reinventing in standard form of "New Fuel Alternative".

## 8.2. NEW ALTERNATIVE FUEL IMPROVEMENT

When the answer to fossil fuel replacement is the "Biodiesel", there still several improvement needed in this new alternative fuel. The manpower or machinery necessary in harvesting vegetables suitable for biodiesel is very high, mass plantation and high maintenance of crops production leads to the need of more efficient type of crops to increase the productivity, faster yield and less manpower. The next reinventing shows the analysis and new solutions to alternative fuel of the world.

### The Prototype – Biodiesel

In February 2010, EPA reported that biodiesel from soy oil results, on average, in a 57% reduction in greenhouse gases compared to fossil diesel, and biodiesel produced from waste grease results in an 86% reduction. The problem of Biodiesel from soy, sunflower, palm is that these plants need to be planted in vast area to have high production, we are talking hundreds of hectares to support the energy usage of the world and the yield time is quite long.

*Example:* We use the same example from 8.1. Biodiesel fuel.

### The Result of Contradiction Analysis

The "plus-state" side of "Biodiesel" can be determined as in A-Matrix as number *02 universality, adaptability*, because the fuel is available anywhere and affordable to everyone. The "minus-state" of the "Biodiesel" can be determined with A-Matrix number *26 quantity of material*, understood as the need of natural resources of the biodiesel is very high to supply in needed time.

#### Inventing Idea

To have a new natural sources, that needed plantation (for increasing oxygen) but need higher and faster yield, less plantation area and the result of alternative fuel is environmentally safe. These factors can contribute to reducing $CO_2$. Through references between A-Matrix number 02 and 26, the results are:

- *Navigator 07. Dynamization* – according to this navigator, dynamization, the new natural resources can be planted in various way, even the plantation can be dynamized horizontally, by using moveable frames.
- *Navigator 10. Copying* – as additional navigator, which uses similar way to produce fuel by natural resources such as vegetable/plant oil.
- *Navigator 01. Change in the aggregate state of an object* - Algae plantation uses more water than soil. This is different than other vegetable plantations. Algae can use waste water as their food with some treatment.
- *Navigator 12. Local property* - Algae can produce oil just by pressing it by screw press. The method of producing the oil is faster, simpler and cheaper than the method of producing oil from soy bean, sunflower or palm oil.
- *Navigator 19. Transition into another dimension* – considering water as a replacement to soil.

### 4.2.2 The Artefact (solution) – Algae Fuel

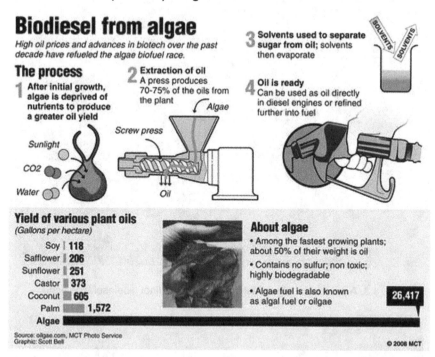

# Biodiesel from algae

High oil prices and advances in biotech over the past decade have refueled the algae biofuel race.

## The process

**1** After initial growth, algae is deprived of nutrients to produce a greater oil yield

Sunlight

CO2

Water

Screw press

Oil

**2 Extraction of oil**
A press produces 70-75% of the oils from the plant

Algae

**3** Solvents used to separate sugar from oil; solvents then evaporate

SOLVENTS

**4 Oil is ready**
Can be used as oil directly in diesel engines or refined further into fuel

**Yield of various plant oils**
(Gallons per hectare)

Soy | 118
Safflower | 206
Sunflower | 251
Castor | 373
Coconut | 605
Palm | 1,572
Algae |

**About algae**
• Among the fastest growing plants; about 50% of their weight is oil
• Contains no sulfur; non toxic; highly biodegradable
• Algae fuel is also known as algal fuel or oilgae

26,417

Source: oilgae.com, MCT Photo Service
Graphic: Scott Bell

© 2008 MCT

fig. 8.7. Graphic from oilgae.com sources, indicates that Algae yield is far higher than any other biodiesel plantation (2008 statistic).

Algae fuel is a type of biodiesel. Algae fuel is very suitable for the purpose of bio fuel which is faster in yield and less plantation area.

Algae works well with area full of $CO_2$ - the goal of all scientist in reducing air pollution is to decrease $CO_2$, because $CO_2$ is hard to vanish unless with natural way by planting trees. If we provide concentrated $CO_2$ to the algae, it will turbocharge the production. Algae can produce up to 300 times more oil per acre than conventional crops, such as rapeseed, palms, soybeans, or jatropha.

As Algae has a harvesting cycle of 1–10 days, it permits several harvests in a very short time frame, a differing strategy to yearly crops. Algae can also be grown on land that is not suitable for other established crops, for instance, arid land, land with excessively saline soil, and drought-stricken land. This minimizes the issue of taking away pieces of land from the cultivation of food crops.

Algae can grow 20 to 30 times faster than food crops. The result of re-inventing in standard form is represented at the figure 8.13.

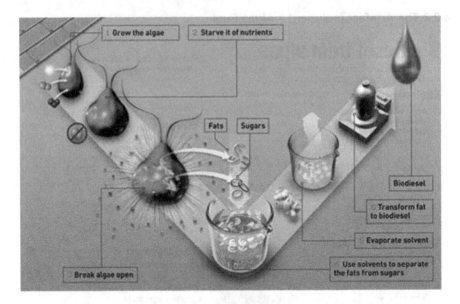

**fig. 8.8.** A process of algae fuel from growing to final biodiesel type fuel.

**fig. 8.9.** An Algae fuel in the form of oil, its finished product is in green color.

**fig. 8.10 (left).** Algae found in swamp naturally but now is a multimillion-industry item.

**fig. 8.11 (right).** Vertical farming of algae for fuel purposes.

**fig. 8.12** Hydrogenase: Bio-Hydrogen airship – an imagination illustration of future algae farming with collection of $CO_2$ from air and on water for algae's essential food.

Then, some questions should be asked to confirm that the solution is right.

- **Is contradiction eliminated**? = Yes
- **Are there some super-effects?** = Algae treat $CO_2$ as food. This is very good for reducing $CO_2$ in the air.

## TARGETING

Biodiesel can be used alone, or blended with petro diesel. The problem of Biodiesel which is mainly from soy, sunflower or palm (vegetable based) is that these plants need to be planted in mass area to achieve high production (to support world's energy request) and the yield time is quite long.

## REDUCING

**FIM:** X-resource, without producing the inadmissible negative effects, provides together with other existing resources obtaining *[to have less area of plantation and faster yield]*.

## Standard (Technical) Contradiction Model

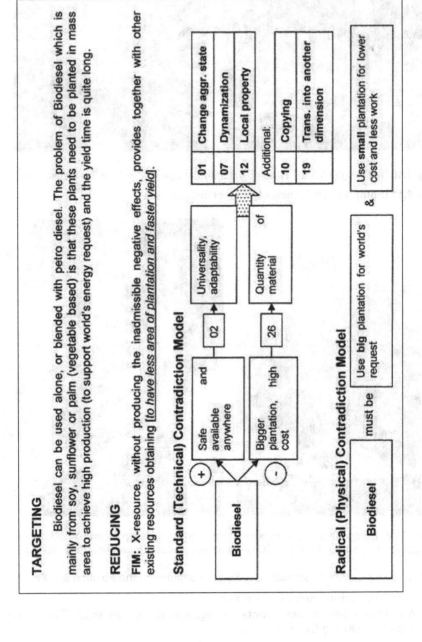

## Radical (Physical) Contradiction Model

## INVENTING

According to navigators: **07. Dynamization:** the new natural resources can be planted in various ways (vertical or horizontal) even the plantation can be dynamized, by using moveable frames, **01. Change in the aggregate state of an object:** Algae plantation uses more water than soil, **12.** Local **property:** Algae can produce oil just by pressing it by screw press, faster, simpler and cheaper. Additional navigators: **10. Copying:** which uses similar way as vegetable/plant oil and **19. Transition into another dimension:** to use water to plant algae (considering a replacement with soil).

## ZOOMING

Are contradictions removed?: (Yes) – No.

Super-effects: The algae treat CO2 as their food and consumed used water (recycling).

Negative effects: No

## Picture

## Brief description

To have alternative fuels besides planting with mass area according to navigators 07, 01, 12, 10 and 19. The target is to have also a new natural source, that needed only small plantation area with higher and faster yield and produce safe fuel for the environment still can support world's oil request.

**fig. 8.13.** Reinventing in standard form of "New Alternative Fuel Improvement".

## 8.3.  ELECTRICAL ENERGY SUPPLY IMPROVEMENT

### The Prototype – EV Battery, Lithium Ion

Batteries for electric cars have become more interesting and attractive with new technology such as Li-ion (lithium ion) battery and the current development is "Li-ion Polymer" battery that have higher power and energy density. Electric vehicle battery requires driver's alert to charge the battery or switching charge-sustaining mode when the battery is exhausted. Most "Plug-in Hybrids" (PHEV) operates at startup with charge-depleting mode and switched to charge-sustaining mode after the battery reached its minimum "state of charge" (SOC). The total mass weight of Li-ion battery can reach up to 364 kg. Still the charging method and weight of batteries are the main issues with the battery gets heavier for purposes more than operating electrical items in the car.

**Example.** Lithium ion consists of cobalt oxide positive electrode (cathode) and graphite carbon in negative electrode (anode). One of main advantages of cobalt-based battery is because of its high energy density. In electric car or hybrid cars, it requires connecting Li-ion batteries in parallel circuit because it is more efficient than connecting single large battery. Li-ion Polymer battery on the other hand has low cost of manufacture and high adaptability to wider varieties of packaging shapes. Lithium-ion Polymer batteries appear in consumer electronics items starting around 1996, does not require strong metal casing so it is lightweight about 20% lighter than Li-ion battery.

**fig. 8.13.** An example of Li-ion Polymer battery used for electric vehicles.

**fig. 8.14.** Installation of Li-ion Polymer battery. Such size could only run electrical items in the car and not yet capable to move the whole car on the road.

The major drawbacks of Li-ion Polymer battery installed in a hybrid car is two: charging method and weight, although it is lighter than Li-ion battery. Li-ion Polymer battery has to be protected from overcharging, by applying limitation no more than 4.24V per use. Overcharging this type of battery will result to fire and worst explosions but in December 2007, Toshiba Company developed Li-ion Polymer battery with just 5 to 6 minutes charging time.

Compared to Li-ion battery, Li-ion Polymer battery disadvantages are that it have greater life-cycle degradation rate. On the weight matter, typically a single cell of Li-ion Polymer battery needs 300 watt-hour per kilogram for full recharge. For an average car, it requires 30 horsepower (hp) for one hour daily commute, with 1hp equivalent to 750 watts; resulting 22,500 watt-hour per kilogram.

## The Result of Contradiction Analysis

The "plus-state" side of "Li-ion Polymer Battery" can be determined as in A-Matrix as number **37; energy used of the moveable object**, because the use of the battery is for moveable object, the electric vehicle. The "minus-state" of the "Li-ion Polymer Battery" can be determined with A-Matrix number **32; weight of the moveable object**, understood as the main problem for elimination, weight of several Li-ion Polymer makes the weight consume more electric power supplied by it.

The other problem; "recharging method" considered minor as it is essential to recharge any type of energy supplied to a car.

## Inventing Idea

To have a new method or material to store and discharge electrical energy. The idea is to use the car body as the electrical storage so that uses of battery will be less. The car will be lighter and consume less power.

- **Navigator 04. Replacement of mechanical matter** – according to this navigator, recommendation of using material that can store energy, replacing battery which is heavy and bulky.

- **Navigator 06. Use of mechanical oscillations** – according to this navigator, the body captures energy from electricity supplied to it, ions moves (like oscillations) within material to electrodes and producing voltage across itself and keeps recycling.

- **Navigator 37. Equipotentiality** – according to this navigator, rather than having battery placed inside the car's system and electricity runs in a normal way, this new method doesn't require electricity to work under the chassis or car's engine but on the body itself, going straight to electrical items such as headlamps, radio antenna, rear lamps etc.

## The Artefact (solution) – Body Work Powered Car

A study made by a group of engineers from the department of Aeronautics at Imperial College London lead by Dr. Emile Greenhalgh in collaborations with Volvo carmakers on a 3.4 million euro project made a revolutionary findings. They are developing a prototype material that stores and discharges electrical energy, allowing parts of a car's bodywork to double up as its battery.

The material which is a new type of composite, can store and discharge electrical energy work like battery, suggested placement at hood, roof and door panels. Cars equipped with this material use very less battery and sustain longer on the road without frequent charging. By developing this idea, the super effect is decreasing numbers of charging and charging stations. Lightweight electrical car too contribute less road wear. This idea is also capable to be used in consumer products such as mobile phones with no batteries or laptop that can run power by its casing, and importantly auto-charging.

The material studied could be made up from carbon fibers and polymer resin, potential in electrical storage and capable of discharging quicker than Li-ion Polymer battery. The technology too can be charged with normal plugging into household power supply.

Further development of the capacity of storing and supply electrical energy for greater use can be realized.

The result of re-inventing in standard form is represented at the figure 8.19.

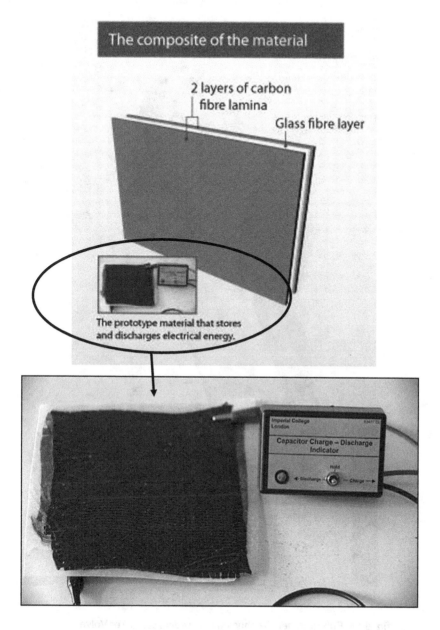

**fig. 8.15.** Early study and testing of the conceptual "Body Work Powered" technology which uses new composite material..

In the future could make the bonnet, car doors and roof from the material. As the material develops it can become more structural creating a bigger saving in weight.

The material could also be charged when the car is on the move, recycling the energy created when the car brakes.

Where it would be used

A composite wheel well is being developed to replace the metal flooring. Giving as much as a 15% reduction in weigth.

For example a Tesla Roadster's weigth would be reduced from 1.2tonnes to 750kg

Leading to even greater performance return

**fig. 8.16.** Future Battery Technology - Swedish carmaker Volvo teaming up with Imperial College London.

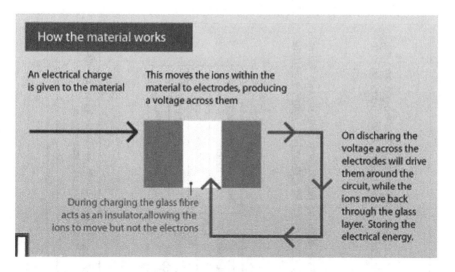

**fig. 8.17.** Diagram of how the material works.

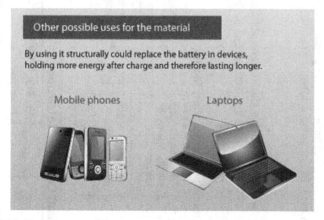

**fig. 8.18.** There are possibilities of applying this technology to other products. (Pictures above are from EU Infrastructure at www.euinfrastructure.com).

Then, some questions should be asked to confirm that the solution is right.

- **Is contradiction eliminated?** = Yes.
- **Are there some super-effects?** = the weight is light and need not to re-charge.
- **Need development?** = Yes, for higher capacity of electrical energy and sustainability.
- **Change in environment?** = No

## TARGETING

Electric car's batteries have a weight between 33kg to 50kg per average car. The weight and the size of, for example lithium ion polymer battery consume spaces and become heavy. If the purpose is to run the car on the road, bigger amount of battery needed.

## REDUCING

FIM: X-resource, without producing the inadmissible negative effects, provides together with other existing resources obtaining *[to have a new way of storing energy which is lightweight, recyclable and environmental friendly]*.

## Standard (Technical) Contradiction Model

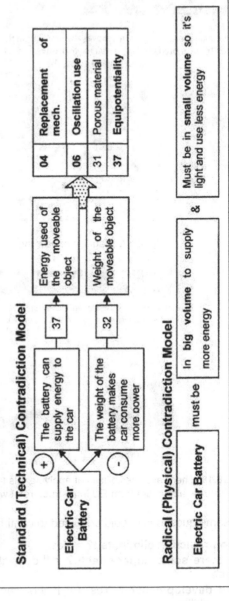

| 04 | Replacement of mech. |
| 06 | Oscillation use |
| 31 | Porous material |
| 37 | Equipotentiality |

## Radical (Physical) Contradiction Model

| Electric Car Battery | must be | In big volume to supply more energy | & | Must be in small volume so it's light and use less energy |

## INVENTING

According to navigators: **04. Replacement of mechanical matter:** by using material that can store energy, **06. Use of mechanical oscillations:** the body captures energy from electricity supplied to it and keeps recycling and **37. Equipotentiality:** rather than having battery placed inside the car's system and electricity runs in a normal way, this new method doesn't require electricity to work under the chassis or car's engine but on the body itself.

## ZOOMING

Are contradictions removed?: (Yes). – No.

Super-effects: The weight is drastically reduced and need no more battery recharging.

Negative effects: No

## Brief description

The revolutionary idea is to use the body of the car as storage and produce electricity according to navigators 04, 06, 37.

## Picture

**fig. 8.19.** Reinventing in standard form of "Electrical Energy Supply Improvement".

## 8.4. OCEAN TRANSPORT ENERGY ALTERNATIVE

### The Prototype – Cargo Ship

Cargo ship or freighter is a type of ship or vessel that carries cargo, goods and materials across ocean in very low speed. Although cargo ships produce only 1.75% greenhouse gas compared to road transport which is 10.5%, cargo ships uses the cheapest fuel, which they call "bunker fuel"; the remains of gases and high-grade fuels refined from crude oil. Bunker fuel contains up to 5000 times more sulfur than diesel, a single container ship emits more pollution than 2000 diesel trucks. This is really a disturbing environment effect on the ocean and air.

**Example:** Cargo ship latest engine model – Wärtsilä-Sulzer RTA96-C produced by Swiss company Wärtsila-Sulzer is a turbocharged two-stroke diesel engine and the most powerful and most efficient engine in the world today. It comes with 6 cylinder in-line through a whopping 14 cylinder version, produces 7780 horsepower(hp) and the measurements of the engine are 89 feet length by 44 feet height with total weight of 2300 tons. Its fuel consumption is about 0.260 lbs/hp/hour compared to automotive and small aircraft that consume 0.40 to 0.60 lbs/hp/hour range.

**fig. 8.20.** The14 cylinder of Wärtsilä-Sulzer RTA96-C produces an incredible 108,920 hp at a speed of 102t/min.

**fig. 8.21.** Emma Maersk owned by A.P.Moller-Maersk Group, operation since 2006. Sister ships of Emma are Ebba, Edith, Eleonora, Elly, Estelle, Eugen and Evelyn.

One of very famous cargo ship, Emma Maersk is powered by this engine, claims that the engine protect environment with exhaust heat recovery and co-generation. But there are occasions that Emma Maersk burns bunker fuel, which produces high sulfur with 2.5% to 4.5%, estimated over 2000 times more than allowed in current automotive fuel.

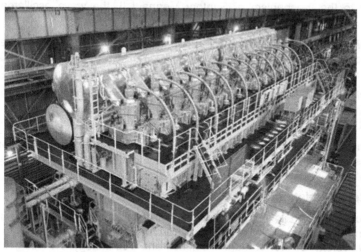

**fig. 8.22.** An overall look of the engine Wärtsilä-Sulzer RTA96-C.

## The Result of Contradiction Analysis

During solving this contradiction process, writer selects several A-Matrix, known as MICO (Multiple Input Cluster Out) for finding the most exact solutions. The results are: the "plus-state" side of "cargo ship" can be determined as in A-Matrix as number *29; stabile structure of the object and 37; energy use of the moveable object*, because the ship is a stable body that can carry heavy weight and floats on water. The ship is too uses less fuel (bunker fuel), very energy efficient. The "minus-state" of the "cargo ship" can be determined with A-Matrix number *13; external damaging factor, 19; volume of the moveable object and 32; weight of the moveable object* understood as the ship disadvantages are releasing greenhouse gases, the gigantic sizes and weight.

## Inventing Idea

To have an alternative or a stages replacement of the big cargo ship engine's, to use ocean waves as the driving power for the ship. The speed is not an issue but travelling in slow speed with free emissions of pollutions is the main target.

- *Navigator 01. Change in the aggregate state of an object* – according to this navigator, changing the level of complexity from complex big engine to a simpler mechanism.

- *Navigator 06. Use of mechanical oscillations* – the oscillations here is the wave from the ocean itself. By applying wave as the main energy supplier to the mechanism of Suntory Mermaid II, the wave oscillates and moves the boat.

- *Navigator 04. Replacement of mechanical matter* – the replacement of big complicated mechanism of ship's engine to a very simple propulsion that suits ocean waves to move the water transportation.

- *Navigator 23. Use of inert media* – the Suntory Mermaid II mechanism need no motor and pump for moving the boat, meaning the boat is using inert parts but managed well so that the inert part can move the boat.

- **Navigator 13. Inexpensive short-life object as a replacement for expensive long-life one** – uses of simple fins that can be replaced easily and less cost.

## The Artefact (solution) – "Suntory Mermaid II"

The "Suntory Mermaid II" is a wave powered boat designed by Dr. Yutaka Terao from the department of naval architecture and ocean engineering from Tokai University School of Marine Science and Technology. He designed the boat's propulsion system. It is a promising technology for cargo ship because it runs in low speed and by application of multiple fins in higher scale, it can support greater weight.

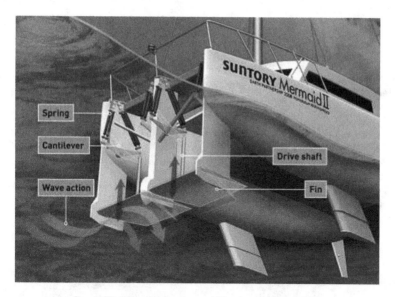

**fig. 8.23.** The mechanism of "Suntory Mermaid II"
that uses wave as energy provider.

"Suntory Mermaid II" propulsion mechanism is mounted under the bow, designed to pull the boat rather than pushing it forward, regardless of weather,
wave height or direction. The mechanism of Suntory Mermaid II consists of two
horizontal fins placed side by side that moves up and down according to motions of the waves, creating a dolphin-like tail kick that propels the boat.

His previous design was Malt's Mermaid I, II and III (1996-2002) and Suntory
Mermaid I (2004).

According to Yutaka, "water wave are a negative factor for a ship and slows it
down but the Suntory can transform wave energy into propulsive power regardless of where the wave comes from." Some improvement in the wave propulsion technology combined with wind sailing can be the best alternative to fossil
fuel. The idea of using wave as the power for moving ships started since 1895.
One design is from a German engineer R.J. May design a wave powered ship
with 9 spouts, a suction pump and an electric motor.

The result of re-inventing in standard form is represented at the figure 8.28.
Then, some questions should be asked to confirm that the solution is right.

- **Is contradiction eliminated?** = Yes

- **Are there some super-effects?** = Reduce pollution drastically and new
  wave energy development.

- **Need development?** = further development for bigger and heavier ships.

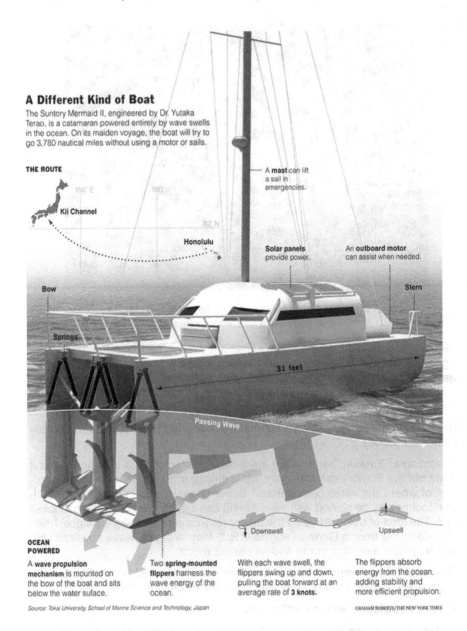

**A Different Kind of Boat**
The Suntory Mermaid II, engineered by Dr. Yutaka Terao, is a catamaran powered entirely by wave swells in the ocean. On its maiden voyage, the boat will try to go 3,780 nautical miles without using a motor or sails.

THE ROUTE

150° E
180°
30° N

Kii Channel

Honolulu

A **mast** can lift a sail in emergencies.

**Solar panels** provide power.

An **outboard motor** can assist when needed.

Bow

Stern

Springs

31 feet

Passing Wave

Downswell

Upswell

**OCEAN POWERED**

| | | | |
|---|---|---|---|
| A **wave propulsion mechanism** is mounted on the bow of the boat and sits below the water suface. | Two **spring-mounted flippers** harness the wave energy of the ocean. | With each wave swell, the flippers swing up and down, pulling the boat forward at an average rate of **3 knots**. | The flippers absorb energy from the ocean, adding stability and more efficient propulsion. |

Source: Tokai University, School of Marine Science and Technology, Japan

GRAHAM ROBERTS/THE NEW YORK TIMES

**fig. 8.24.** The look of "Suntory Mermaid II" and his testing journey be-tween Hawaii and Japan (source from The New York Times, 2008).

Another cargo ship company, Wallenius Wilhelmsen has designed a concept vessel for environmentally sound ocean transport, equipped with fin technology as the Suntory Mermaid ship has with additional solar technology. The conceptual ship is named "Orcelle" (fig. 8.25-8.27).

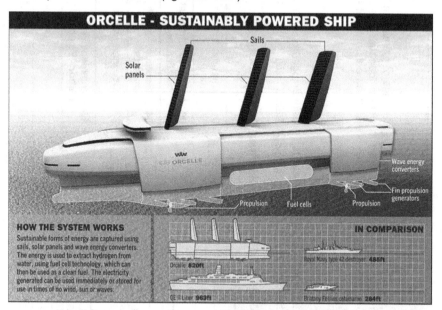

**fig. 8.25.** The overall look of "Orcelle" vessel and its system. Notice the fin propulsion on both side front and rear under the ship generates power form ocean waves.

**fig. 8.26.** E/S "Orcelle" from Wallenius Wilhelmsen Company.

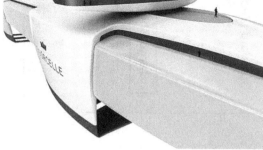

**fig. 8.27.** The size of "Orcelle" comparing to human.

## TARGETING

Cargo ships use the cheapest fuel, which they call 'bunker fuel'; the remains of gases and high-grade fuels refined from crude oil. Bunker fuel contains up to 5000 times more sulfur than diesel, a single container ship emits more pollution than 2000 diesel trucks. This is really a disturbing environment effect on the ocean and air.

## REDUCING

FIM: X-resource, without producing the inadmissible negative effects, provides together with other existing resources obtaining [to have more sustainable ships uses less dangerous energy].

## Standard (Technical) Contradiction Model – MICO (Multiple Input Clusters Out)

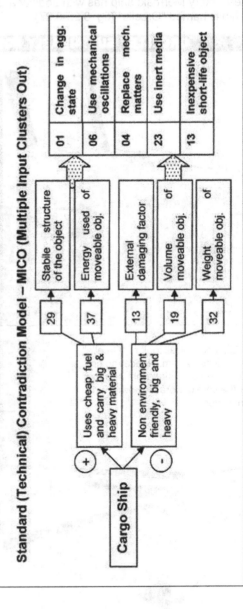

## INVENTING

According to navigators: **01. Change in the aggregate state of an object**: from complex big engine to a simpler mechanism, **06. Use of mechanical oscillations**: The oscillation here is the wave from the ocean itself, **04. Replacement of mechanical matter**: simple propulsion that suits ocean waves to move the water transportation, **23. Use of inert media**: the boat is using inert parts but can move the boat and **13. Inexpensive short-life object as a replacement for expensive long-life one**: uses of simple fins that can be replaced easily and less cost.

## ZOOMING

Are contradictions removed?: (Yes.) – No.

Super-effects: Saves cost of fuel drastically

Negative effects: No

## Picture

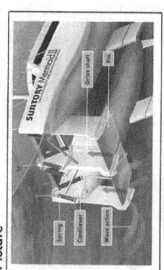

## Brief description

To have an alternative or a stages replacement of the big cargo ship engine's, to use ocean wave's special propulsion construction according to navigators 01, 06, 04, 23 and 13. The speed is not an issue but travelling in slow speed with free emissions of pollutions is the main target.

**fig. 8.28.** Reinventing in standard form of "Ocean Transport Energy Alternative".

## 8.5. SOLAR TECHNOLOGY IMPROVEMENT

The TRIZ application on this example is done with "Radical Contradiction" method. "Radical Contradiction" occurred when both "plus-state" and "minus-state" are experiencing the same problem elements. As for this "Hybrid Car Solar Roof Panels", the main target in improvements is the factor "flexibility". Let us go through the analysis.

### The Prototype – Hybrid Car Solar Roof Panels

It is common now that the position of solar panel placed on roof of a car. But in the recent years, there are various pattern designs of roof solar panels. Such conceptual car designed with way more attractive solar panel appearance for example from the TV series "Viper" (1994), with honeycomb pattern and moveable.

**Example.** The solar panel system collects clean renewable solar energy and utilizes it to charge the supplemental and HV battery pack but there are some solar cars uses direct power straight to an electric motor. Solar panels built on the rooftop absorb energy just enough to operate electrical items in a car such as air-conditioning.

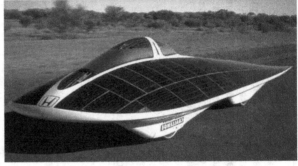

**fig. 8.29.** Some of existing prototype of solar car, which can only run on limited kilometers.

**fig. 8.30.** Artist impression of a solar motorcycle that has flexible solar panels to charge the battery.

The sun is abundant source of energy and able to deliver 10,000 times more energy than we can consume. In current situation, it is very difficult to generate enough power to move a vehicle from the sun energy, and right now it doesn't have any impact on car's fuel efficiency. The solar panels currently made up of semiconductors, made of silicon, are just 17% efficient when the weather condition is good with the sun is shining and the cost is expensive for automakers. Presently, the place for solar panel on car is placed on rooftop where it is always exposed to the sky, irrelevant on weight and limited size of solar absorption.

**fig. 8.31 (left).** A solar panel on car's rooftop, built by the Kyocera Corporation.
**fig. 8.32 (right).** The Toyota "Prius" car with enhancement on the solar panel roof, with convertible design.

### The Result of Contradiction Analysis

The "plus-state" side of "Hybrid Car Solar Roof Panels" can be determined as; solar panel is **not flexible** as the panel need to be exposed to the sun, usually placed on the roof surfaces of a car, flat surfaces. The "minus-state" of the "Hybrid Car Solar Roof Panels" can be determined with; solar panel is **flexible** placed other than roof, side body has to be designed so that it can fix the solar panel's flat surface.

### Inventing Idea

To have a type of "paint" that can work as photovoltaic cells, absorbing solar energy and emit it through electrical power. This can be great in hybrid or full electric car that uses battery to operate the engine. The particles of the solar spray are called "Buckyballs" - which is actually "fullerene" molecule composition. Like paint, the composite can be sprayed onto other materials and used as portable electricity. A sweater coated in the material could power a cell phone or other wireless devices. A hydrogen-powered car painted with the film could potentially convert enough energy into electricity to continually recharge the car's battery.

- *Navigator 19. Transition into another dimension* – transition of solar panel which is rigid, non flexible and limited space, now transformed in paint form, flexible, painted on all automotive body and absorbs solar benefits to supply more energy.

- *Navigator 25. Use of flexible covers and thin films* – using flexible covers, the artefact is in liquid form that covers the body of a car (which all exterior body of a car exposed to the sun.

- *Navigator 34. Matryoshka (nested doll)* – the paint nested on a surface of car's body, replacing solar panel nested on the roof.

- *Separation in material* – according to Af-Catalogue, the separation of material suits, from a solid material to a liquid form.

## The Artefact (solution) – Solar-On Spray

Nano technology can realize this idea. Studies by Prof. Somenath Mitra from the New Jersey's Insitute of Technology says, the spray on solar that uses "buckyballs" concept is inexpensive solar material that can be sprayed on surfaces or printed on plastic with for example ink-jet printer. "Fullerene" is a single wall carbon nanotube for polymer bulk that interfaces photovoltaic cells, is featured as the June 21st, 2007 cover story of the journal of "Materials Chemistry" published by the Royal Society of Chemistry. The solar cell developed uses a carbon nanotube complex. Nanotubes can conduct current better than any electrical wires and significantly better conductor than copper, combined with tiny carbon buckyballs (known as fullerenes) and formed in snake-like structures. Buckyballs doesn't make electrons flow but trap them. Add sunlight to excite the polymers, and the buckyballs will grab the electrons. Nanotubes, behaving like copper wires, will then be able to make the electrons or current flow.

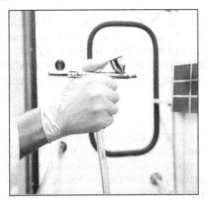 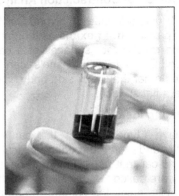

**fig. 8.31 (left).** A test of the solar paint sprayed onto sheets of plastic or stainless steel done in laboratory of University of Texas.
**fig. 8.32 (right).** The University of Texas also doing tiny photovoltaics in inks is made of Copper Indium Gallium Selenide (CIGS), sun-absorbing particles 10,000 times thinner than a strand of human hair.

Current development of solar cells which is by organic solar cells is cheaper as to silicon cells. We can foresee a great deal of interest in solar spray because solar cells can be inexpensively printed or simply painted on exterior building walls or roof tops. Imagine someday driving in your hybrid car with a solar panel painted all over the car, which produce more power to drive the engine. The opportunities are endless. It is also noted that this method by far is non-toxic to human health. By using this unique combination of organic solar cells, it will increase efficiency of the future paint solar cells.

**fig. 8.33.** The solar spray ink with the size of a
grain of rice can generate 7.8 volts of electricity.

The result of re-inventing in standard form is represented at the figure 8.36. Then, some questions should be asked to confirm that the solution is right.

- **Is contradiction eliminated?** = Yes

- **Are there some super-effects?** = The solar power collected are higher than normal solar panel

- **Need development?** = need long lasting photovoltaic paint

- **Change in environment?** = more environmental friendly

**fig. 8.34 (left).** The molecular structures of Buckyballs in a planetary shape, absorbs light from distant.
**fig. 8.35 (right).** Prof. Somenath Mitra, PhD from New Jersey Institute of Technology is the inventor of solar spray.

## TARGETING

Solar panel normally placed on the rooftop of a car. The system collects clean renewable solar energy and utilizes it to charge the supplemental and HV battery pack. The problem of this solar panel approach is that the energy sources collected can only be used for electrical items in a vehicle, not moving the vehicle itself.

## REDUCING

**FIM:** X-resource, without producing the inadmissible negative effects, provides together with other existing resources obtaining [to have more access to solar energy].

## Standard (Technical) Contradiction Model

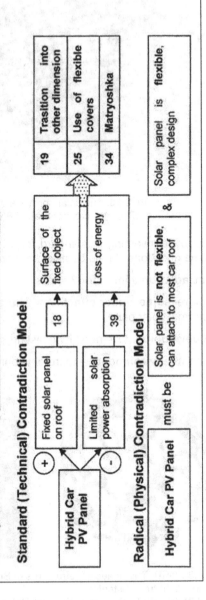

| 19 | Trasition into other dimension |
| 25 | Use of flexible covers |
| 34 | Matryoshka |

Surface of the fixed object — 18 — Fixed solar panel on roof

Loss of energy — 39 — Limited solar power absorption

Hybrid Car PV Panel (+) (−)

## Radical (Physical) Contradiction Model

Hybrid Car PV Panel | must be | Solar panel is **not flexible**, can attach to most car roof | & | Solar panel is flexible, complex design

## INVENTING

According to navigators: **19. Transition into another dimension:** transition of rigid, non flexible to a flexible form, **25. Use of flexible covers and thin films:** flexible covers, the artefact is in liquid form that covers the body of a car, **34. Matryoshka (nested doll):** the paint nested on a surface of car's body, replacing solar panel nested on the roof and the primary fundamental model is **Sp. Separation in Space:** instead of just on the roof, it is now can be applied to all body part.

## ZOOMING

Are contradictions removed?: (Yes.) – No.

Super-effects: The solar paint collected power higher than normal solar panel, can apply to other

electrical product as well.

Negative effects: No

## Picture

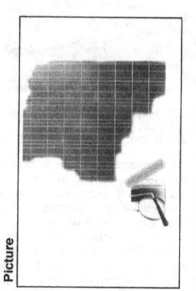

## Brief description

To have a "paint" according to navigators 19 25 and 34; that can work as photovoltaic cells, absorbing solar energy and emit it through electrical power. This can be great in hybrid or fully electric car that uses battery to operate the engine. The paint on all vehicles body and can absorb sun light at any side of the body.

**fig. 8.36.** Reinventing in standard form of "Solar Technology Improvement"

## 8.6. ENERGY SUPPLY FOR RAIL TRANSPORT IMPROVEMENT

### The Prototype – Electric Multiple Unit Trains

An "Electric Multiple Unit" train or EMU train is a unit consisting of more than one passenger carriages which carry passengers, using electricity as the motive power. To run the EMU, electric cables are build along the commuter rail, (for as long as the commuter runs, many kilometers) and build overhead of the train so that the electricity flows on the electric rod (the electric rod is on top of the commuter). Although the EMU trains uses no fuel, but electricity consumptions is very high. Electricity power too uses fossil fuel, but on EMU case, it's not direct. The additional problem is building the electricity supports along train routes are very high in cost and maintenance, plus visual pollution resulting disturbed city attractiveness.

*Example:* There are some famous electric multiple units (EMU) in the world and well-known EMU used very frequently is in the country of Japan, the "Shinkansen" and in Germany which they call ICE.

**fig. 8.37 (left).** Shinkansen "Kodama" 8-car Type 500, built by West Japan Railway Company (JR West).
**fig. 8.38 (right).** A German EMU - ICE 3 (Inter City Express,) built by Siemens AG.

The "Shinkansen" of Japan is built by a Japanese company named Japan Railways Group, starting in 1964. Shinkansen uses advanced technologies, operate in high speed punctually and separated from conventional rail lines. It also equipped with Automated Train Control (ATC) signal system, a different approach than the trackside signals. Shinkansen uses a 25,000V AC overhead power supply, air sealed to ensure stable air pressure when entering tunnels at high speed. The punctuality of Shinkansen is no doubt the most accurate, with average arrival time is within 6 seconds including natural and human accidents. The environmental impact of this train is excellent with only 16% $CO_2$ emissions from Tokyo to Osaka compared to journey by car.

The well known company Siemens and Bombardier built the "ICE 3" to achieve the goal of designing a higher-powered, lighter train than its predecessors. Travelling by this train gives you an experience of smooth ride, quiet and very fast journey. Licensed for 330km/h and has reached 368km/h on first trial, the ICE 3 is EMU unit with under floor motors throughout, not more than 400m and compatible to the electricity, signaling and communication systems. ICE 3 is equipped with "eddy-current" brake that ensures free wear braking. Achieving lightweight, ICE 3 uses air as cooling medium instead of air conditioning, similar to airplanes cooling mechanism.

## The Result of Contradiction Analysis

The "plus-state" side of "Electric Multiple Unit Trains" can be determined as in A-Matrix as number *29; stable structure of the object* - because the structure alongside train rail can give stable continuous electricity to the train while moving. The "minus-state" of the "Electric Multiple Unit Trains" can be determined with A-Matrix number *37; energy use of the movable object* - understood as the train is a movable object, and the electrical energy used need to be active 24 hours when the metal rod touching the cable alongside rail.

## Inventing Idea

An idea of an electric trains that consumes no fuel, gets electrical charges while in motion without plugging in – and then goes "anywhere". High in safety because of the cables are in the ground.

- *Navigator 37. Equipotentiality* – the cables that are build give electric charges to the train need not be constructed overhead of the train. It can be built on the ground and still gives electric charges.

- *Navigator 11. Inverse action* – the cables can be built under the train, 12 cm from the ground surface but still the train can absorb the electrical power to run. This way it uses less material, less cost of cable holders, constructions, weather exposures and safe for the human.

- *Navigator 18. Mediator* – mediator needed to supply electricity to the train. In this new design, they uses magnet.

- *Navigator 04. Replacement of mechanical matter* – the use of magnetic mechanism as the key energy supply when charging.

## The Artefact (solution) – OLEV (Online Electric Vehicle)

Online Electric Vehicle (OLEV) - a technology developed and deployed by Korean institute, Korea Advanced Institute of Science and Technology (KAIST) - a prototype implementation of OLEV technology that picks up electricity from power cables buried underground through a non-contact magnetic charging method. This revolutionary electric vehicle absorbs energy from cables beneath the road, about 12cm, rather than relying on batteries. This public transportation made available for full electric run. Facing significant problems such as overcoming the limitations involving lithium battery in terms of power capacity, weight, raw material price, recharging time and preparation of charging stations make OLEV is the new method of reducing those problems.

A German research company "Ingenieurgesellschaft Auto und Verkehr" or IAV GmbH also studied the same technology and applied to a tram manufactured by Bombardier Transportation.

**fig. 8.39 (left).** Prototype car for testing from Korean Advanced Institute of Science and Technology (KAIST).
**fig. 8.40 (right).** The launching of the OLEV technology.

**fig. 8.41.** A German branch of Bombardier Transportation produces tram that uses the same technology, researched by IAV GmbH naming it PRIMOVE.

**fig. 8.42.** The energy flow of the train's system – eliminates overhead wires and increase city's attractiveness. Safe transfers of inductive power, eliminates wear on parts and components plus operates in all weather and ground conditions.

The result of re-inventing in standard form is represented at the figure 8.46. Then, some questions should be asked to confirm that the solution is right.

- **Is contradiction eliminated?** = Yes

- **Are there some super-effects?** = Safer for the environment, less construction cost (12cm from ground surface) and less material.

- **Need development?** = magnetic electric recharge technology.

**fig. 8.43.** A pick-up coil underneath the tram turns the magnetic field created by the cables in the ground into an electric current, feeds the vehicle traction system.

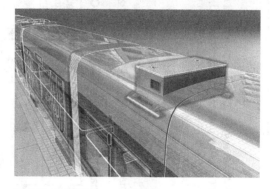

**fig. 8.44.** Uses MITRAC Energy Saver for breakdown, stores energy during braking.

**fig. 8.45.** South Korea's OLEV at an amusement park south of Seoul, runs along blue lines underneath is power strips.

## TARGETING

An electric multiple unit (EMU) is a multiple unit consisting of more than one passenger carriages, using electricity as the motive power. To run the EMU, electric cables are building along the commuter rail, (for as long as the commuter runs) and build high on the top of the train so that the electricity flows on the electric rod. High cost and maintenance.

## REDUCING

FIM: X-resource, without producing the inadmissible negative effects, provides together with other existing resources obtaining [to have other energy source, less visible, safe].

## Standard (Technical) Contradiction Model – MICO (Multiple Input Clusters Out)

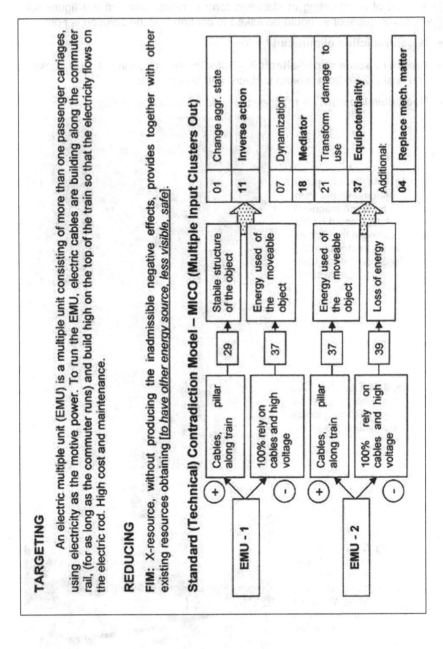

## INVENTING

According to navigator: **37. Equipotentiality:** the cables build underground give electric charges to the train, **11. Inverse action:** the cables can be build under the train, 12 cm in the ground but still the train can absorb the electrical power to run and **18. Mediator:** needed to supply electricity to the train. In this new design, the mediator is magnet. The additional navigator is **04.** Replacement of mechanical matter: the uses of magnetic mechanism as the key energy supplier when charging.

## ZOOMING

Are contradictions removed?: (Yes) – No.

Super-effects: ___ Safe for environment, very less or little construction.

Negative effects: ___ No

## Picture

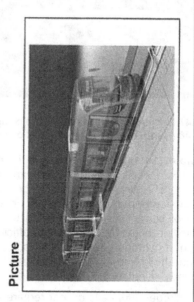

## Brief description

An idea of an electric train/car that consumes no fuel and gets electrical charges with the use of magnetic pulses planted underground – charging the car's power supply while in motion, according to navigators 37, 11, 18 and 04.

**fig. 8.46.** Reinventing in standard form of "Energy Supply for Rail Transport Improvement".

## 8.7. NEW TECHNOLOGY OF ENGINE PERFORMANCE

### The Prototype – Hydrogen Vehicle

Hydrogen vehicle uses Hydrogen ($H_2$) as its onboard fuel for motive power. It converts chemical energy of $H_2$ to mechanical energy either by burning $H_2$ in an internal combustion engine (ICE), or by reacting $H_2$ with $O_2$ in a fuel cell to run electric motors. $H_2$ fuels currently is most frequently made from methane, however, it can also be produced from wind, solar or nuclear sources. The attraction of using $H_2$ as an energy currency is that, if $H_2$ is prepared without using fossil fuel inputs, vehicle propulsion would not contribute to carbon dioxide emissions.

For the next twenty years, hybrid-electric ICE powered vehicle will be beneficial for a life-cycle perspective as fuel cell. Massachusetts Institute of Technology (MIT) researchers found that fuel cell vehicles do not have significant advantages in terms of energy consumption or $CO_2$ emissions but will accomplish significant reductions in energy usage by transportation sector with $CO_2$ emissions reductions.

**fig. 8.47 (left).** A hydrogen car from the company Ford.
**fig. 8.48 (right).** Fuelling H2 in the tank of hydrogen powered car.

**Example.** Hydrogen storage is still the main issue and it is still about the cost of the liquid, costing four times as expensive as gasoline. Some institutions are creating "effective hydrogen storage" because of inadequate application on most customer demands. They too create "affordable hydrogen fuel cells" because of current fuel cells are ten times more expensive than ICE (internal combustion engines). The overall efficiency is not more than 30% when production of hydrogen from hydrocarbon resources and the challenges of manufacturing hydrogen are the obstacles of operating in cold weather, packaging, reliability and safety.

The hydrogen fuel cells do not require any combustion of carbon-based fuels to run the car but generates electricity using chemical reactions. Although this automotive green technology approach is very good in reducing $CO_2$ the drawbacks quantity is more than normal fossil fuels. Hydrogen has very low density that equals to one-third of methane so some researchers have done special crystalline materials to store hydrogen at greater densities and at lower pressures. *"Experts say it will be 40 years or more before hydrogen has any meaningful impact on gasoline consumption or global warming, and we can't afford to wait that long. In the meantime, fuel cells are diverting resources from more immediate solutions."* [152]

## The Result of Contradiction Analysis

The "plus-state" side of "Hydrogen Vehicle" can be determined as in A-Matrix as number *01; productivity* - because the production of energy produced by fuel cell to convert to electricity is very efficient and $H_2$ can be easily renewed. The "minus-state" of the "Hydrogen Vehicle" can be determined with A-Matrix number *32; weight of the moveable object* - understood as the storage of hydrogen tank is heavy, and consumes spaces, because of its high flammable nature and will be extremely hard to store in a tank within cars, because any slight bump may cause the hydrogen to explode.

## Inventing Idea

To have a new type of mechanism that will run better, safe, environmental friendly and economical. The mechanism may use very little chemical or no chemical at all.

- *Navigator 27. Full use of thermal expansion* - this navigator is the driving navigator for the solution of replacing the Hydrogen car which is quite dangerous when compacted. Different than hydrogen, compressed air can be compressed, lead to compact sizes & lightweight.

- *Navigator 01. Change in the aggregate state of an object* - according to this navigator, instead of using chemical, the use of air that is compressed will move the vehicle. Instead of using chemical, it uses air.

- *Navigator 10. Copying* - according to this navigator, the engine still runs with internal combustion engine (ICE). The medium of energy supply is the differences. Both are environmental friendly, but compressed air are a lot easier to produce & economical plus safer.

- *Navigator 14. Use of pneumatic or hydraulic constructions* – uses pneumatic machine idea into the car's conceptual system.

- *Navigator 32. Counter-weight* – usage of light material and have aerodynamics effects.

## The Artefact (solution) – Compressed Air Car

Compressed-air powered car have the thermodynamic efficiency by a pneumatic engine and considers higher advantages between compressed air and chemical storage of potential energy. Competition between hybrid electric vehicles and pneumatic-combustion hybrid is technologically feasible and inex-

---

[152] Excerpts from the online technology news website "Wired News", May 2008.

pensive, opposed to the gas & oxygen as fuel cells car explosions of internal-combustion models. It takes only few minutes for the model "CityCAT", designed by Guy Negre, a former F1 engineer and the inventor of the "Compressed Air Vehicle", to refuel at gas stations equipped with custom air compressor units, costing around US2 dollars to fill the car's carbon-fiber tanks with 340 liters of air at 4350 psi. Drivers also will be able to plug into the electrical grid and use the car's built-in compressor to refill the tanks in about 4 hours.

**fig. 8.49 (left).** CityCAT, designed by Guy Negre, CEO of Motor Development International (MDI).

**fig. 8.50 (right).** Compressed air car from TATA Motors India – four-cylinder, 800 cc boxer-style engine with 25hp, sufficient for the 750kg car a top speed of 68 mph. The car has a range of 125 miles on the 4400 psi air held in carbon-fiber tanks.

Compressed-air car can run 35mph entirely on air, uses a trickle of petrol to heat and compress more air to reach higher speeds up to 90mph, and the car cost is just around US 20,000 dollars. It is super-efficient, 100mph plus vehicles to help mankind from its oil addictions. The Compressed Air Technology System (CATS) have significant step for zero-emission transport, safe, quiet, top speed of 110km/h and range of 200km with cost next to nothing. A French company together with TATA Motors India had an agreement in 2007 to develop the engine.

Although current research still consumes time, it is the cheapest, environmentally-neutral compressed-air engine and such proposal as "AIRPod" was created, to cater urban and industrial transport. The "AIRPod" is a personal 3 to 4 seater city commuter, airport, train station and workers in factory transporter, with joystick maneuvering. Weighing around 220kg, it only needs 5 and half horse power from its air engine to reach top speeds as little as 90 seconds with only 1.10 euro fueling cost.

Then, some questions should be asked to confirm that the solution is right.

- **Is contradiction eliminated?** = Yes
- **Are there some super-effects?** = It can go high in speed, lightweight and economical.

The result of re-inventing in standard form is represented at the figure 8.55.

**fig. 8.51 (left).** AIRPod: tiny air-powered commuter costs half a Euro per 100km.
**fig. 8.52 (right).** A 3D presentation of the seat locations in the "AIRPod".
The single seat is for the driver.

**Compressed Air Engine (CAE)**

**fig. 13.53.** A 3D presentations of the compressed-air engine for the "AIRPod".

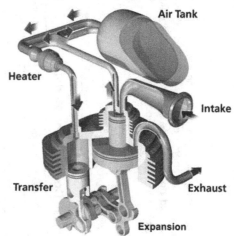

Air Tank

Heater

Intake

Transfer

Exhaust

Expansion

Courtesy of www.zeropollutionmotors.us

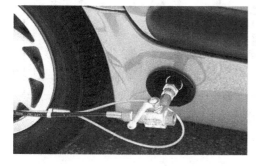

**fig. 8.54.** Fueling port, took only few minutes to refuel at MDI-supplied air compressor unit gas stations. Users can also refill its tank using the car's built-in compressor and plug into electric point at home for 4 hours.

## TARGETING

Hydrogen car uses hydrogen as its onboard fuel for motive power with electric motors. The attraction of using H2 as an energy currency is that, if H2 is prepared without using fossil fuel inputs, vehicle propulsion would not contribute to carbon dioxide emissions. The drawback is that it needs big tanks to store the liquid and explosion exposure is possible.

## REDUCING

**FIM:** X-resource, without producing the inadmissible negative effects, provides together with other existing resources obtaining [to have less chemical or no chemical, small tank].

## Standard (Technical) Contradiction Model

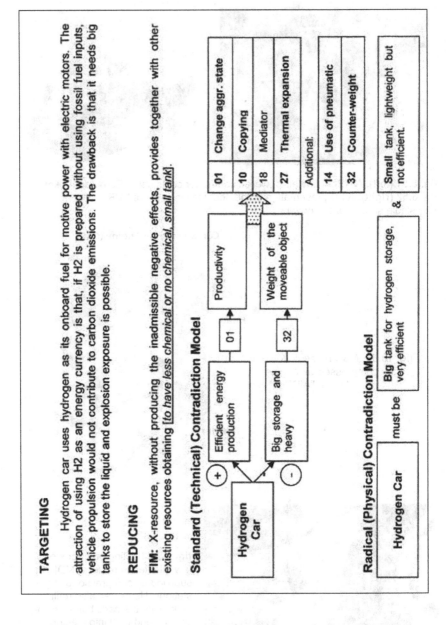

## Radical (Physical) Contradiction Model

## INVENTING

According to navigator 27. **Full use of thermal expansion:** the driving navigator. Using thermal expansion to produce compressed air. Compact sizes & lightweight, **01. Change in the aggregate state of an object:** instead of using chemical it uses air, and **10. Copying:** the engine still runs with ICE (Internal Combustion Engine). The medium of energy supply is the differences. Additional navigator is **14. Use of pneumatic or hydraulic constructions:** uses pneumatic machine idea into the car's conceptual system and **32. Counter-weight:** the use of light material and aerodynamics effect.

## ZOOMING

Are contradictions removed?: (Yes.) – No.

Super-effects: The speed is far better and lightweight

Negative effects: No

## Picture

## Brief description

As a replacement of chemical based energy supply (benzene, petrol, and diesel) according to navigators 27, 01, 10, 14 and 32. To have an engine that uses compressed air as a replacement of hydrogen based energy. It is lightweight, saves cost and environmental friendly.

**fig. 8.55.** Reinventing in standard form of "New Technology of Engine Performance".

## 8.8. PUBLIC TRANSPORTATION IMPROVEMENT – BUS

### The Prototype – QR Code Route Bus

The technology is very reliable but the route is fixed and non-flexible. It is well planned buses; for passenger taking & going out all over Berlin areas.

**Example.** Buses that uses QR Code Route is a company named BVG (Berliner Verkehrsbetriebe) in Berlin, Germany. The company provides linkage to all area in Berlin by S-Bahn train, U-Bahn train, tram and buses. BVG buses are an example of very systematic, flexible mode of transportation because it can move along planned destination or different maneuver when emergency occurred. QR Code or "Quick-Response Code" is originated in Japan and widely used in the mobile communication.

**fig. 8.56 (left).** An example of QR Code. With detection software such as "QR Code Reader", scanned code can redirect you to stored URL and access buses timetable.
**fig. 8.57 (right).** BVG Buses around Berlin urban and suburban areas that exercise the QR Code system.

**fig. 8.58.** The QR code on every bus stop.

QR Codes is originated from Japan by a company name "Denso-Wave" starting in 1994. In Japan, this code is used very commonly in mobile services. It was created originally for tracking parts in vehicle manufacturing. It works as "physical world hyperlinks", which with some applications or software installed in mobile or public transport that scans correctly and redirect to the programmed URL. Each bus has a GPS which continuously updates the bus company server with its position. These results to buses giving announcements of stops arrivals when the bus reached within 250m from the bus stop provided there are QR codes nearby and the bus detected it. It is also a tool for buses route planning.

Bus passengers can browse mobilized page when they scan QR code at bus stop which shows a list of approaching buses, their location and estimated time of arrivals and if there are any alter-native buses going in the same direction, so they no longer have to wait and wondering when is the bus is coming. BVG has applied QR Code system, starting with Wireless Application Protocol (WAP), that brings the BVG-mobile internet with i-mode and currently using Extensible Hypertext Markup Language (XHTML).

### The Result of Contradiction Analysis

The "plus-state" side of "QR Code Route Bus" can be determined as in A-Matrix as number *04; reliability* - because the system is efficient and nearly accurate. The "minus-state" of the "QR Code Route Bus" can be determined with A-Matrix number *02; universality, adaptability* - understood as the destination routes is too planned and non-flexible. If flexibility exists, the number of transportation needed is very high.

### Inventing Idea

To have a bus concept for public transportation that can deliver passengers faster, comfortable and runs in flexible route. The vehicles drive at normal speeds on existing roads but can speed up to 250km/h on special tracks.

*"The basic idea is to offer comfortable, demand driven, point-to-point transport that is able to compete with cars and high speed trains"* [153].

By means of ICT-technology (comparable to SMS and internet), one can "order" a vehicle. The vehicle will pick you up close to your point of departure and will bring you close to your point of destination. It will minimize the number of stops and changing means of transport will not be necessary anymore.

- *Navigator 01. Change in the aggregate state of an object* – changes of the degree of flexibility is applied. The route is not 100% planned and can be changed. As long as the driver have the help of the route navigator and from the help of ICT, picking up and leaving passenger is fast and comfortable.

**fig. 8.59.** The uses of QR code are now applicable all over Europe's public transportation.

---

[153] Technical University of Delft, Netherland project of "Superbus", lead by Prof. Wubbo Ockels.

- **Navigator 18. Mediator** – the ICT Technology is the mediator to the Superbus driver. Compared to QR Code, when destination arrived, the code response that the destination had arrived. Different from QR Code, the ICT not just announcing arrivals but too recommends faster route of destination and pick-up calls from passenger from any places that is not in the planned route.

- **Navigator 32. Counter-weight** – the design of Superbus is hard to recognize as a bus. This is because the shape is more aero-dynamic for speed & comfortability. The interior is as the same level as a limousine. Divided into two sections, single travelers and by group, Superbus gives maximum travel comfortability.

- **Additional Fundamental Ti. Separation in time** – according to Af-Catalogue, the separation of time is definitely describing the Superbus. Concerning waiting time and travelling time for passengers, Superbus uses ICT to pick-up and arrives faster.

- **Additional Fundamental Sp. Separation in space** – according to Af-Catalogue, Superbus are best on the Supertrack route, which is separated from the normal or planned route. The speed on this track can reach up to 250km/h. Superbus can travel on normal road and "Supertrack" road.

**The Artefact (solution) - ICT Technology Route: Superbus**

Build by a team of designers and engineers from Technical University of Delft, Netherlands; Superbus is a conceptual bus for sustainable mobility, new logistics with dedicated infrastructure called Supertrack. The intelligence of Superbus is that the route and schedule are not fixed, by-demand transport, via central routing optimization system that does not fixed schedule and the logistics allow high flexibility. One of Superbus features is when in summer, the Superbus will store heat as the heating mechanism for winter; it will be geothermically heated in order to prevent icing. In realizing point-to-point transport possible, ordering system through internet and SMS is developed, combining as many passengers at the same point and same destination as possible, decreasing stops and waiting time. In future, Superbus is planning to have its vehicle on the Supertrack every 7 seconds.

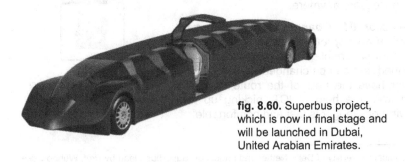

**fig. 8.60.** Superbus project, which is now in final stage and will be launched in Dubai, United Arabian Emirates.

Superbus equipped with navigation system, obstacle detection, communication system, fail safe system and control system. There are two driving mode, by which driver assisted controlled on existing roads, autopilot on Supertrack. With length of 15,000 mm, weight distribution of 34/66, the energy consumption is less because it's lightweight. Electrically powered vehicle, it is sustainable with replacement after 1.5million kilometers. The result of re-inventing in standard form is represented at the figure 8.64.

**fig. 8.61 (left).** The frame of actual Superbus.
**fig. 8.62 (right).** The testing of Superbus chassis on real road.

**fig. 8.63.** The finished body of Superbus. This will be installed on the tested chassis.

Then, some questions should be asked to confirm that the solution is right.

- **Is contradiction eliminated?** = Yes

- **Are there some super-effects?** = Usage of internet can widen to other functions for Superbus.

- **Change in environment?** = The construction of Supertrack for faster Superbus journey.

## TARGETING

Application of QR Codes system on public transportation (bus): the technology is very reliable but the flexibility of the route is fixed. The need of several buses to cater the area for passenger taking & going out of the bus is already planned. Some destination requires inter-changing from one bus to another and requires waiting time.

## REDUCING

**FIM:** X-resource, without producing the inadmissible negative effects, provides together with other existing resources obtaining [*to have a bus route- flexible, fast and universal*].

## Standard (Technical) Contradiction Model

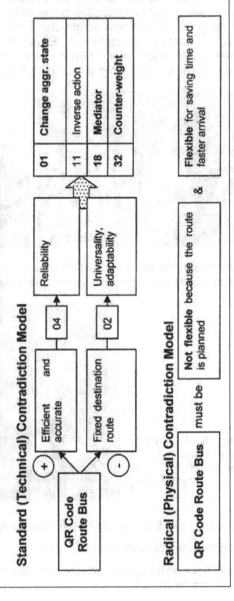

## Radical (Physical) Contradiction Model

## INVENTING

According to navigators: **01. Change in the aggregate state of an object**: The route is not 100% planned and can be changed, **18. Mediator**: the ICT Technology is the mediator, **32. Counter-weight**: the design of Superbus is hard to recognize as a bus. This is because the shape is more aero-dynamic for speed & comfortability.

## ZOOMING

Are contradictions removed?: (Yes) – No.

Super-effects: ___The usage of ICT not only on ordering route, but also ordering meals, etc.___

Negative effects: ___No___

## Picture

## Brief description

By means of ICT-technology, one can "order" a vehicle by ideas from navigators 01, 18 and 32. The vehicle will pick you up close to your point of departure and will bring you close to your point of destination. It will minimize the number of stops and changing means of transport will not be necessary anymore.

**fig. 8.64.** Reinventing in standard form of "Public Transportation Improvement – Bus".

## 8.9. RAIL AND ROAD TRANSPORT ALTERNATIVE

### The Prototype – Train and Automobiles.

Analysis between "Rail Transport" and "Road Transport" resulted that rail transport gives higher speed but less maneuverability. On the other hand, road transport gives maximum maneuverability but if high in speed, it will cause great danger to safety. Both transportation works on land and require additional construction of rail road with strong foundation to support the weight of the train and for road transport, construction of tar road is essential for automobiles travelling on it. Both are higher in cost but are the major type of transportation nowadays. After comparing with both means of traffic, the best selection of improvement is rail transport as speed is the demand in changing a complete new nature of traffic but also considering high in safety and environmentally friendly.

**Example.** Rail transport is specialized to transport passengers and goods by way of wheeled vehicles running on rail tracks. Tracks usually consist of steel rails installed on sleepers and ballast. New type of train, Electric Multi Unit (EMU) have additional rail of electricity cables along the rail for overhead electrical supply to the train. Being very high in cost for construction and operation, only limited numbers of train available especially in developing and under-development countries.

**fig. 8.65.** An EMU train, KTM Komuter from Malaysia

**fig. 8.66.** Putra LRT travelling across Kuala Lumpur city.

There is need of simple, fast and lower cost of transportation in the city areas, which have bad congestion every day. The quality of air and time spent travelling with road transport is getting lower and travelling with train requires waiting time and many stops. A very light construction above the ground with advantages of extreme level, straight for faster speed and relatively independent from topography of construction site is needed.

**fig. 8.67 (left).** The Kuala Lumpur city scenery, the capital city of Malaysia.
**fig. 8.68 (right).** The twin tower of Petronas is one of Malaysia's prides.

**fig. 8.69.** The city of Kuala Lumpur have limited space for commute area, unavailable to built new traffic roads or train constructions.

## The Result of Contradiction Analysis

The "plus-state" side of "Rail Transport" can be determined as in A-Matrix as number *03; level of automation, 04; reliability, 29; stabile structure of the object,*, because the rail transport; train, have high automation level, high reliability for smooth transport, stabile in structures and stable in movement. The "minus-state" of the "Rail Transport" can be determined with A-Matrix number *07; complexity of construction, 37; energy use of the moveable object, 39;loss of energy*, understood as the train and facilities construction is quite complex, the electrical energy used have disadvantage – of high cost, maintenance and complex construction.

## Inventing Idea

The target is to have a high speed transport, level rail, high in safety and environment friendly new type of transport. It also fulfils universality and versatility that can go across cities and cross countries. The new transport can travel without so many disturbances of congestions and very safe mode of public transportation.

- *Navigator 03. Segmentation* – small high speed unit is required instead of heavy long train with high consume on electrical energy.

- *Navigator 04. Replacement of mechanical matter* – replace mechanical structures of huge beams and metal tracks with extreme stretchable cable constructed for example at high rise buildings support pillars.

- *Navigator 05. Separation* – the new transport is separated from land, elevated and positioned in air.

- *Navigator 11. Inverse action* – it uses cable string, replacing train tracks.

## The Artefact (solution) – String Transportation Unit (STU)

A typical problem facing most major cities around the world is the increasing congestion with land transport. The solution to decrease congestion and still can give excellent travel efficiency is the String Transport Unit (STU), invented by Russian engineer Anatoly Unitsky. The unique aspects of STU are the efficiency that is far greater than cars, monorail, planes, trains or virtually any other transport system. First, the steel wheel or track interface has been developed to be twice as efficient as that of a train wheel (and 10 times better than a car wheel on a concrete road). Second, In the case of single rail passenger STU systems, the anchors support or pillars of STU cables can be a medium to high-rise buildings and the elevation or position above the ground removes a major source of wind resistance thus increases speed smoothly. The pick-up and drop point of passengers are mostly on high rise buildings with elevators straight to the roof top of buildings. Third, the aerodynamics shape of the vehicle has been refined to produce an incredibly low wind-drag vehicle. STU also caters transportation in large numbers of passengers, freight and mining transportation.

The result of re-inventing in standard form is represented at the figure 8.77.

Then, some questions should be asked to confirm that the solution is right.

- **Is contradiction eliminated?** = Yes

- **Are there some super-effects?** = doesn't involve in road congestion and smooth ride.

- **Development** = construction of cable connections.

**fig. 8.70.** Inventor and author Anatoliy Unitsky and his key monographs (https://rsw-systems.com)

**fig. 8.71.** Integration of String SkyWay System with urban landscape

**fig. 8.72.** The interior look of STU public transporter.

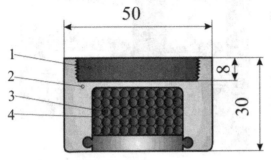

**fig. 8.73.** The cross section of the "track" – combines the flexibility of a string with the rigidity of a steel beam.

1.  Rail head - steel
2.  Body - aluminum composite
3.  String - high strength steel wire, 50 wires of 3 mm each
4.  Filler - epoxy resin type

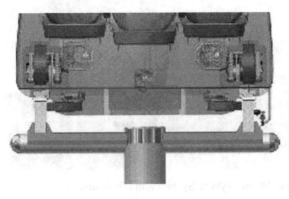

**fig. 8.74.** STU moves along cable which is very stable and bump free, very different than cable cars

**fig. 8.75.** Today there are several projects developing
for United Arab Emirates, India and other countries

**fig. 8.76.** One of the Cargo Projects for Australia

## TARGETING

Rail transport is the means of transporting passengers and goods by way of wheeled vehicles running on rail tracks, run on a prepared surface, rail vehicles, directionally guided by the tracks they run on. Then road transport as we all know is the main way of transportation in the world.

## REDUCING

**FIM:** X-resource, without producing the inadmissible negative effects, provides together with other existing resources obtaining [to have a new rail transport that do not require big land, less maintenance, less cost and manpower].

## Standard (Technical) Contradiction Model

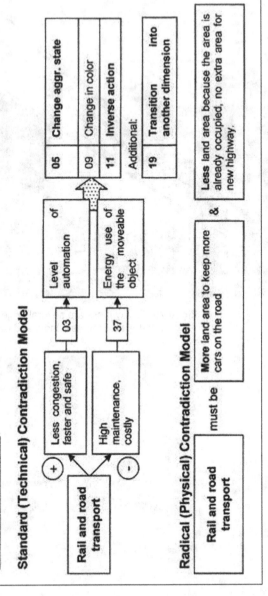

## Radical (Physical) Contradiction Model

| Rail and road transport | must be | More land area to keep more cars on the road | & | Less land area because the area is already occupied, no extra area for new highway. |

## INVENTING

According to navigators: 03. **Segmentation:** small high speed unit instead of heavy long train, 04. **Replacement of mechanical matter:** replacing mechanical structures of huge beams and metal tracks with extreme stretchable cable, 05. **Separation:** separated from land, elevated and positioned in air (z-axis), 11. **Inverse action:** it uses cable string, replacing train tracks and additional navigator 19. **Transition into another dimension:** Z-axis lift, the transport is lifted from the ground with string construction.

## ZOOMING

Are contradictions removed? (Yes) – No.

Super-effects: Doesn't involve in road congestion and smooth ride.

Negative effects: No

## Picture

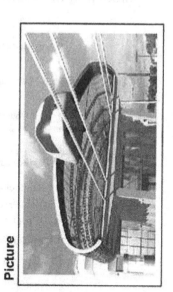

## Brief description

To have a new type of transport - String transportation, universal and versatile that can go across cities and cross countries. By using navigators 03, 04, 05, 11 and 19 safe and fast with extremely low cost both in operation and construction. The new transport can travel without so many disturbances of congestions and certainly an environmentally friendly public transportation.

**fig. 13.77.** Reinventing in standard form of
"Public String Transportation System SkyWay" by inventor Unitsky.

# 9 Managing Value Across Industries: Modern TRIZ Modeling of Jerome Lemelson's Inventions [154]

The American dream is that if the average American invents something novel and worthy of patenting, he'll find someone to license it. However, for most contemporary inventors, it hasn't worked out that way. The independent inventor today still has an extremely difficult time convincing corporations that he has a product which deserves to be on the market. Most companies have a tremendous resistance to ideas and technology developed on the outside. [155]

*Jerome Lemelson*

## Technische Universität Berlin

*"Wir haben die ideen für die zukunft"*

Global Production Engineering
International Masters Program

## Managing Value Across Industries: Modern TRIZ Modeling of Jerome Lemelson's Inventions

by

**PRABHAKARAN ADHINAMILAGI**

A thesis submitted in partial fulfillment of the requirements for the degree of
'Master of Science in Global Production Engineering'

Under the guidance of
**Professor, Dr. Dr. Sc. techn. Michael Orloff**

**Fakultät V – Verkehrs- und Maschinensysteme**
Produktionstechnisches Zentrum

**Technische Universität Berlin**

**Global Production Engineering**

**Dean: Prof. Dr.-Ing. Günther Seliger**

**Department of Assembly Technology and Factory Management**

**Berlin, Germany**

---

[154] Part 9 is based on the edited and supplemented fragments of the master thesis by Prabhakaran Adhinamilagi (India), a certified graduate of MTRIZ course at GPE Program, TUB
[155] quoted from https://en.wikipedia.org/wiki/Jerome_H._Lemelson

© Springer Nature Switzerland AG 2020
M. A. Orloff, *Modern TRIZ Modeling in Master Programs*,
https://doi.org/10.1007/978-3-030-37417-4_9

# 9.1. JEROME H. LEMELSON

## 9.1.1. Biography [156]

Jerome Lemelson, also called "Jerry", was born in New York on July 18, 1923.

Having the role model as Thomas Edison, he earned himself the fame as second most prolific inventor after Edison. He has to his name, to date, 606 patents and counting in a myriad of fields from medical and industrial technologies to gadgets and toys. A consummate inventor, Lemelson it seems used to sleep with a notebook beside him to wake up and jot down his endless ideas that he dreamt.

Airplanes and technology were his prime passion as a child. This would come to influence the later part of his career eventually. He had a bachelor's degree and master's degree in aeronautical engineering and another master's in industrial engineering. He then went ahead to serve in the Army Air Corps during World War II.

Soon after Lemelson began filing his first patents in 1950, he was eventually producing one patent a month. Lemelson's patents after some time spanned various fields – including many for industrial automation – and have played an integral part in products and technologies we use every day.

A champion of the independent inventor, Lemelson has challenged many companies for patent infringement or incorporating certain inventions of his into new designs without licensing them. After facing several challenges, in 1975, he joined the Patent and Trademark Office Advisory Committee to improve the way patenting is performed.

In the 1993, the Lemelsons established the Lemelson Foundation to inspire the next generation of inventors, innovators and entrepreneurs in the United States. The Foundation has developed several programs intended to motivate and prepare young people to create, develop and commercialize new technologies; and to increase public awareness of the critical role that inventors, innovators and entrepreneurs play in sustaining and strengthening America's economic vitality and shaping and improving our daily lives.

A black day for Invention, Lemelson died on October 1, 1997 a year after he was diagnosed with liver cancer. Lemelson continued to invent after his diagnosis, concentrating even more on medical technologies, particularly those that could be used for cancer treatment. His innovative ideas never ceased, nor could his illness hinder his drive to invent – he submitted almost 40 patent applications during his last year.

Lemelson used to say *"I am always looking for problems to solve. I cannot look at a new technology without asking: How it can it be improved?"*

---

[156] The description above is an adoption from *Lemelson MIT, 2012*

His Son Rob Lemelson used to quote about him that *"He was so busy invent-ing 24 hours a day"*. These talk about the passion he had to invent.

Arthur Molella, the director of the Lemelson Center for the Study of Invention and Innovation at the Smithsonian Institution also once praised Lemelson say-ing "He had a fruitful, creative mind that has made him a great role model to independent inventors."

### 9.1.2. List of Patents for Modeling [157]

As of October 2009, Lemelson has to his name, 606 patents. A very few of his renowned inventions include fax transmission; VCRs; portable tape players; camcorders; and the bar code scanner. Others like illuminated highway mark-er; a talking thermometer; a video telephone; a credit verification device; an automated warehouse system; and a patient monitoring system have also been in prominence around the world.

Owing to the scale of the list of patents, it is not included in this Thesis work.

1. Advancement in Bar Codes (U.S. Patent No. 6,543,691)
2. Durable Valves and Valve Components (U.S. Patent No. 5,040,501)
3. Composite Reinforced Foam Sheet (U.S. Patent No. 3,556,918)
4. Extensive Forest Fire Detection System (U.S. Patent No. 5,832,187)
5. Container with Inbuilt Dispenser Straw (U.S. Patent No. 4,226,356)
6. Photosensitive Contact Lens (U.S. Patent No. 4,681,412)
7. Facsimile – Transmission & Error Proofing (U.S. Patent No. 3,084,213)
8. Catheter and Implant Method (U.S. Patent No. 4,588,395)
9. Modern Fences (U.S. Patent number: 3,933,311)
10. Autonomous Assembly Robot (U.S. Patent No. 6,898,484)
11. Magnetic Tape Transducing System (U.S. Patent No. 3,842,432)
12. Advanced Reflectors (U.S. Patent No. 4,127,693)
13. Improved Roller Skate (U.S. Patent No.4,273,345)
14. Towed Watercraft Steering Method (U.S. Patent No. 5,462,001)
15. Wheel Inspection Device (U.S. Patent No. 3,148,535)
16. Reflective Thread (U.S. Patent No. 3,050,824)

---

[157] It could be seen at the reference *Lemelson Patents, 2012*

## 9.2. Information Management:

### Advanced Bar Codes (U.S. Patent No. 6,543,691)

**Prototype (Engineering Level)**

Bar codes are widely used today to mark objects in order to provide rapidly readable codes containing information relating to the object, such as, its identity or its price. It is also widely used in manufacturing plants to identify the parts.

Bar codes usually consist of a horizontal series of printed vertical parallel lines of varying thickness and spacing juxtaposed to produce a particular reflectivity pattern when the bars are scanned sequentially in a straight line with a light beam, such as, a laser beam, in a direction transverse to the orientation of the bars.

**fig. 9.1.** A Conventional Bar Code

**Problems with the Prototype**

Normally, a bar code may be used to store primary information relating to the product's identification. However it is not known to provide secondary information storage capability on the same barcode.

The need for this secondary information storage can specifically be seen in manufacturing operations where several barcodes are placed on single product – each containing one primary information. A problem with using multiple bar codes on a product has been that they occupy substantial space and reading is complicated. They are also able to store only limited amount of information.[158]

There have been however, several new inventions like the following format to increase the information density. But none of the known prior art is designed to be downward compatible with the conventional Bar Code technology while at the same time increasing the storage density (Batterman et al.,1992).

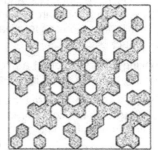

**fig. 9.2.** A format of the information storage cell to increase the storage density (Batterman et al.,1992)

**Result of Contradiction Analysis (Creative Level)**

The contradiction analysis could be structured in this problem of barcodes as follows. The "plus" in the system would be and "Increased information storage" but by doing so, we are faced with a "minus" contradiction of "Necessity to make the bar code bigger". To convert it in the language of TRIZ, the "plus" transforms into "Universality or Adaptability" because of the required adaptabi-

---

[158] Above sections are adapted from – Lemelson et al., 2003

lity to increased information storage and the "minus" transforms into "Length of the fixed object.

The Altshuller Matrix yields the navigators to be used as – 1. Change in the aggregate state, 3. Segmentation and 16. Partial or excess effect.

### Inventing Idea

The idea for the resultant artifact could be obtained with the navigator number 3 that is, Segmentation.

Each vertical stripe of the bar code is segmented horizontally. Every original vertical line now serves as miniature bar code storing secondary information. This way the information storage density is drastically increased.

### Psychological Level

Knowledge of barcode system itself could help us attain the Artifact. An equivalent barcode is embedded in each stripe at a right angle.

The new artifact could also be comprehended from the exposure to day to day storage systems, which one may come across to which this barcode system of information storage could be analogized.

A long row of library rack which is segmented horizontally to accommodate more books or Information stored in each spiral track of a compact disc which is further divided into "pits" and "lands". A mind which compares things not related to each other could also help in the transformation process. For example, a comparison of transition of CDs to DVDs with the Barcode system could also lead to the answer. CDs contain small pits which are read with laser as a 0 or 1. DVDs have finer pits and hence more information could be stored in it. Hence making the Barcode lines significantly smaller and trying to accommodate it in the same base area could generate the same result as the artifact (Compact Disc, 2012).

Looking for a solution at a microscopic level could yield the answer rather that thinking from the same level as the prototype.

The new system of Barcodes seems very practical to be used in products that need high information storage. This guarantees minimal space utilization with maximum possible data storage.

### Resultant Artifact

US Patent No. 6,543,691                    Date of Patent: April 8, 2003

**Method and apparatus for encoding and decoding bar codes
with primary and secondary information and method of
using such Bar Codes**

**Inventors: Jerome H. Lemelson; John H. Hiett**

**Abstract:**

A bar code has primary information encoded in one direction (e.g., horizontally) and secondary information encoded in another direction (e.g., vertically) in single or multiple tracks in selected ones of the vertical bars of a bar code.

Using a non-linear, variable amplitude scanner, all of the primary bars are scanned in the one direction to obtain all of the primary information and all of those vertical bars having secondary information are scanned in the other direction to obtain all of the secondary information.

The one direction which is perpendicular to the vertical primary bars, is determined by first rotating the scan path axis until both start and end code bars are read thereby placing the scan path entirely within the total bar code, and, then, further rotating the scan path to determine the direction of the minimum crossing width of the total bar code.

Secondary information is scanned in planes orthogonal to the one direction after those vertical bars having such information are first identified and selected. In a preferred embodiment, the decoded secondary information may be used to control selected station process operations for selected products in a continuous manufacturing assembly line (Lemelson et al., 2003).

**fig. 9.3.** Two formats of rationalized Bar Codes (Lemelson et al., 2003)

**Extracting** of invention Bar Codes is shown in fig. 14.4. The symbol "LC" in the table means "Level of Compliance" – estimation of adequacy of selected transformation model to real change by transition from prototype to the artifact under studying. Two pluses means high level of compliance, one plus – satisfactory adequacy.

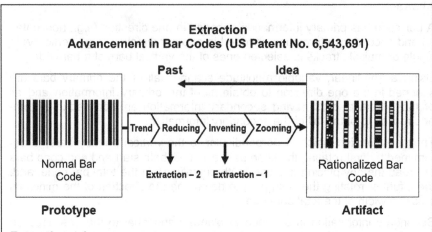

**Extraction – 1**

| LC | No. | Navigator | Substantiation for Extraction |
|----|-----|-----------|-------------------------------|
| ++ | 03 | Segmentation | Segment the individual vertical stripes further into horizontal stripes. |
| + | 05 | Separation | Take out the functional barcode and analyze where it could |
| + | 10 | Copying | Place a copy of the barcode into each vertical |
| ++ | 12 | Local property | Change the property of each stripe so that it acts like a barcode on its own. |
| ++ | 19 | Transition into another dimension | Rather than orienting the barcode only horizontally, use also |
| ++ | 34 | Matryoshka | Nest miniature barcodes into each stripe. |
| + | 35 | Unite | Unite two barcodes perpendicular to each |

**Extraction – 2**

**Standard Contradiction**: A Barcode should have increased information storage capability but it makes the length of the Barcode itself longer.

**Radical Contradiction**: A Barcode should be long to hold extra information but it should be short to look elegant and be rationalized.

**FIM**: X- resource, together with available or modified resources, and without making the object more complex or introducing any negative properties guarantees attainment of the following Ideal Final Result (Multiple information can be stored in single Barcode)

**fig. 9.4.** Extracting of invention Bar Codes (Lemelson et al., 2003)

## Advancement in Bar Codes (U.S. Patent No. 6,543,691)

**TREND:** Bar codes are widely used today in most Objects to mark and to provide rapidly readable codes containing the identity information of the Object. Essentially they are made up of vertical bars of varying lengths depending on the information to be stored and the gaps produce a reflectivity pattern, dependent of the width of the line when a light beam like a laser is incident on the same.

*Standard Bar*

**PROBLEM:** Only short and specific information can be stored in Bar codes. Storage of additional secondary information is not possible.

## REDUCING

**FIM:** X- resource, together with available or modified resources, and without making the object more complex or introducing any negative properties guarantees attainment of the following IFR ( Multiple information can be stored in single Barcode)

Standard (Technical) Contradiction Model

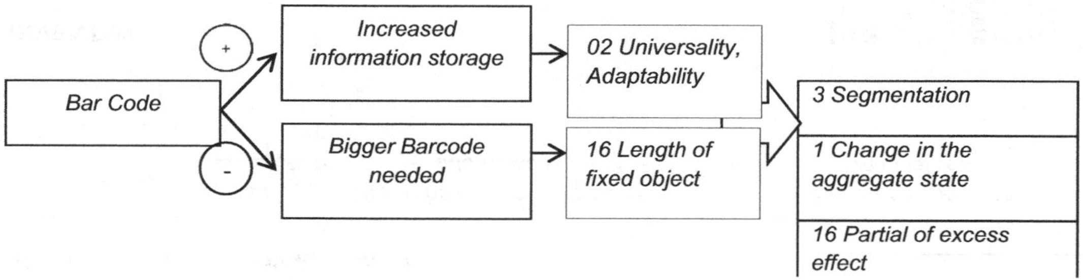

fig. 9.5 - beg. Reinventing in standard form of
Advanced Bar Codes (U.S. Patent No. 6,543,691)

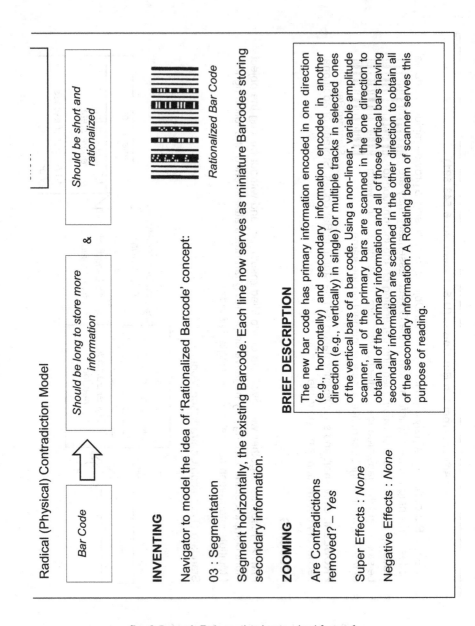

Radical (Physical) Contradiction Model

Bar Code

*Should be long to store more information* & *Should be short and rationalized*

*Rationalized Bar Code*

## INVENTING

Navigator to model the idea of 'Rationalized Barcode' concept:

03 : Segmentation

Segment horizontally, the existing Barcode. Each line now serves as miniature Barcodes storing secondary information.

## ZOOMING

Are Contradictions removed? – Yes

Super Effects : None

Negative Effects : None

## BRIEF DESCRIPTION

The new bar code has primary information encoded in one direction (e.g., horizontally) and secondary information encoded in another direction (e.g., vertically) in single) or multiple tracks in selected ones of the vertical bars of a bar code. Using a non-linear, variable amplitude scanner, all of the primary bars are scanned in the one direction to obtain all of the primary information and all of those vertical bars having secondary information are scanned in the other direction to obtain all of the secondary information. A Rotating beam of scanner serves this purpose of reading.

**fig. 9.5 - end.** Reinventing in standard form of Advanced Bar Codes (U.S. Patent No. 6,543,691)

## 9.3. Precision Technology:

### Durable Valves and Valve Component (U.S. Patent No. 5,040,501)

### Prior Art (Engineering Level)

Poppet valves are predominantly used in Internal Combustion Engines. Their function is to shut and give way in the intake and exhaust ports when necessary. Also called a "Mushroom Valve", they control the flow of vaporous fuel during each combustion cycle of the engine (Dyke Gasoline Engine, 1920).

### Problems with the Prototype

As mentioned by Lemelson (1991), Valves such as exhaust valves for internal combustion engines and other valves are subject to wear and corrosion. In certain instances, the impact forces may cause degradation in structure and structural failure.

A select portion of the surface of a valve or valve seat is normally subject to degradation during use.

Typically, it is due to the corrosive effects of fluid particles passing through the valve.

**fig. 9.6.** A poppet Valve (Uploads 2012)

### Result of Contradiction Analysis

The positive requirement from the existing valve and valve component is that. It should have an improved wear resistance property. The negative that arises by having such a system is that a harder material could end up being brittle. By introducing such a valve made of harder material, the whole valve system could break up due to its increased brittleness. TRIZ would define the positive requirement as "Functional time of moving object" and the negative "Strength". The Contradiction analysis gives the following three navigators – inexpensive short life object replaces expensive long life Object, Local property and Preliminary Action

### 2.4. Inventing Idea

The Inventing idea could be achieved with the navigator 'Local property'. Only the local property that is, the surface of the valve and valve seat are hardened by coating the surfaces with synthetic diamond material

### 2.5. Psychological Level

The inventive idea for the artifact could be achieved through the knowledge of natural materials and their basic properties. The chipping off of the Valve and its seating area is mainly attributed to the chemical corrosiveness or high forces involved in the region of fuel delivery.

Hence, knowledge of the basic physics and chemistry involved in the valve region is required for the generation of the solution.

A Valve made up of a hard and chemically non-reactive substance could solve the problem. Hence, nature's Diamond, which is the hardest material known to man would be the ideal material to be used. Anything else would come second to the ideal. So a Valve made of Diamond would be very expensive and impractical to use.

This brings the requirement down to the smallest possible usage of Diamond. On the other hand, the valve surface is the only region that gets chipped off. We therefore have favorable requirements from either side. Locating Diamond only on the surface seems to be the only choice. Through this, we could achieve the final Artifact of Valve coated with microscopic Diamond particles.

Observing and relating with nature could generate the artifact from another angle too. The Valve could be considered as "soft" for the gaseous vapors and harsh environment inside. Looking for similar cases in nature where "soft" parts are protected by the harshness of nature could also help. A fruit for example is protected by its comparatively harder Shell. Soft molluscs are also protected in nature by a similar hard Shell. Looking further for soft things that exist in nature could yield a similar or same analogy to our required Artifact.

The general requirement for solving this problem seems to be a basic knowledge of nature and a mind to relate unrelated things. Not thinking purely from the perspective of an Engineer could help at this stage of inventive creativity.

### Resultant Artifact

#### US Patent No. 5,040,501        Date of Patent: Aug 20, 1991

#### Valves and Valve Components

#### Inventors: Jerome H. Lemelson

**Abstract:**

Improved valves, such as exhaust valves for internal combustion engines and other valves subject to wear and corrosion and, in certain instances, impact forces which may cause degradation in structure and structural failure.

In one form, a select portion of the surface of a Valve component or components subject to degradation during use such as erosive and/or corrosive effects of fluid particles and liquid or vaporous fluid passing through the Valve, is coated with a synthetic diamond material which is formed in situ thereon.

In another form, the entire surface of the valve component is so coated. The component may be a movable poppet member for an exhaust valve for a combustion chamber of an internal combustion piston engine.

The valve seat or insert may also be coated with synthetic diamond material, particularly the circular tapered inside surface thereof against which a portion of the underside of the head of the valve poppet which engages the seat when the valve is spring closed.

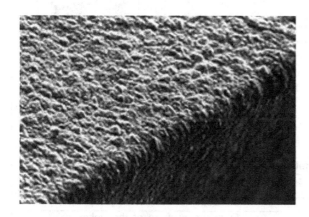

**fig. 9.7.** Microscopic View of Valve surface coated with synthetic Diamond (Nova Diamant 2012)

By coating the entire head and stem of the valve poppet with synthetic diamond and overcoating or plating a solid lubricant, such as chromium on the outer surface of the diamond coating a number of advantages over conventional valve construction are derived including better heat and corrosion resistence, reduced wear resulting from seat and valve head impact contact and a reduction in the enlargement of surface cracks.

Similar improvement are effected for the valve seat when so coated and protected. In a modified form, the entire interior or selected portions of the wall of the valve body or the combustion chamber containing the valve may be coated with synthetic diamond material with or without a protective overcoating (Lemelson, 1991).

**Extracting** of invention Bar Codes is shown in fig. 9.8.

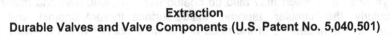

## Extraction
## Durable Valves and Valve Components (U.S. Patent No. 5,040,501)

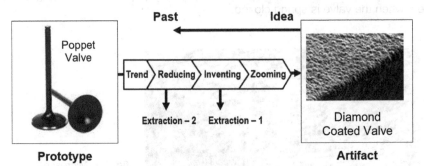

**Prototype**　　　　　　　　　　　　　　　　　　**Artifact**

### Extraction – 1

| LC | No. | Navigator | Substantiation for Extraction |
|----|-----|-----------|-------------------------------|
| + | 01 | Change in the aggregate state of an object | The hardness is increased on the surface. |
| ++ | 12 | Local property | Only the surface of the valve is hardened with Synthetic Diamond |
| ++ | 17 | Use of composite materials | A layer of Diamond is deposited on the surface of the metal Valve. |
| + | 18 | Mediator | Diamond layer acts as a mediator which prevents the inner layer from chipping off. |
| + | 34 | Matryoshka (nested doll) | The metal valve is completely located inside the diamond case |
| + | 35 | Unite | Diamond and metal are united to give the valve the property of both. |
| ++ | 39 | Preliminary counter-action | The harmful action of heat and pressure is mitigated by coating the valve with Synthetic Diamond in advance. |

### Extraction - 2

**Standard Contradiction:** The valves and valve components should possess improved wear resistance but this could change the endurance strength to the existing pressure and heat in the valve zone.

**Radical Contradiction:** The valves and valve components should be hard to withstand the wear but it should not be hard as it could make the valve brittle.

**FIM:** X- resource, together with available or modified resources, and without making the object more complex or introducing any negative properties guarantees attainment of the following Ideal Final Result. (No wear/corrosion of the valve)

**fig. 9.8.** Durable Valves and Valve Component (Lemelson, 1991)

**Wear resistant Valves and Valve components (U.S. Patent No. 5,040,501)**

*A Poppet Valve*

**TREND:** Poppet valves are mainly used in internal combustion engines and other engines operating on similar principle to open and close the intake and exhaust port when it is required. Normally, a portion of the surface of the valve or valve seat is normally subject to degradation during use. Typically, it is due to the erosive and/or corrosive effects of fluid particles and liquid or vaporous fluid passing through the valve.

PROBLEM: Frequent wear and corrosion results in the failure of Valve.

**REDUCING**

**FIM:** X- resource, together with available or modified resources, and without making the object more complex or introducing any negative properties guarantees attainment of the following IFR (No wear or corrosion of the Valve)

Standard (Technical) Contradiction Model

Valves and valve components

(+) Improved wear resistance for a long time → 23 Functional time of moving object

(−) A harder material could end up being brittle → 28 Strength

13 Inexpensive short life object replaces expensive long life Object

12 Local property

02 Preliminary Action

**fig. 9.9 - beg.** Reinventing in standard form of Durable Valves and Valve Component (U.S. Patent No. 5,040,501)

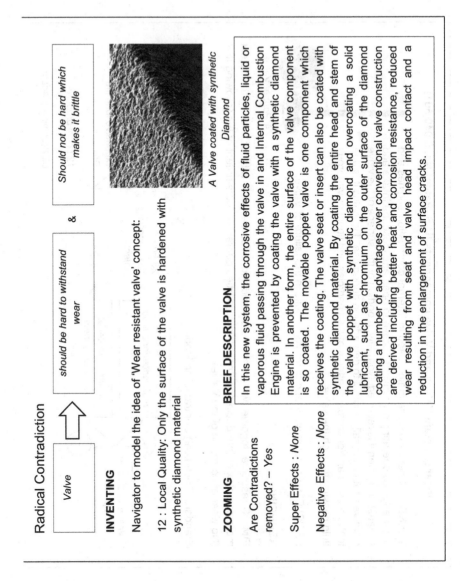

**Radical Contradiction**

| Valve | ⟹ | should be hard to withstand wear | & | Should not be hard which makes it brittle |

**INVENTING**

Navigator to model the idea of 'Wear resistant valve' concept:

12 : Local Quality: Only the surface of the valve is hardened with synthetic diamond material

*A Valve coated with synthetic Diamond*

**ZOOMING**

Are Contradictions removed? – *Yes*

Super Effects : *None*

Negative Effects : *None*

**BRIEF DESCRIPTION**

In this new system, the corrosive effects of fluid particles, liquid or vaporous fluid passing through the valve in and Internal Combustion Engine is prevented by coating the valve with a synthetic diamond material. In another form, the entire surface of the valve component is so coated. The movable poppet valve is one component which receives the coating. The valve seat or insert can also be coated with synthetic diamond material. By coating the entire head and stem of the valve poppet with synthetic diamond and overcoating a solid lubricant, such as chromium on the outer surface of the diamond coating a number of advantages over conventional valve construction are derived including better heat and corrosion resistance, reduced wear resulting from seat and valve head impact contact and a reduction in the enlargement of surface cracks.

**fig. 9.9 - end.** Reinventing in standard form of
Durable Valves and Valve Component (U.S. Patent No. 5,040,501)

## 9.4. Industrial Hardware:

### Composite Reinforced Foam Sheet (U.S. Patent No. 3,556,918)

### Prior Art (Engineering Level)

A foam sheet reinforced with abrasive particle has been widely used as cleaning article since some time now. It finds its use from household needs to industrial applications. Whatever specific purpose it might be used for, the primary function remains the same – Cleaning!

### Problems / Requirements in the Prototype

fig. 9.10. Foam Sheet (University Products 2012)

"In the cleaning of utensils such as pots and pans or articles containing greases and the like, it is necessary to abrade or scour as well as rinse the surface of said articles in order to effect proper cleaning. The use of a sponge and water may absorb or wipe up some of the grease but the conventional soft and flexible sponge or pad made of flexible cellular plastic tends to spread a good deal of the adhered substances rather than remove it. It is often necessary to combine the absorbing action of the sponge with an abrading action by utilizing a scraper pad of steel wool or the like. However, the alternate use of two cleaning implements is time consuming and requires a considerable amount of additional effort and wasted movement in switching from one cleaning implement to the other" (Lemelson et al,1968).

### Result of Contradiction Analysis

A sponge scattered with abrasives which is laminated to hold back the cellular mass together is pretty easy to manufacture. But such a product loses its lamination pretty soon because of its own abrasives acting on the laminate. These positives and negatives could be represented as "Ease of Manufacture" and "Stability of the Object" respectively in TRIZ. The Altshuller Matrix yields the following results for this contradiction – Previously installed cushion, Inverse action & Segmentation. The following section talks about the inventing idea using one of these navigators.

### Inventing Idea

The Idea for inventing the new and improved Abrasive Foam Sheet could be obtained using the navigator "Inverse Action". This advises us to do things in reverse. Having abrasive outside, say as a wire mesh rather than embedding abrasives particles inside the cellular mass solves the contradiction. "The wire mesh holds the cellular mass together and also acts as an abrasive to perform the scrubbing action" (Lemelson et al. 1971).

### Psychological Level

To attain the Artifact's inventive idea, no extra knowledge of any field is required in this case. A blend of slight creativity and knowledge of TRIZ would generate this result. Modern TRIZ gives the Navigator "Inverse Action" to be worked with. It recommends something to be done in reverse. Either the cleaning process has to be done in reverse (part rubbed on fixed Foam Sheet rather than Foam Sheet on the Part), which does not sound practical owing to the variety of sizes of Articles that need cleaning.

Since the Foam Sheet only consists of two parts that is Foam and the Abrasive, putting things in reverse order that is foam inside the abrasive mesh rather than abrasive particle inside foam sheet does not seem hard to conceive.

This Artifact could be achieved also by observing and imitating the nature. TRIZ leads us to the coast and sailing to the horizon depends on this observing power of the person.

Soft Foam and Hard Abrasive is certainly not a good combination. But such a combination exists in nature and often soft core is covered by a hard cover (Eg. Coconut or Fish and its Scales), Inverse of the Prototype at our hand, which has hard abrasives covered by Foam Sheet. Hence and Inverse action yields the required Artifact.

### Resultant Artifact

**US Patent No. 3,556,918**                      **Date of Patent: Jan 19, 1971**

**Title: Composite Reinforced Plastic Foam Sheet**

**Inventors: Jerome H. Lemelson et al.**

**Abstract:** A composite, sheet-like article of manufacture is provided comprising a base sheet made of flexible, cellular 15 plastic material and a plurality of reinforcing elements secured to one or both major surfaces of the sheet or extending through the sheet.

Where the composite sheet is to be utilized as a cleaning and scouring device, the parallel extending reinforcing elements are surface se- 20 cured and serve to scour grease while the sponge- like base element serves to retain and dispense a cleaning liquid such as water.

The composite sheet may be formed by welding strip like or filamental elements to the surfaces of the sheet 25 or by co-extruding said elements with the base material (Lemelson et al., 1971).

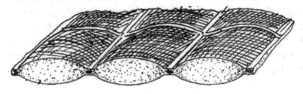

fig. 9.11. A composite reinforced foam sheet (Lemelson et al. 1971).

**Extracting** of "Composite reinforced foam sheet" is shown in fig. 9.12.

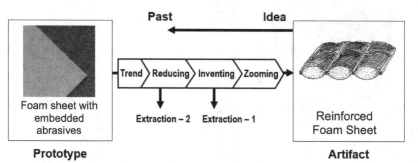

**Extraction**
**Composite Reinforced Foam Sheet (U.S. Patent No. 3,556,918)**

Foam sheet with embedded abrasives

**Prototype**

Reinforced Foam Sheet

**Artifact**

**Extraction – 1**

| LC | No. | Navigator | Substantiation for Extraction |
|----|-----|-----------|-------------------------------|
| + | 01 | Change in the aggreg. state ... | The sheet is made a little rigid in flexibility by wrapping it with the mesh. |
| + | 05 | Separation | The abrasive action is completely separated from within the sheet and located outside it. |
| ++ | 11 | Inverse action | Rather than embedding abrasives inside the cellular mass, the abrasive mesh surrounds it |
| + | 18 | Mediator | The mediator 'wire mesh' is used to perform scouring function. |
| + | 20 | Universality | The artifact is used for both scouring (mesh) and holding the cleaning liquid (cellular mass). |
| ++ | 21 | Transform damage into use | The abrasive mesh is used to hold the cellular mass together. |
| + | 31 | Use of porous materials | The abrasive sheet is porous to enable it to hold and deliver the cleaning liquid. |
| ++ | 34 | Matryoshka | The cellular mass is nested within the mesh. |

**Extraction - 2**

**Standard Contradiction:** Laminated sponge with scattered abrasives is easy to perceive and manufacture but the abrasives weaken the cellular mass and it disintegrates eventually.

**Radical Contradiction:** Cellular mass should not contain abrasives as it weakens it but it should also contain abrasives for it should have scouring ability.

**FIM:** X-resource, together with available or modified resources, and without making the object more complex or introducing any negative properties guarantees attainment of the following Ideal Final Result (Cellular mass strength is retained overtime.)

**fig. 9.12.** A composite reinforced foam sheet (Lemelson et al. 1971).

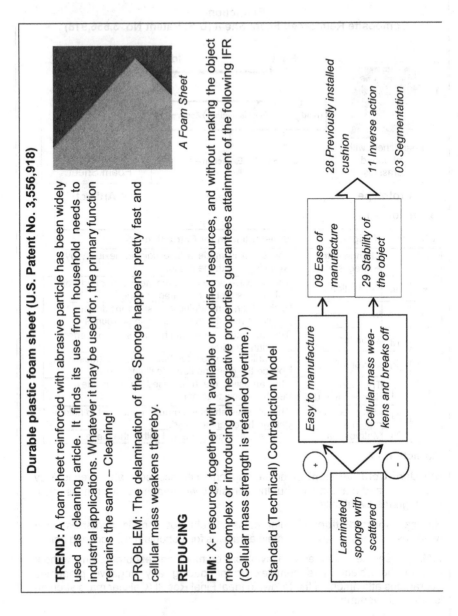

**Durable plastic foam sheet (U.S. Patent No. 3,556,918)**

A Foam Sheet

**TREND:** A foam sheet reinforced with abrasive particle has been widely used as cleaning article. It finds its use from household needs to industrial applications. Whatever it may be used for, the primary function remains the same – Cleaning!

**PROBLEM:** The delamination of the Sponge happens pretty fast and cellular mass weakens thereby.

**REDUCING**

**FIM:** X- resource, together with available or modified resources, and without making the object more complex or introducing any negative properties guarantees attainment of the following IFR (Cellular mass strength is retained overtime.)

Standard (Technical) Contradiction Model

Laminated sponge with scattered

(+) Easy to manufacture → 09 Ease of manufacture → 28 Previously installed cushion

11 Inverse action

(–) Cellular mass weakens and breaks off → 29 Stability of the object → 03 Segmentation

**fig. 9.13 - beg.** Reinventing in standard form of Composite Reinforced Foam Sheet (U.S. Patent No. 3,556,918)

Radical (Physical) Contradiction Model

| Cellular mass | ⟸ | Should not contain abrasive as it weakens cellular mass | & | Should contain abrasive as it should have scouring ability |

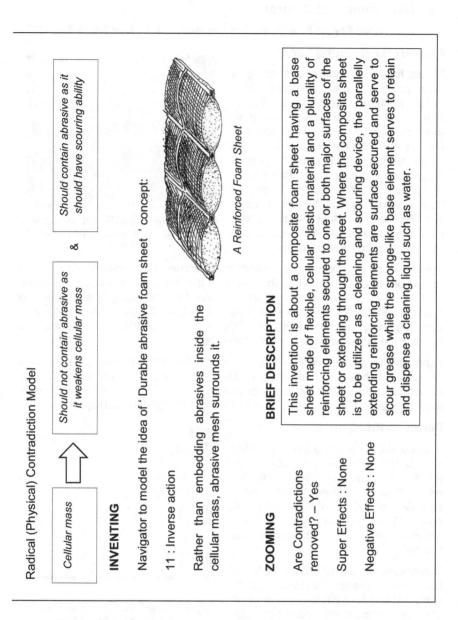

*A Reinforced Foam Sheet*

## INVENTING

Navigator to model the idea of 'Durable abrasive foam sheet' concept:

11 : Inverse action

Rather than embedding abrasives inside the cellular mass, abrasive mesh surrounds it.

## BRIEF DESCRIPTION

This invention is about a composite foam sheet having a base sheet made of flexible, cellular plastic material and a plurality of reinforcing elements secured to one or both major surfaces of the sheet or extending through the sheet. Where the composite sheet is to be utilized as a cleaning and scouring device, the parallelly extending reinforcing elements are surface secured and serve to scour grease while the sponge-like base element serves to retain and dispense a cleaning liquid such as water.

## ZOOMING

Are Contradictions removed? – Yes

Super Effects : None

Negative Effects : None

**fig. 9.13 - end.** Reinventing in standard form of Composite Reinforced Foam Sheet (U.S. Patent No. 3,556,918)

## 9.5.   Monitoring and Control:

### Extensive Forest Fire Detection System (U.S. Patent No. 5,832,187)

**Prior Art (Engineering Level)**

Forest fires can destroy everything in their paths in mere minutes. Also called Wildfires, Land fire or Forest Fire, 100,000 of these clear out on an average 4 million to 5 million acres of land in the U.S. every year. In recent years, wildfires have burned up to 9 million acres of land. A wildfire moves at speeds of up to 23 kilometers an hour, consuming everything in its path. The fire triangle of fuel, oxygen, and a heat source talks about the conditions that are needed for a fire to start.

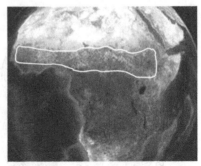

**fig. 9.14.** Regions of Forest fire occurrence in Africa (Environmental Graffiti 2006)

Fuel could be any flammable material in the immediate vicinity of the fire. The Air we breathe supplies enough oxygen to the fire and it is always available. Since we find several articles around us that burn off easily, the fuel is also readily available for a Forest Fire. Now, a spark is all it takes for a Wildfire to start. Lightning, burning campfires or cigarettes, hot winds, and even the Sun can provide sufficient heat to spark a wildfire.

Although four out of every 5 Forest fires are started by humans, a prolonged dry spell and heat could also spark off a fire. That spark is enough for Wildfire to consume thousands of acres of forest and the  fire  could rage  up  to weeks together (Above sections  are adapted from Wildfire, 2012).

**Detection Problems**

A quick detection of the forest fire after it is started has been a challenge. Often the fire goes unnoticed until the expanse of the flames reach kilometers. One of the methods that used to prevail for detecting a forest fire before this invention was as in the representation below.

There were several pillars on the forest floor which were mounted with Infrared cameras and controlling devices which sent the signals to the nearest fire station.

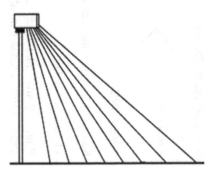

**fig. 9.15.** Past method of Forest fire detection on the forest floor

The problem however was that this system was a little far from practicality. In order to cover the vast expanse of the forest floors, one needed to install these controlling pillars in great numbers, which would involve a lot in maintenance cost apart from the installation challenge and its prohibiting cost (Colstoun et al., 1986).

## Result of Contradiction Analysis

The positive requirement of the contradiction existing in the current Forest fire detection system could be defined as "A complete coverage of the forest land". The negative that arises by having such a system is that "several pillars mounted with Infrared Camera would be required" to achieve this.

This is not practical as it would involve high cost of installation, running and maintenance. TRIZ defines the positive requirement as "Surface of fixed object" and the negative "Quantity of Material".

The Contradiction analysis gives the following four navigators - Separation, Use of mechanical oscillation, Use of composite materials and Asymmetry.

## Inventing Idea

The Inventing idea could be achieved with the navigator "Separation". Separating out the functional part which is the Infrared Camera and placing it at a highest possible location, i.e. on a satellite revolving the Earth.

## Psychological Level

The thought process to generate the inventive idea for Forest Fires detection could take place with the knowledge of basic devices and systems that existed during the time of the prototype's existence.

One could still come up with the inventive idea in this case with the knowledge of the prototype alone. For example, Sensors that detect the physical quantity of something and alert the observer already exists in the prototype device.

The only possible option that seemingly remains is to separate out the sensor and place it as high as possible to ensure maximum monitoring capability.

Again, stepping out of Engineer's shoes could solve this problem too. A practical thought for highest possible place could be a tall Pillar or the peak of a nearby mountain. Initially, the thought of a satellite being the highest possible place does not cross the mind.

Psychological stereotyping doesn't let us think beyond a Mountain.

Hence a simple set of questions could help one reach the state of artifact. "What is the highest place that one could reach?" If the answer is still a Mountain, then a question of "What is higher than that?" could help.

Posing that question again without thinking about the problem could help us achieve the Ideal Final Result of placing the Sensors in a Satellite.

Resultant Artifact

US Patent No. 5,832,187                         Date of Patent: April 8, 2003

### Title: Fire detection systems and methods

### Inventors: Robert D. Pedersen, Jerome H. Lemelson

**Abstract:**
The invention is a system for automatically detecting fires in select areas, and reacting thereto to put out the fires. A stationary, earth orbit satellite, pilotless drone aircrafts or piloted aircraft contains one or more infrared detectors and optical means for detecting small fires when they first occur in fields and wooded areas, preferably where man made campfires and trash dumping are prohibited.

A computer in the satellite, drone or piloted aircraft and/or on the ground receives and analyzes image signals of the earth area or areas being monitored and, upon detecting infrared radiation of varying intensity and variable shape, indicative that a fire has started, generates coded signals which are (a) indicative of the coordinate locations of the fire, (b) the extent of the fire, (c) the shape of the area(s) burning, (d) the direction of movement of the fires (e) the speed(s) of the flame fronts, (f) smoke condition, (g) intensity of the fire, (h) fire ball location(s), etc.

In one form, a conventional and/or infrared television camera(s) generates image or video signals which are digitized and recorded in memory and/or on tape or disc. Expert system logic, such as fuzzy logic, is used to prioritize dangerous areas to assist in directing firefighting. Such information, or select portions thereof, is automatically transmitted to one or more earth bound and/or aircraft contained receivers where fire-fighting equipment such as helicopter and aircraft containing drainable water or other fire-fighting agents is available for immediate dispatch to the fire zone (Lemelson et al. 1971).

**fig. 9.16.** An image of the landscape from the infrared camera, with "white" regions signifying high heat signature. (Miller McCune 2010)

**Extracting** of "Extensive Forest Fire Detection System" is shown in fig. 9.17.

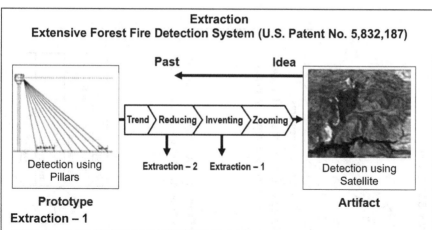

**Extraction**
**Extensive Forest Fire Detection System (U.S. Patent No. 5,832,187)**

**Prototype**                                                           **Artifact**

**Extraction – 1**

| LC | No. | Navigator | Substantiation for Extraction |
|---|---|---|---|
| ++ | 05 | Separation | Separate out Infrared Camera) and place it in the highest possible location, i.e. on a Satellite revolving the Earth. |
| + | 07 | Dynamization | Cameras are made more dynamic by placing it on the Satellite. |
| + | 10 | Copying | A copy of the forest cover is reproduced by placing the Camera on a Satellite. |
| ++ | 18 | Mediator | Satellite serves as a mediator holding and controlling the infrared Camera. |
| + | 19 | Transition into another dimension | Camera is located in a higher dimension of Space. |
| + | 20 | Universality | Several adjacent Forest lands could be covered by the same Camera. |
| + | 28 | Previously installed cushions | The impact of a Forest Fire is prevented by tracking the Forest beforehand. |
| + | 40 | Uninterrupted useful function | Uninterrupted tracking is performed irrespective of Cloud cover and other climatic variations. |

**Extraction - 2**

**Standard Contradiction:** The fire detection Pillar should have a range that covers the entire forest but several Pillars would be required to do the same.

**Radical Contradiction:** The Sensing Pillars should be as many as possible to increase the coverage it should also be as less as possible to reduce cost and complication.

**FIM:** X-resource, together with available or modified resources, and without making the object more complex or introducing any negative properties guarantees attainment of: *Easy fire detection over a long range*.

**fig. 9.17.** Fire detection systems and methods (Lemelson et al. 2003).

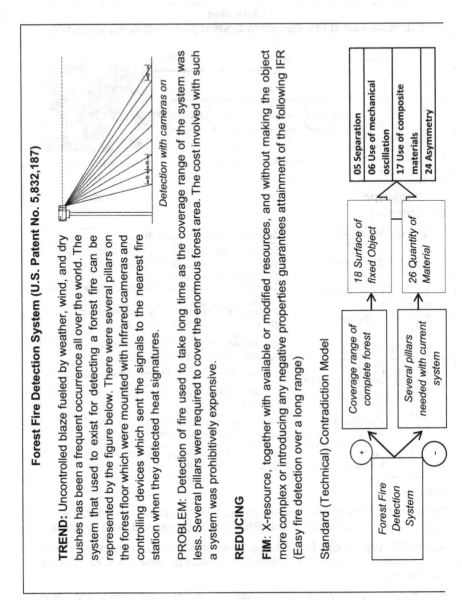

**Forest Fire Detection System (U.S. Patent No. 5,832,187)**

*Detection with cameras on*

**TREND:** Uncontrolled blaze fueled by weather, wind, and dry bushes has been a frequent occurrence all over the world. The system that used to exist for detecting a forest fire can be represented by the figure below. There were several pillars on the forest floor which were mounted with Infrared cameras and controlling devices which sent the signals to the nearest fire station when they detected heat signatures.

**PROBLEM:** Detection of fire used to take long time as the coverage range of the system was less. Several pillars were required to cover the enormous forest area. The cost involved with such a system was prohibitively expensive.

**REDUCING**

**FIM:** X-resource, together with available or modified resources, and without making the object more complex or introducing any negative properties guarantees attainment of the following IFR (Easy fire detection over a long range)

Standard (Technical) Contradiction Model

| Coverage range of complete forest | 18 Surface of fixed Object |
| Several pillars needed with current system | 26 Quantity of Material |

Forest Fire Detection System

05 Separation
06 Use of mechanical oscillation
17 Use of composite materials
24 Asymmetry

**fig. 9.18 - beg.** Reinventing in standard form of
Extensive Forest Fire Detection System (U.S. Patent No. 5,832,187)

Radical (Physical) Contradiction Model

Fire Detection

⇧

The Sensing Pillars should be as many as possible to increase the coverage

&

The sensing pillars should be as less as possible to reduce cost and complication

## INVENTING

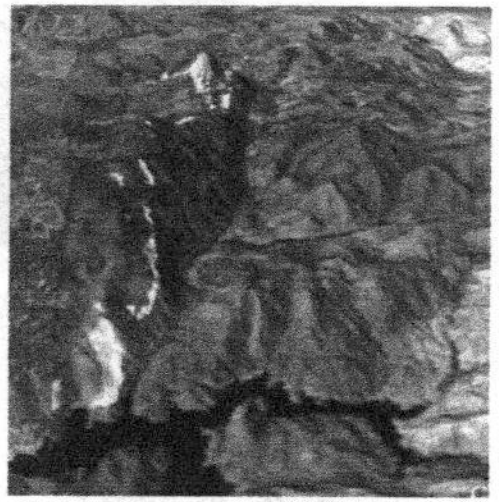

Navigator to model the idea of 'Forest Fire Detection' concept:

05: Separation: Separate out the functional part (Infrared Camera) and place it in the highest possible location, i.e. on a satellite revolving the Earth.

## ZOOMING

Are Contradictions removed? – Yes

Super Effects : None

Negative Effects : None

## BRIEF DESCRIPTION

The system automatically detects fire in a forest and puts it out. A stationary, earth orbit satellite, pilotless drone aircrafts or piloted aircraft contains one or more infrared detectors and optical means for detecting small fires when they first occur. A computer in the satellite, drone or piloted aircraft and/or on the ground receives and analyzes image signals of the earth area or areas being monitored and, upon detecting infrared radiation of varying intensity and variable shape, indicative that a fire has started, generates coded signals precisely defining the features of the outbreak and transmitting the codes to the nearest fire station.

**fig. 9.18 - end.** Reinventing in standard form of Extensive Forest Fire Detection System (U.S. Patent No. 5,832,187)

## 9.6. Beverage:

### Container with Inbuilt Dispenser Straw (U.S. Patent No. 4,226,356)

### Prior Art (Engineering Level)

A beverage can, originally introduced to eradicate the practice of paying deposit for the beer bottles can be seen universally today. It today is used to pack virtually all kinds of beverages. It is primarily made of aluminium and sealed air tight before its dispatch to the shops and supermarkets (Aluminium Can Beverage, 2012).

The "cling on" opener clip of the Aluminium Can is very user friendly and the drink is normally consumed directly or with a Straw.

**fig. 9.19.** A Beverage Can (Colour box 2011)

### Detection Problems

The consumption of the drink with a straw has several advantages. It prevents the drink from spilling onto the clothes apart from keep the tooth from decay. However, providing a straw along with the beverage Can has been a problem both in terms of handling and appearance.

Various proposals have been made for providing tubes or drinking straws in containers, such as metal cans containing carbonated and non-carbonated drinks, to provide means for drawing the contents of the container into the mouth of the person.

These proposals have not been commercially successful for a number of reasons including the fact that a conventional drinking straw disposed in such a container can be no longer than the length of the container and, as a result, when the upper end of the straw floats upwardly through the opening in the container when the closure is removed or partially removed, the lower end of the straw cannot provide access to within more than about an inch of the bottom of the container as at least a similar amount is required to protrude from the upper end of the container to permit it to be placed between the lips of the person drinking the contents (Lemelson, 1980).

### Result of Contradiction Analysis

The contradiction analysis for this problem could be performed as here below. The Beverage Can should be of simple construction but the risk of spilling increases without a straw and with simple construction. Hence the "plus" could be defined as avoiding "Complexity of construction" and the "minus" as "Ease of use".

The Altshuller matrix gives away the following navigator recommendations - 13. Inexpensive short life object as a replacement for expensive long life object, 39.Preliminary counter action, 10.Copying and 18.Mediator.

## Inventing Idea

The inventing idea could be obtained by using the navigator 39. Preliminary counteraction – that is the unavailability of straw during consumption is counteracted by placing straw inside, in advance and 18. Mediator – that is the beverage can itself acts as a mediator carrying the straw during storage and transportation. Both these navigators could be used to invent the same efficient concept. The equivalent of model 39 in this situation can be model 02 Preliminary action (see extraction).

## Psychological Level

To perceive the inventive idea of this artifact, no special knowledge of a field is seemingly required. After TRIZ matrix yields us the navigator "Preliminary counteraction", the psychological level of approaching the solution could take place as following.

Simple questioning could generate great ideas for this. What counteracts spilling? - A Straw. Hence providing the Straw at a preliminary stage is TRIZ's recommendation. Hence placing the Straw along with the Can during the manufacture serves the purpose.

When thought from out-of-the-box perspective, it could also be approached by analogizing the problem. Thinking of an analogy, in which way and in which system is the future risk mitigated, by taking preliminary counteraction?

A Ship's future risk of sinking for example is counteracted by placing inside it, the lifesaving systems like lifeboats and jackets.

Now it could very well lead to counteracting the future spilling of Beverage by placing inside it, a system that prevents it. For instance, the same straw that is offered separately could be placed inside the Can.

Another highly probable answer due to psychological stereotyping could be to place the straw outside the container so that it could be separated and used during consumption.

This is not ideal for transportation and is certainly not the best idea from cleanliness perspective too.

This could however be easily solved and artifact state be reached by random positioning trials of the Straw around the container.

**Resultant Artifact**

US Patent No. 4,226,356                    Date of Patent: Oct 7, 1980

### Title: Container and Dispenser Straw
### Inventors: Jerome H. Lemelson

**Abstract:**

A liquid container such as a metal or plastic canister or can having a drinking straw disposed therein which straw becomes available for drinking the liquid contents of the container when the container is opened.

The container top wall is provided with a finger pulled tab secured to or extending from a removable portion of the top wall and, when pulled, the tab not only provides an opening in the top wall but exposes or pulls out of the container the upper portion of a drinking straw.

The straw is a plastic or impregnated paper tube containing a pleated or bendable portion thereof and the upper portion of the straw is deformed either downwardly along the remaining body of the straw or conforms to that portion of the container top wall which is removed so that, upon removal of the removable portion, the upper portion of the straw will be pulled thereby through the opening so that it may be engaged by mouth and used.

The lower portion of the straw may be secured or frictionally held against the side or bottom wall of the container to maintain it in place (Lemelson, 1980).

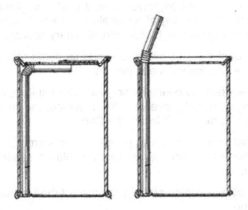

**fig. 9.20.** Containers with dispenser straw (Lemelson, 1980)

**Extracting** of "Containers with dispenser straw" is shown in fig. 9.21.

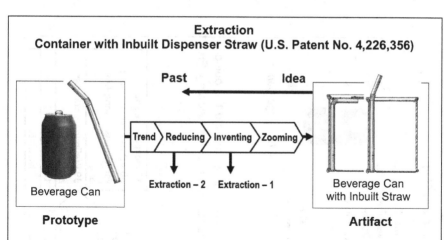

### Extraction
### Container with Inbuilt Dispenser Straw (U.S. Patent No. 4,226,356)

### Extraction – 1

| LC | No. | Navigator | Substantiation for Extraction |
|----|-----|-----------|-------------------------------|
| + | 02 | Preliminary Action | Straw is placed inside the Can in advance during the manufacture of the Can. |
| + | 07 | Dynamization | Straw is made to pop out when the beverage can is opened. |
| + | 12 | Local Property | The local property of the Straw is changed to make it flex when the Can is sealed or opened. |
| ++ | 18 | Mediator | Beverage Can itself acts as a mediator carrying the straw during storage and |
| ++ | 34 | Matryoshka | The Straw is nested within the Can and dispatched. |
| + | 35 | Unite | The Can and straw are united during manufacture. |

### Extraction – 2

**Standard Contradiction:** Beverage Can should be of simple construction and so exist in the current basic shape. But without a straw along with it, there is a risk of spilling the Drink on oneself.

**Radical Contradiction:** The Beverage Can should not contain the straw during storage and transport but it should contain the straw during its consumption.

**FIM:** X-resource, together with available or modified resources, and without making the object more complex or introducing any negative properties guarantees attainment of: *Can having a Straw ready to use.*

**fig. 9.21.** Containers with dispenser straw (Lemelson, 1980).

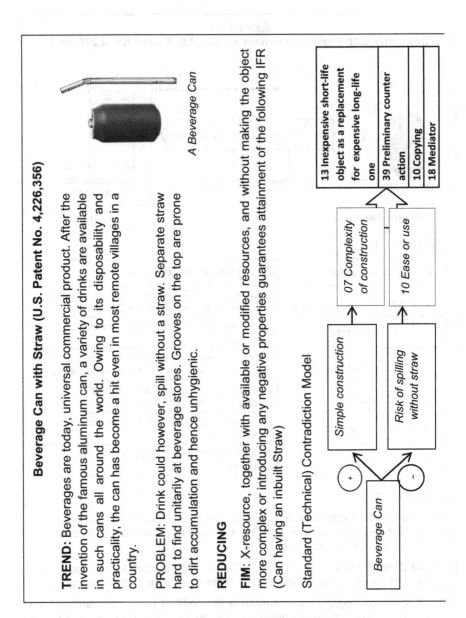

**Beverage Can with Straw (U.S. Patent No. 4,226,356)**

**TREND:** Beverages are today, universal commercial product. After the invention of the famous aluminum can, a variety of drinks are available in such cans all around the world. Owing to its disposability and practicality, the can has become a hit even in most remote villages in a country.

**PROBLEM:** Drink could however, spill without a straw. Separate straw hard to find unitarily at beverage stores. Grooves on the top are prone to dirt accumulation and hence unhygienic.

**REDUCING**

**FIM:** X-resource, together with available or modified resources, and without making the object more complex or introducing any negative properties guarantees attainment of the following IFR (Can having an inbuilt Straw)

Standard (Technical) Contradiction Model

*A Beverage Can*

13 Inexpensive short-life object as a replacement for expensive long-life one

39 Preliminary counter action

10 Copying

18 Mediator

07 Complexity of construction

10 Ease or use

Simple construction

Risk of spilling without straw

+

−

Beverage Can

**fig. 9.22 - beg.** Reinventing in standard form of Container with Inbuilt Dispenser Straw (U.S. Patent No. 4,226,356)

Radical (Physical) Contradiction Model

| Can | $\Rightarrow$ | *Straw should not be there during storage and transport* | & | *Straw should be there during consumption* |

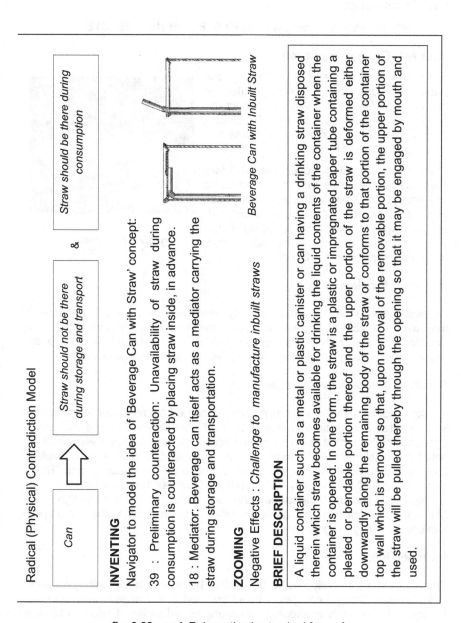

*Beverage Can with Inbuilt Straw*

## INVENTING

Navigator to model the idea of 'Beverage Can with Straw' concept:

39 : Preliminary counteraction: Unavailability of straw during consumption is counteracted by placing straw inside, in advance.

18 : Mediator: Beverage can itself acts as a mediator carrying the straw during storage and transportation.

## ZOOMING

Negative Effects : *Challenge to manufacture inbuilt straws*

## BRIEF DESCRIPTION

A liquid container such as a metal or plastic canister or can having a drinking straw disposed therein which straw becomes available for drinking the liquid contents of the container when the container is opened. In one form, the straw is a plastic or impregnated paper tube containing a pleated or bendable portion thereof and the upper portion of the straw is deformed either downwardly along the remaining body of the straw or conforms to that portion of the container top wall which is removed so that, upon removal of the removable portion, the upper portion of the straw will be pulled thereby through the opening so that it may be engaged by mouth and used.

**fig. 9.22 - end.** Reinventing in standard form of
Container with Inbuilt Dispenser Straw (U.S. Patent No. 4,226,356)

## 9.7.   Cosmetics:

### Photosensitive Contact Lens (U.S. Patent No. 4,681,412)

**Prior Art (Engineering Level)**

A contact lens typically is a flexible and transparent lens which is gently located on the surface of the eye, mainly to correct the problems with vision. These days the contact lenses have evolved into several forms and for various different functions.

**fig. 9.23.** A contact lens (My lot 2006)

They have advantage over the standard spectacles in a sense that they tend to be tiny and fit great within the eye. They also help the wearer by not giving an uncomfortable feeling when worn. By providing better vision peripherally, contact lenses clearly have an edge over the Spectacles.

**Detection Problems**

Contact lenses were very user friendly. However, during the bright daylight they were not able to protect the eyes against harmful UV rays or the extra ambient light. Although it was previously inconceivable for a lens to have light filtering ability, an improvement in the system of lenses was needed.

**Result of Contradiction Analysis**

The contradiction analysis in this problem is carried out as follows. The Contact lens requires a dark tinted material on itself that can reduce the intensity of light passing through it to filter out the bright light. Doing so, we also block the light transmission through it in the normal room light conditions. TRIZ explains the positive and negative aspect of this problem as "Adaptability" and "Brightness of Lighting" respectively. Adaptability is taken as the "positive" as the lens is required to be adaptable to the conditions of both low and bright lights. The Matrix has the following navigators to offer – Universality, Transform damage into use, Copying, Segmentation.

**Inventing Idea**

The idea for the invention could be derived by brainstorming with the navigators provided. Several navigators could be used in this case to solve the contradiction. Universality is adopted by making the same lens flexible to be used in both bright and low light. Segmenting the Lens also could help in deriving the invention idea.

The Lens is divided into two segments. The photosensitive front part and transparent back part. The transparent back part acts as a normal vision correction lens and the front part controls the light passing through by adjusting the photosensitive material in it automatically depending on the intensity of incident light. The navigator of 'Transforming damage into use' also helps invent the solution. The damaging bright light by itself controls the light passage through the lens. This is achieved by making it with photosensitive material (Lemelson, 1987).

**Resultant Artifact**

**US Patent No. 4,681,412**                    **Date of Patent: Jul. 21, 1987**

### Title: Photosensitive Contact Lens

### Inventors: Jerome H. Lemelson

**Abstract:**

The improved contact lens has a unitary molded central portion containing a light sensitive material which varies in light transmission with variations in the intensity of ambient light. One embodiment has an assembly of components secured to the lens by molding. The assembly also contains a light sensitive material (Lemelson, 1987).

The photosensitive chemical within the lens orients itself to allow or block light when the conditions are dull and luminous respectively, automatically. This way a single lens is able to guard the eye against the bright light and also correct the vision).

**fig. 9.24.** Photosensitive contact lenses – sectional view (Lemelson, 1987)

**Extracting** of "Photosensitive contact lenses" is shown in fig. 9.25.

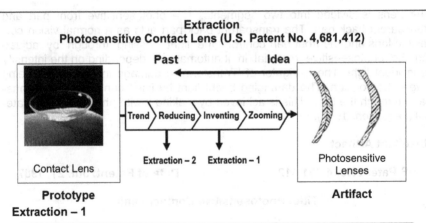

**Extraction – 1**

| LC | No. | Navigator | Substantiation for Extraction |
|----|-----|-----------|-------------------------------|
| ++ | 03 | Segmentation | Divided into two segments. Front - photosensitive part and back - transparent part |
| + | 07 | Dynamization | The photosensitive particle inside the lens orients itself depending on the intensity of incident light |
| + | 17 | Use of Composite Materials | The lens is made out of composite of photosensitive particle and transparent base. |
| + | 18 | Mediator | The photosensitive crystal acts as a mediator controlling the amount of light. |
| ++ | 20 | Universality | The same lens is made flexible to be used in both bright and low light. |
| ++ | 21 | Transform damage into use | The bright light by itself controls the light passage through the lens. This is achieved by making it with photosensitive material. |
| + | 40 | Uninterrupted useful function | By automatically orienting itself, the photosensitive part helps the lens perform a non-stop function. |

**Extraction - 2**

Standard Contradiction: Dark tinted contact lenses would serve the purpose of blocking the extra ambient light but it blocks too much light during normal room conditions

**Radical Contradiction:** The contact lens should allow less light to pass through when outdoors but it should allow more light to pass through when indoors.

**FIM:** X-resource, together with available or modified resources, and without making the object more complex or introducing any negative properties guarantees attainment of: *Lens filters UV light by itself*.

**fig. 9.25.** Photosensitive Contact Lens (Lemelson, 1987).

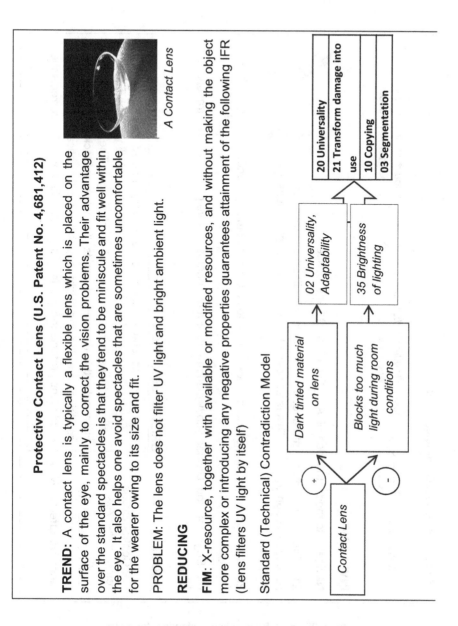

## Protective Contact Lens (U.S. Patent No. 4,681,412)

A Contact Lens

**TREND**: A contact lens is typically a flexible lens which is placed on the surface of the eye, mainly to correct the vision problems. Their advantage over the standard spectacles is that they tend to be miniscule and fit well within the eye. It also helps one avoid spectacles that are sometimes uncomfortable for the wearer owing to its size and fit.

PROBLEM: The lens does not filter UV light and bright ambient light.

### REDUCING

**FIM**: X-resource, together with available or modified resources, and without making the object more complex or introducing any negative properties guarantees attainment of the following IFR (Lens filters UV light by itself)

Standard (Technical) Contradiction Model

fig. 9.26 - beg. Reinventing in standard form of Photosensitive Contact Lens (U.S. Patent No. 4,681,412)

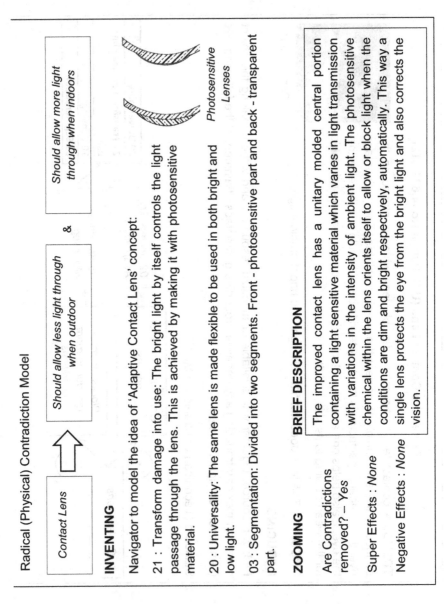

Radical (Physical) Contradiction Model

| Contact Lens | Should allow less light through when outdoor | & | Should allow more light through when indoors |

*Photosensitive Lenses*

## INVENTING

Navigator to model the idea of 'Adaptive Contact Lens' concept:

21 : Transform damage into use: The bright light by itself controls the light passage through the lens. This is achieved by making it with photosensitive material.

20 : Universality: The same lens is made flexible to be used in both bright and low light.

03 : Segmentation: Divided into two segments. Front - photosensitive part and back - transparent part.

## ZOOMING

Are Contradictions removed? – Yes

Super Effects : *None*

Negative Effects : *None*

## BRIEF DESCRIPTION

The improved contact lens has a unitary molded central portion containing a light sensitive material which varies in light transmission with variations in the intensity of ambient light. The photosensitive chemical within the lens orients itself to allow or block light when the conditions are dim and bright respectively, automatically. This way a single lens protects the eye from the bright light and also corrects the vision.

**fig. 9.26 - end.** Reinventing in standard form of Photosensitive Contact Lens (U.S. Patent No. 4,681,412)

## 9.8. Communication & Electronics:

### Facsimile – Transmission & Error Proofing (U.S. Patent No. 3,084,213)

### Prior Art (Engineering Level)

Fax (abbreviation for Facsimile), is a device used for the transmission of scanned printed material to another receiving device which is connected to a printer. Traditionally, two fax machines (located only in telegraphic offices) were connected with dedicated lines.

**fig. 9.27.** A Transmitter in operation at the 'New York Herald' Office (Forgotten futures 1900)

A scanner scanned the printed material and converted the data into electronic codes. These codes were then transmitted through these dedicated lines to the specific telegraphic offices as required by the customer. The receiver used to collect his fax from the fax receiving office, where the electronic codes are translated back into a printed material.

Although fax system found its effective use during its age, the system itself was inherently ineffective for the following main reasons.

### Detection Problems

1. The users had to go to telegraphic offices for sending and receiving fax. Universal connectivity was required. But such a system was inconceivable as very high connectivity would have been required between every user. This would have been very expensive apart from the complicated controls required.

2. Several errors and omissions used to pop up at the receiver end. Sometimes this would be so drastic that the whole document had to be scanned again and resent.

There were several other problems with the facsimile system like its bulky size, lack of automatic transmission, slow printing & requirement of manual attendance for loading and unloading the photosensitive paper from the drum, we here will focus only on the two problems we discussed. (The above sections are adapted from Lemelson, 1963).

## Result of Contradiction Analysis

Transmission – The transmission of fax messages used to happen only between telegraphic offices. Hence to solve such a problem, one needs a universal connectivity. But by having such a universal connectivity, the high level of network control and automation required for that is certainly a problem that would be introduced in the system. These positives and negatives could be represented as "Complexity of construction" and "Level of Automation" respectively in TRIZ. The Altshuller Matrix yields the following results for this contradiction – Dynamization, Segmentation and/or Mediator.

Error Proofing –The data loss was a problem prevailing widely in those days. Hence "an improvement of data loss during transmission was required" as a positive improvement. But to attain that quality of reproducibility, the transmission distance had to be short (negative of contradiction). These positives and negatives could be represented as "Loss of Information" and "Length of fixed object" respectively in TRIZ. The Altshuller Matrix yields "Copying" as the navigator to solve this problem.

### Inventing Idea

Transmission: The Idea for inventing the new and improved connectivity in the Fax transmission system could be obtained using the navigator 'Mediator'. It advises to look for some mediator devices that could transmit the fax signals universally. Hence the already existing telephone lines are used as a mediator for data transfer. The voice signals of the telephone and the visual signals of the fax are transmitted through a common shared telephone wire.

Error Proofing: Lossless data transmission of data in the fax system could be attained by the navigator - Copying. "A Recording strip is in place which records a copy of the image. This verifies the image transmitted with the actual image and eliminates data loss in transmission" (Lemelson, 1963).

### Resultant Artifact

**US Patent No. 3,084,213**              **Date of Patent: April 2, 1963**

**Title: Facsimile Apparatus**

**Inventors: Jerome H. Lemelson**

**Abstract:**

1. The new facsimile system includes a master or transmitting station from which picture signals originate and are transmitted over one or more wires of a conventional telephone switching system to one or more receiving stations by semiautomatic means, without the need for manual attention of the apparatus at the receiving station.

A video camera is provided at the transmitter station having a slow enough scanning system to permit the wire transmission of the video picture signal derived from a single scan or frame sweep over said conventional phone lines.

As a result, the operator at source may transmit picture information to any business or location having one or more telephone line outlets and having the necessary receiving and picture printing equipment.

2. A recording member fragment is utilized in the apparatus which contains a magnetic recording strip running parallel to the picture strip on which signals may be recorded for controlling apparatus in the selection of one or more circuits over which the associated picture signal is to be transmitted.

This provides a means for recording any errors or omissions in the transmission of a signal and is corrected by the process of feedback (Lemelson, 1963).

**fig. 9.28.** Facsimile Apparatus (Lemelson, 1963)

**Extracting** of "Facsimile Apparatus" is shown in fig. 9.29.

## Extraction
### Facsimile Apparatus – Transmission (U.S. Patent No. 3,084,213)

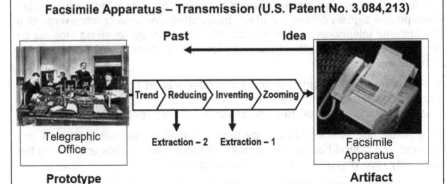

**Past**      **Idea**

Trend ⟩ Reducing ⟩ Inventing ⟩ Zooming

Telegraphic Office     Extraction – 2    Extraction – 1     Facsimile Apparatus

**Prototype**                 **Artifact**

### Extraction – 1

| LC | No. | Navigator | Substantiation for Extraction |
|---|---|---|---|
| + | 01 | Change in the aggregate state of an object | The narrow network connections are made broad by using the existing telephone lines |
| + | 05 | Separation | The fax data transmission function in taken out of conventional line and placed in the telephone lines. |
| + | 10 | Copying | The networking in telephone network is copied and used for facsimile system too. |
| ++ | 18 | Mediator | Already existing telephone lines are used as a mediator for data transfer |
| ++ | 20 | Universality | The facsimile system which used to serve only between two telegraphic offices can now be used by all users |
| + | 35 | Unite | The facsimile signals and telephone signals are sent through the same wire |

### Extraction - 2

**Standard Contradiction:** The transmission of fax should be universal i.e. from every user to every other user but a high level of networking and control is required for that.

**Radical Contradiction:** The density of network required should be high for universal usage but also low to reduce the complexity.

**FIM:** X-resource, together with available or modified resources, and without making the object more complex or introducing any negative properties guarantees attainment of: *User to user direct transmission of data possible.*

**fig. 9.29.** Facsimile Apparatus – Transmission (Lemelson, 1963).

## Extraction
### Facsimile Apparatus – Error Proofing (U.S. Patent No. 3,084,213)

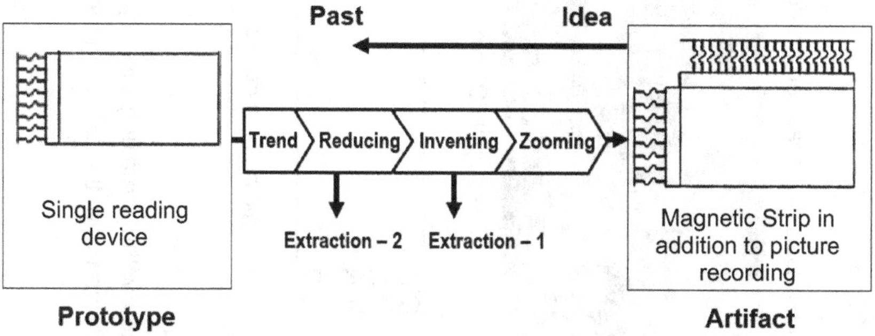

**Past**       **Idea**

Trend > Reducing > Inventing > Zooming >

Single reading device

Extraction – 2     Extraction – 1

Magnetic Strip in addition to picture recording

**Prototype**           **Artifact**

### Extraction – 1

| LC | No. | Navigator | Substantiation for Extraction |
|----|-----|-----------|-------------------------------|
| + | 02 | Preliminary action | The recording member corrects errors in the transmission |
| + | 03 | Segmentation | The scanning is segmented into scanner and another magnetic strip member which tracks and corrects the difference in data |
| ++ | 10 | Copying | Recording member is in place which records a copy of the image. This verifies the image transmitted with the actual image and eliminates data loss in transmission. |
| + | 15 | Discard and renewal of parts | The old data is taken off the magnetic strip when new scan is made. |
| + | 18 | Mediator | The mediator recording strip ensures reproducibility |
| + | 24 | Asymmetry | Another recording strip is placed offset to the main recording member for error correction |
| ++ | 29 | Self-servicing | The system itself correct itself by matching the data recorded in two strips and correcting errors |
| + | 35 | Unite | A second recording strip is united with main member |
| ++ | 36 | Feedback | One magnetic strip constantly gives |

### Extraction - 2

**Standard Contradiction:** The data loss during transmission has to be low for good reproducibility but for that, the distance of transmission should not be long.

**Radical Contradiction:** Distance of transmission has to be long because it is required to connect all people but it should also be short to eliminate data loss.

**FIM:** X-resource, together with available or modified resources, and without making the object more complex or introducing any negative properties guarantees attainment of: *There is no data loss during transmission.*

**fig. 9.30.** Facsimile Apparatus – Error Proofing (Lemelson, 1963).

## Facsimile Apparatus (U.S. Patent No. 3,084,213 )

**TREND**: Fax (abbreviation for facsimile), is a device used for the transmission of scanned printed material to another receiving device which is connected to a printer. Traditionally, two fax machines (located only in telegraphic offices) were connected with dedicated lines. A scanner scanned the printed material and converted the data into electronic codes. These codes were then transmitted through these dedicated lines to the specific telegraphic offices as required by the customer. The receiver used to collect his fax from the fax receiving office, where the electronic codes are translated back into a printed material.

*A Telegraphic Office*

*A Single reading device*

PROBLEM:

1. Transmission is possible only between telegraphic offices which are connected by dedicated cable.

2. Data loss during transmission.

**REDUCING**

**FIM**: X-resource, together with available or modified resources, and without making the object more complex or introducing any negative properties guarantees attainment of the following Ideal Final Results

**fig. 9.31 - beg.** Reinventing in standard form of
Facsimile – Transmission & Error Proofing (U.S. Patent No. 3,084,213)

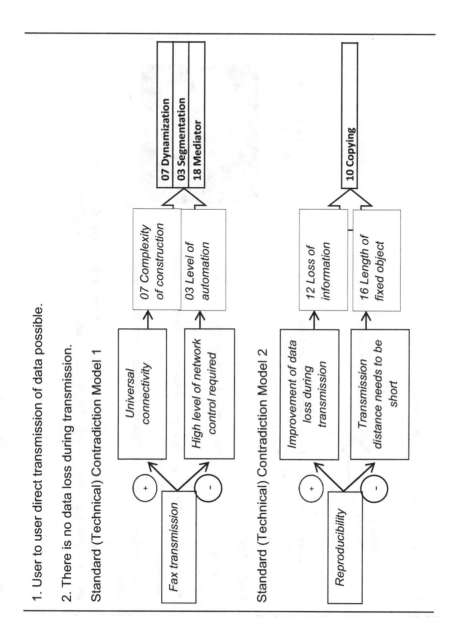

**fig. 9.31 – cont. 1.** Reinventing in standard form of
Facsimile – Transmission & Error Proofing (U.S. Patent No. 3,084,213)

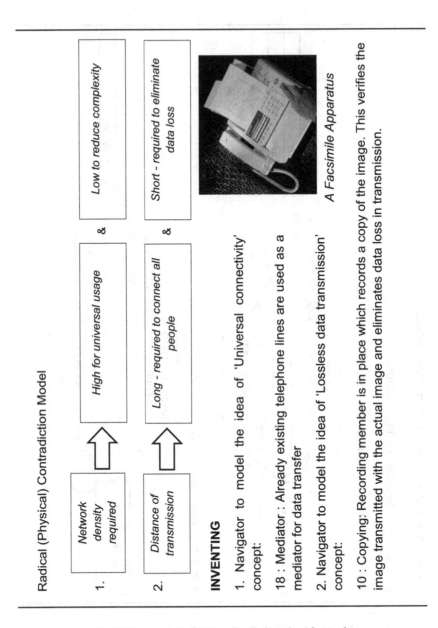

Radical (Physical) Contradiction Model

| 1. | Network density required | ⟹ | High for universal usage | & | Low to reduce complexity |
| 2. | Distance of transmission | ⟹ | Long - required to connect all people | & | Short - required to eliminate data loss |

A Facsimile Apparatus

**INVENTING**

1. Navigator to model the idea of 'Universal connectivity' concept:

18 : Mediator : Already existing telephone lines are used as a mediator for data transfer

2. Navigator to model the idea of 'Lossless data transmission' concept:

10 : Copying: Recording member is in place which records a copy of the image. This verifies the image transmitted with the actual image and eliminates data loss in transmission.

**fig. 9.31 – cont. 2.** Reinventing in standard form of
Facsimile – Transmission & Error Proofing (U.S. Patent No. 3,084,213)

## ZOOMING

Are Contradictions removed? – Yes

Super Effects : None

Negative Effects : None

*A Magnetic Strip in addition to picture recording strip*

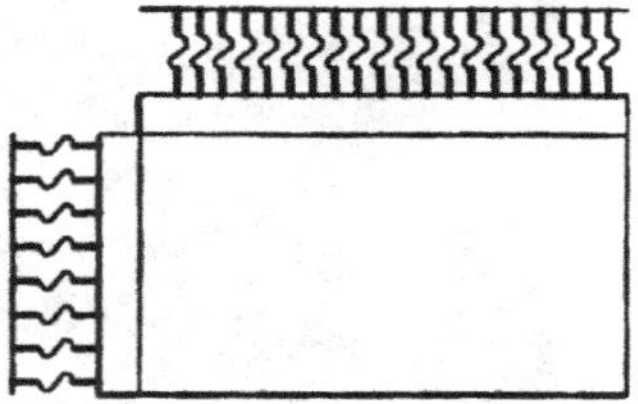

### BRIEF DESCRIPTION 1 – Universal connectivity

The new facsimile system includes a master or transmitting station from which picture signals originate and are transmitted over one or more wires of a conventional telephone switching system to one or more receiving stations by semiautomatic means, without the need for manual attention of the apparatus at the receiving station. This could be received by anyone who has a suitable facsimile machine at the receiver end and the data can be retrieved using a device like printer connected to the said facsimile system.

### BRIEF DESCRIPTION 2 – Lossless data transmission

A recording member fragment is utilized in the apparatus which contains a magnetic recording strip running parallel to the picture strip on which signals may be recorded for controlling apparatus in the selection of one or more circuits over which the associated picture signal is to be transmitted. This provides a means for recording any errors or omissions in the transmission of a signal and is corrected by the process of feedback.

**fig. 9.31 - end.** Reinventing in standard form of
Facsimile – Transmission & Error Proofing (U.S. Patent No. 3,084,213)

## 9.9.  Instruments / Medical:

### Catheter and Implant Method (U.S. Patent No. 4,588,395)

### Prior Art (Engineering Level)

In the olden days, drainage of unwanted bodily fluids, administration of fluids, gases, or medicines into the body cavity was possible only through surgery.

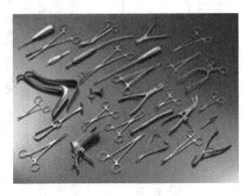

**fig. 9.32.** Surgical Instruments
(Shree enterprisers 2012)

The precision was great because that specific section of the skin was surgically cut to access the body cavity, which enabled the forceps to deliver the medicine in the required area. The surgical cuts usually had to be big to enable easy passage of tools and give good visibility into the cavity.

After the delivery of the medicine or implant at the defined location, the cuts are stitched back and a dressing is placed on the wound. The healing process normally takes weeks together. A scar is also unavoidable post-surgery in this case. The patient is also mostly bed ridden during this stage of wound healing (Lemelson, 1986).

### Result of Contradiction Analysis

The contradiction posed by the surgical procedure to implant or deliver medicine locally could be defined as following. The surgical procedure offers a clear view of the location of implant or medicine and hence the positioning becomes easy. This is the positive. Having this positive feature in our process poses the negative of having "Incisions in the body" and "Healing process and time". The positive, hence could be defined in the language of TRIZ as "Reliability". The negative clearly could be said as an "Internal damaging factor".

### Inventing Idea

The idea for solving the abovementioned contradiction could be achieved by the navigators offered by the Altshuller matrix when the positives and negatives are looked up in the table. It says to – Change the aggregate state of the object; Perform separation – that is taking out required features and eliminating unnecessary ones; Use of composite materials and Copying.

Separation helps the invention process by separating out the necessary small incision for the passage of medicine from the unnecessary big surgical incisions.

The system is further improved by changing the system's aggregate state i.e. the Inflexible forceps are replaced by flexible catheter which delivers the implant or medicine internally.

A camera on the tip of the catheter enables perfect positioning. This goes with the navigator "Copying".

**Resultant Artifact**

**US Patent No. 4,588,395**          **Date of Patent: May 13, 1986**

**Title: Catheter and Implant Method**

**Inventors: Jerome H. Lemelson**

**Abstract:**

A device and method are provided for disposing a quantity of matter at a select location within an animal or human body. In one form, the matter is a solid material, such as medication which dissolves with time, a source of radiation such as gamma radiation, a sensor or transducer in a housing which includes a short wave transmitter or other material or device which is operable to beneficially effect human tissue or body function when it is inserted into the body.

The material is retained within a housing located at an end of a flexible tube or catheter which is inserted into a body cavity and is manipulated from the other end thereof to a predetermined location within the body cavity.

Suitable actuating means located at the external end of the catheter is operated, when the head end thereof is at a predetermined location within the human body, to cause the material supported within the head to be ejected therefrom' (Lemelson, 1986).

fig. 9.33. Cross section view of catheter with material held in housing (Lemelson, 1986)

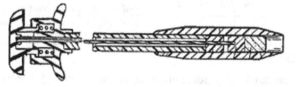

fig. 9.34. Cross section view of catheter with material just before the delivery (Lemelson, 1986).

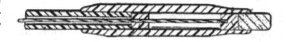

**Extracting** of "Catheter and Implant Method" is shown in fig. 9.35.

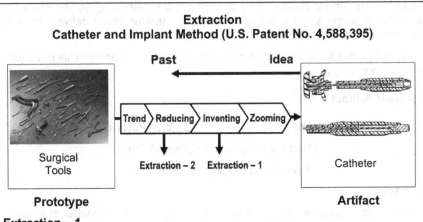

**Extraction**
**Catheter and Implant Method (U.S. Patent No. 4,588,395)**

Past                    Idea

Trend  Reducing  Inventing  Zooming

Surgical Tools          Extraction – 2    Extraction – 1          Catheter

Prototype                                          Artifact

**Extraction – 1**

| LC | No. | Navigator | Substantiation for Extraction |
|----|-----|-----------|-------------------------------|
| ++ | 01 | Change in the aggregate state of an Object | Inflexible Forceps are replaced by flexible Catheter which delivers the implant or medicine internally. |
| ++ | 05 | Separation | The necessary small incision is separated out from the unnecessary big surgical incisions. |
| + | 07 | Dynamization | The device (Insertion) is made dynamic relative to surgeon's hands compared against the Forceps. |
| + | 10 | Copying | The functional copy (from Forceps) of grabbing and releasing the implant is placed on the tip of Catheter. |
| + | 18 | Mediator | Mechanical wiring are used as mediator between the implant and surgeon's hands |
| ++ | 34 | Matryoshka | The implant is placed within the Catheter before it is delivered at its position. |
| + | 35 | Unite | The functions of several surgical instruments are given to the Catheter. |

**Extraction - 2**

**Standard Contradiction:** In conventional surgical procedure, the positioning of the implant is precise but it involves a lot of surgical incisions and healing.

**Radical Contradiction:** Surgical cuts should not be there as it requires healing but surgical cuts are required to position the implant properly.

**FIM:** X-resource, together with available or modified resources, and without making the object more complex or introducing any negative properties guarantees attainment of: *Positioning is perfect without any need for surgical cuts.*

**fig. 9.35.** Catheter and Implant Method (Lemelson, 1986).

## Catheter and Implant Method (U.S. Patent No. 4,588,395)

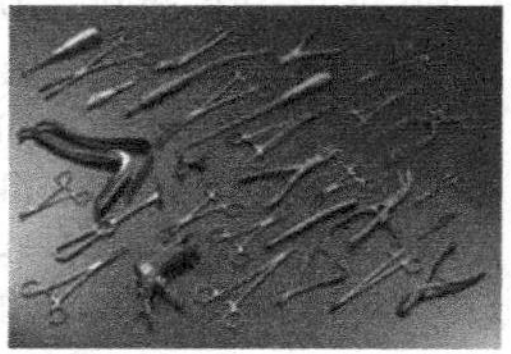

*Surgical Tools*

**TREND:** In the olden days, drainage of unwanted bodily fluids, administration of fluids, gases, or medicines into the body cavity was possible only through surgery. The precision was great because that specific section of the skin was surgically cut to access the body cavity, which enabled the forceps to deliver the medicine in the required area. After the delivery of the medicine or implant at the defined location, the cuts are stitched back and a dressing is placed on the wound.

**PROBLEM:** Even small localized medicine delivery or implants requires surgical incisions.

**REDUCING**

**FIM:** X-resource, together with available or modified resources, and without making the object more complex or introducing any negative properties guarantees attainment of the following IFR (Implant delivered inside body cavity without surgical procedure)

Standard (Technical) Contradiction Model

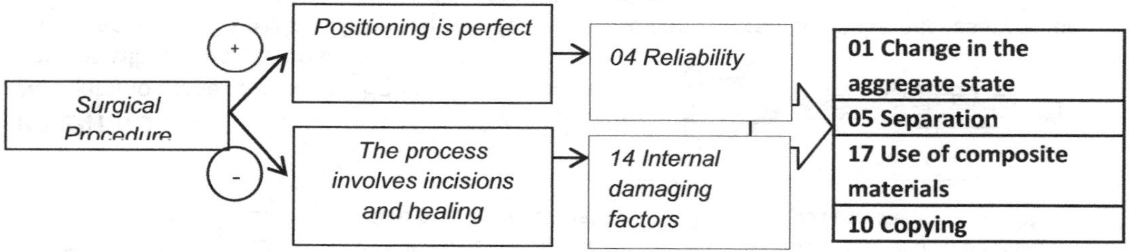

fig. 9.36 - beg. Reinventing in standard form of
Catheter and Implant Method (U.S. Patent No. 4,588,395)

Radical (Physical) Contradiction Model

| Implanting method | Surgical cuts should not be there as it requires healing | & | Surgical cuts should be there to position the implant right |

A cross section of Catheter

## INVENTING

Navigator to model the idea of 'Body implant without surgery' concept:

05 : Separation:  Smallest incision required for implant to pass is created i.e the necessary small incision is separated out from the unnecessary big surgical incisions.

01 : Change in aggregate state : Inflexible forceps are replaced by flexible catheter which delivers the implant or medicine internally.

## ZOOMING

Are Contradictions removed? – Yes

Super Effects :
None

Negative Effects :
None

## BRIEF DESCRIPTION

A Catheter is a device and method for disposing a quantity of matter at a select location within an animal or human body. The matter could be a solid material, such as medication which dissolves with time, a source of radiation such as gamma radiation, a sensor or transducer in a housing which includes a short wave transmitter or other material or device which is operable when it is inserted into the body. The material is retained within a housing located at an end of a flexible tube or catheter which is inserted into a body cavity and is manipulated from the other end thereof to a predetermined location within the body cavity. Actuators located at the external end of the catheter are operated, when the head end is at a predetermined location within the human body, to cause the material supported within the head to be ejected.

**fig. 9.36 - end.** Reinventing in standard form of
Catheter and Implant Method (U.S. Patent No. 4,588,395)

## 9.10. Manufacturing:

### Modern Fences (U.S. Patent No. 3,933,311)

### Prior Art (Engineering Level)

Fences are widely known structures that come in various forms. Mainly used as a demarcation tool to create boundary between two areas, fences serve the purpose of peaceful co-existence among the countries of the world as rightly claimed by Robert Frost (American poet; received four Pulitzer Prizes for his Poetry): "Good fences make good neighbors".

### Manufacturing

Several types of fences exist around the world. Here however, we restrict the study to the so called "Chain-linked Fence" which could be seen around the world in several installations.

The manufacturing of such fences follow slightly complex steps and the process itself is similar to weaving of clothes. Wire is fed into the bending machine which bends it in a zigzag fashion and weaving it to the adjacent patterned wire to form the characteristic diamond pattern.

**fig. 9.37.** A Chain Link fence (Bombay Harbor 2012)

### Problems in Manufacture

Working with each wire to create a continuously woven diamond pattern is time consuming and increases the chances of problems with quality of the Fences produced.

A clear inspiration from the Cloth weaving machine, fence weaving process needed some improvement in terms of speed of production and/or the fence design itself (Chain Link, 2012).

### Result of Contradiction Analysis

The contradiction analysis could be structured in the problem of diamond patterned fences as follows. The "plus" required in the system would be that 'processing would be simplified but by doing so, we are faced with a "minus" contradiction where "Many number of Machines are needed" to manage production with simplified process. To convert it in the language of TRIZ, the "plus" transforms into "Complexity of Construction" and the "minus" transforms into "Quantity of material".

The Altshuller Matrix yields the navigators to be used as – 02 Preliminary action, 11 Inverse action, 12 Local property, 13 Inexpensive short-life object replaced for expensive long-life object.

### Inventing Idea

The idea for the resultant artifact could be obtained with the help of following navigators.

02: Preliminary action – The assembly is performed in advance by the machine, that is, the fence is continuously extruded as a single component using a die in the machine. The inventive idea for preliminary action navigator in fence production could very well trigger and generate the answer to do something in advance that is done later – Assembly, in this case. Assembly at earliest possible stage would be during the manufacturing stage itself. A further preliminary action question would generate the idea of conjoint fence wires exiting the machine, which could further narrow down to the idea of extrusion.

12: Local Property – By producing a continuous undivided fence, the need for welding at the localized zones in the fence is removed.

With the help of these two navigators, the idea for a system as developed by Lemelson could be sparked off.

### Resultant Artifact

**US Patent No. 3,933,311**        **Date of Patent: Jan 20, 1976**

#### Title: Extruded Fence

#### Inventors: Jerome H. Lemelson

### Abstract:

A fence structure is provided which is produced of a plurality of lengths of plastic or metal extrusions joined together to form a tandem array of interconnected sections. In one form, the joining means serves one or more functions in addition to that of connecting the fence sections together. A particular form of joining means for the fence sections is also operable to secure the fence to the ground.

In a particular construction, the fence is formed with one or more tubular formations extending longitudinally thereof and the tubular formations of each section of fence are interconnected with those of the other sections by means of the joining means' (Lemelson,1976).

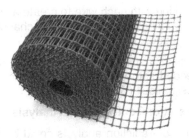

**fig. 9.38.** Extruded Fence Net (Manufacturer 2012)

**Extracting** of "Catheter and Implant Method" is shown in fig. 9.39.

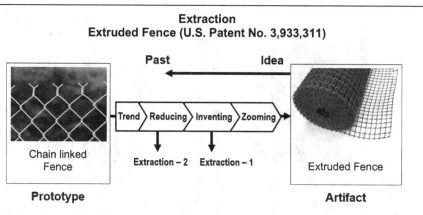

**Extraction**
**Extruded Fence (U.S. Patent No. 3,933,311)**

**Prototype**             **Artifact**

**Extraction – 1**

| LC | No. | Navigator | Substantiation for Extraction |
|---|---|---|---|
| ++ | 02 | Preliminary action | The assembly is made in advance by the machine. |
| + | 12 | Local property | By producing a continuous undivided Fence, the need for welding at the localized zones in the Fence is removed. |
| + | 18 | Mediator | A die is used as a mediator to produce Fences of any shape |
| ++ | 20 | Universality | A single Machine could be used to produce any Fence type. |
| + | 31 | Use of porous materials | Cavities in the Fence's base are used for watering the Garden |
| ++ | 35 | Unite | Various fragments of Fences are united in the extrusion process |
| + | 40 | Uninterrupted useful function | Fence is extruded continuously rather than each link being assembled by the Machine |

**Extraction - 2**

**Standard Contradiction:** Simple process is needed but in that case many number of machines would be needed to manufacture the fence.

**Radical Contradiction:** The manufacturing process must be simple to worl with but at the same time complex to produce all kinds of shapes that fences come in.

**FIM:** X-resource, together with available or modified resources, and without making the object more complex or introducing any negative properties guarantees attainment of: *Fences that need little or no assembly that is – prefabricated fences.*

**fig. 9.39.** Extruded Fence (Lemelson, 1976).

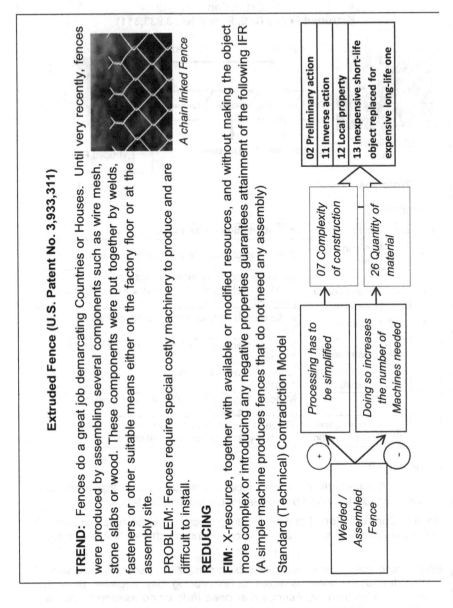

**Extruded Fence (U.S. Patent No. 3,933,311)**

**TREND:** Fences do a great job demarcating Countries or Houses. Until very recently, fences were produced by assembling several components such as wire mesh, stone slabs or wood. These components were put together by welds, fasteners or other suitable means either on the factory floor or at the assembly site.

**PROBLEM:** Fences require special costly machinery to produce and are difficult to install.

**REDUCING**

**FIM:** X-resource, together with available or modified resources, and without making the object more complex or introducing any negative properties guarantees attainment of the following IFR (A simple machine produces fences that do not need any assembly)

Standard (Technical) Contradiction Model

*A chain linked Fence*

Welded / Assembled Fence

(+) Processing has to be simplified

(−) Doing so increases the number of Machines needed

07 Complexity of construction

26 Quantity of material

02 Preliminary action
11 Inverse action
12 Local property
13 Inexpensive short-life object replaced for expensive long-life one

**fig. 9.40 - beg.** Reinventing in standard form of Modern Fences (U.S. Patent No. 3,933,311)

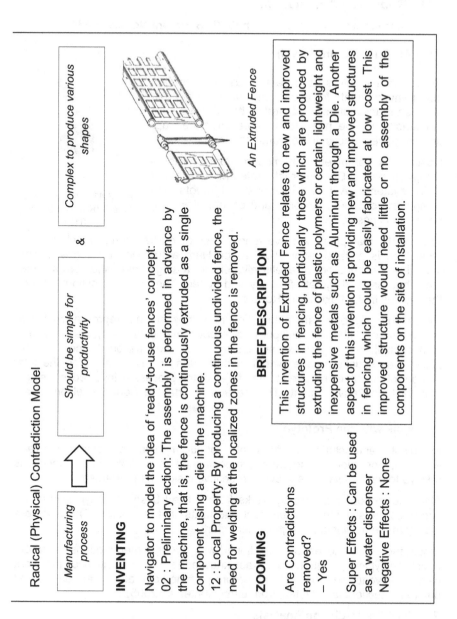

Radical (Physical) Contradiction Model

| Manufacturing process | ⬆ | Should be simple for productivity | & | Complex to produce various shapes |

*An Extruded Fence*

**INVENTING**

Navigator to model the idea of 'ready-to-use fences' concept:

02 : Preliminary action: The assembly is performed in advance by the machine, that is, the fence is continuously extruded as a single component using a die in the machine.

12 : Local Property: By producing a continuous undivided fence, the need for welding at the localized zones in the fence is removed.

**ZOOMING**

Are Contradictions removed?
– Yes

Super Effects : Can be used as a water dispenser

Negative Effects : None

**BRIEF DESCRIPTION**

This invention of Extruded Fence relates to new and improved structures in fencing, particularly those which are produced by extruding the fence of plastic polymers or certain, lightweight and inexpensive metals such as Aluminum through a Die. Another aspect of this invention is providing new and improved structures in fencing which could be easily fabricated at low cost. This improved structure would need little or no assembly of the components on the site of installation.

**fig. 9.40 - end.** Reinventing in standard form of Modern Fences (U.S. Patent No. 3,933,311)

## 9.11. Robotics / Automation:

### Autonomous Assembly Robots (U.S. Patent No. 6,898,484)

**Prior Art (Engineering Level)**

Robotic manipulators have been in use in manufacturing environments for many years now. As the role of these manipulators increased to broader areas of manufacturing, the need for a more structured approach to their functioning became apparent.

Robotics is combined with modem computer technology as part of Computer Integrated Manufacturing (CIM) systems. With the integration of robots with computers, complete flexibility and control of the robotic arm is possible.

**fig. 9.41.** An assembly Robot (Kuka Robotics 2012)

The robotic manipulators serve many different manufacturing functions. The other type is a "hard automation" robot. Hard automation machines are designed and configured to repetitively perform a single task, and normally perform that task with great efficiency and speed. The main distinction between the two lies in the flexibility of Robots.

Most single purpose hard automation machines are not reprogrammable or multipurpose. Although multipurpose robotic manipulators typically have a higher initial cost and slower speeds than hard automation devices they replace, they usually reduce overall manufacturing costs by reducing the total number of machines required to handle all of the processes involved in a single manufacturing environment (This section is adapted from Lemelson et al., 2005).

**Problems with the Prototype**

In typical manufacturing environments, robotic manipulator components (end effectors, work pieces, and parts) are positioned in fixed pre-defined locations relative to each other. This allows a repetitive task to be performed.

Also, the Robot itself typically works from a fixed base. So it typically has a supply of parts made available to it due to its limited mobility. The parts are mostly made available through human effort.

Since the work pieces are positioned with programmed motions, a proper operation typically requires an exact knowledge of the position of the work piece relative to the robot so that preprogrammed operations can be carried out (this section is adapted from Lemelson et al., 2005).

**Result of Contradiction Analysis**

The contradiction analysis could be structured in this problem of autonomous assembly robots as follows. The "plus" in the system would be to "Autonomously handle the part" but by doing so, we are faced with a "minus" contradiction of "Precise calculated positioning". To convert it in the language of TRIZ, the "plus" transforms into "Level of Automation" and the "minus" transforms into "Precision of measurement". The Altshuller Matrix yields the navigators to be used as – 4. Replacement of mechanics, 10.Copying, 2.Preliminary action and 16. Partial or excess effect.

### Inventing Idea

The idea for the resultant artifact could be obtained with the help of following navigators.

2. Preliminary action – The Position Sensors are placed on robotic arm, assembly point and picking shelf in advance and GPS helps the arm track the sensors and parts constantly.

4. Replacement of mechanics – Several sensing devices are used for position determination rather than constant program cycle which operates the arm.

With the help of these two navigators, a comprehensive system as developed by Lemelson could be built.

### Resultant Artifact

**US Patent No. 6,898,484**         **Date of Patent: May 24, 2005**

**Title: Robotic manufacturing and assembly with relative radio positioning using Radio based position determination**

**Inventors: Jerom Lemelson, Dorothy Lemelson et al**

### Abstract:

GPS is a well-established radio navigation technique based on the use of a constellation of twenty-four satellites in carefully placed geo-synchronous orbits. Ground control stations monitor and correct the performance of the satellite broadcasts to maintain a high level of accuracy to triangulate the position of the receiver on earth.

Global positioning system inputs are used in a manufacturing process where location of a work piece relative to a robotic manipulator is input into a control system. The manipulator is located and tracked by using "GPS" signals, as is an associated work piece.

Radio signal based position indicators associated with work pieces transmit work piece location and status. In some embodiments manipulator locations are sensed by position indicators associated with manipulators and signals relating to the position of the manipulators are transmitted to the control system.

The control system controls the manipulator and may also control material handling equipment for the transport of work pieces (Lemelson et al, 2005)

(12) **United States Patent**     (10) Patent No.:     **US 6,898,484 B2**
Lemelson et al.                   (45) Date of Patent:         **May 24, 2005**

(54) **ROBOTIC MANUFACTURING AND ASSEMBLY WITH RELATIVE RADIO POSITIONING USING RADIO BASED LOCATION DETERMINATION**

(76) Inventors: **Jerome H. Lemelson**, deceased, late of Incline Village, NV (US); by **Dorothy Lemelson**, legal representative, 930 Tahoe Blvd., Incline Village, NV (US) 89451-9436; **Robert D. Pedersen**, 7808 Gleneagle, Dallas, TX (US) 75248; **Tracy D. Blake**, 14641 N. 49th Pl., Scottsdale, AZ (US) 85254

(*) Notice: Subject to any disclaimer, the term of this patent is extended or adjusted under 35 U.S.C. 154(b) by 0 days.

(21) Appl. No.: **10/138,064**

(22) Filed: **May 1, 2002**

(65) **Prior Publication Data**

US 2003/0208302 A1 Nov. 6, 2003

(51) Int. Cl.[7] ............................................. G06F 19/00
(52) U.S. Cl. ..................... 700/245; 700/253; 700/257; 700/262; 700/264; 700/302; 318/568.1; 701/23; 901/1; 901/8
(58) Field of Search ...................... 318/568.1; 700/253, 700/257, 262, 264, 302, 245; 901/1, 8; 701/23

(56) **References Cited**

U.S. PATENT DOCUMENTS

4,477,754 A   10/1984  Roch et al. ................. 318/568

| 4,611,292 A | 9/1986 | Ninomiya et al. | .......... 364/559 |
| 5,049,796 A | 9/1991 | Seraji | ........................ 318/568 |

(Continued)

OTHER PUBLICATIONS

Valejo et al., Short–range DGPS for mobile robots with wireless Ethernet links, Jun. 29–Jul. 1, 1998, IEEE, Page(s): 334–339□□.*

Graser Technological Solutions to Autonomous Robot Control, 1998, Internet, pp. 1–8.*

(Continued)

*Primary Examiner*—Thomas G. Black
*Assistant Examiner*—McDieunel Marc
(74) *Attorney, Agent, or Firm*—Edwin A. Suominen; Douglas W. Rudy

(57) **ABSTRACT**

Global positioning system inputs are used in a manufacturing process where location of a work piece relative to a robotic manipulator is input into a control system. The manipulator is located and tracked by using "GPS" signals, as is an associated work piece. Radio signal based position indicators associated with work pieces transmit work piece location and status. In some embodiments manipulator locations are sensed by position indicators associated with manipulators and signals relating to the position of the manipulators are transmitted to the control system. The control system controls the manipulator and may also control material handling equipment for the transport of work pieces.

**29 Claims, 9 Drawing Sheets**

**fig. 9.42.** A GPS based tracking of parts for manufacturing

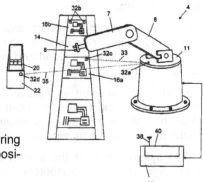

**Extracting** of "Robotic manufacturing and assembly with relative radio positioning" is shown in fig. 9.43.

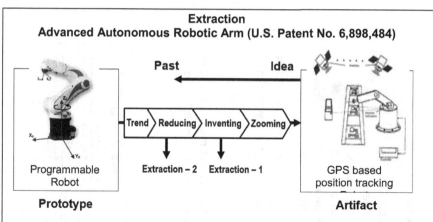

**Extraction**
**Advanced Autonomous Robotic Arm (U.S. Patent No. 6,898,484)**

Prototype                                                       Artifact

**Extraction – 1**

| LC | No. | Navigator | Substantiation for Extraction |
|---|---|---|---|
| ++ | 04 | Replacement of mechanical matter | Sensing devices are used to position rather than constant program cycle which operate the arm. |
| + | 05 | Separation | 'Locating function' is taken out and performed with the help of Sensors. |
| + | 07 | Dynamization | Motion is made with respect to the Position Sensors. |
| + | 12 | Local property | The property of the Work Table is altered to hold Sensor and perform control of Robotic Arm |
| ++ | 18 | Mediator | Position Sensors and GPS Satellites are used as mediating |
| + | 19 | Transition into another dimension | The control is now performed from outside the Robotic system i.e. by sensors on the Shop Floor and Satellites over the Earth. |
| ++ | 36 | Feedback | Constant feedback of position of the arm, raw materials and work station are provided by the GPS satellite. |

**Extraction - 2**

**Standard Contradiction:** The Robotic arm should handle the parts autonomously but in that situation, precise positioning would be a challenge.

**Radical Contradiction:** Robotic Arm's autonomous positioning ability should be there for improvement in the system but the autonomy should not be there as it needs high level of Intelligence and complicated controls.

**FIM (shortly):** *Robot knows the positions and assembles correctly and autonomously.*

**fig. 9.43.** Robotic manufacturing with relative radio positioning (Lemelson, 1976).

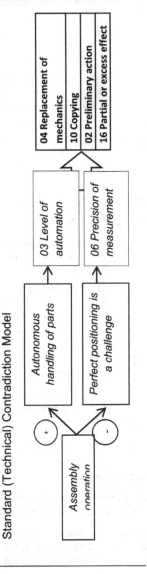

## Advanced Autonomous Robotic Arm (U.S. Patent No. 6,898,484)

**TREND:** Robotic manipulators have been used in automatic manufacturing environments for years, and their use is growing at a rapid pace. They can, in many instances, outperform human labor performing similar operations. Normally the fixed angle of movement of each robotic arm is preprogrammed and this cycle is repeated by the arm. The supporting system like raw material shelf and the assembly spot always need to be in perfect alignment with the arm.

*A programmable Robot*

**PROBLEM:** Autonomous movement of arm, which would offer flexibility, is not possible in the system.

**REDUCING**

**FIM:** X-resource, together with available or modified resources, and without making the object more complex or introducing any negative properties guarantees attainment of the following IFR (Robot knows the positions and assembles correctly and autonomously)

Standard (Technical) Contradiction Model

**fig. 9.44 - beg.** Reinventing in standard form of
Autonomous Assembly Robots (U.S. Patent No. 6,898,484)

Radical (Physical) Contradiction Model

| Robotic assembly operation | Autonomous positioning ability should be there for improvement | & | Autonomous positioning should not be there as it needs high level of Intelligence |

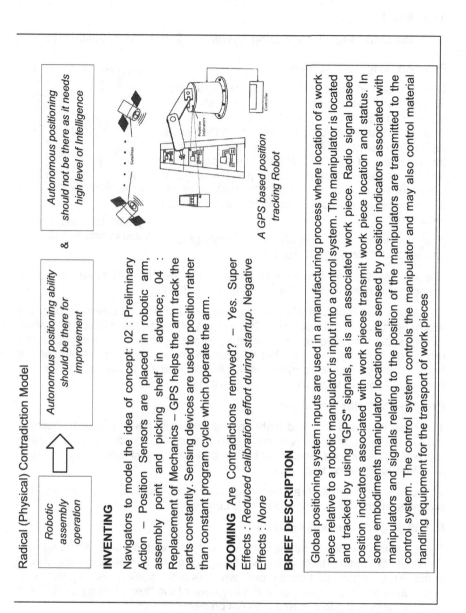

*A GPS based position tracking Robot*

## INVENTING

Navigators to model the idea of concept: 02 : Preliminary Action – Position Sensors are placed in robotic arm, assembly point and picking shelf in advance; 04 : Replacement of Mechanics – GPS helps the arm track the parts constantly. Sensing devices are used to position rather than constant program cycle which operate the arm.

**ZOOMING** Are Contradictions removed? – *Yes*. Super Effects : *Reduced calibration effort during startup*. Negative Effects : *None*

## BRIEF DESCRIPTION

Global positioning system inputs are used in a manufacturing process where location of a work piece relative to a robotic manipulator is input into a control system. The manipulator is located and tracked by using "GPS" signals, as is an associated work piece. Radio signal based position indicators associated with work pieces transmit work piece location and status. In some embodiments manipulator locations are sensed by position indicators associated with manipulators and signals relating to the position of the manipulators are transmitted to the control system. The control system controls the manipulator and may also control material handling equipment for the transport of work pieces

**fig. 9.44 - end.** Reinventing in standard form of
Autonomous Assembly Robots (U.S. Patent No. 6,898,484)

## 9.12. Music:

### Magnetic Tape Transducing System (U.S. Patent No. 3,842,432)

### Prior Art (Engineering Level)

Oberlin Smith in 1878 had a Eureka moment when he proposed the idea of telephone signals being recorded onto Steel Piano Wire. This was the first time when Music or Voice was ever recorded into a replayable format. Electronic amplification during the playback of recorded signals were expensive and hence the concept of recording was going nowhere for the next several years. By 1930, electronics were evolving and in parallel evolved the wire recorders.

**fig. 9.45.** Wire Recorder (Video Interchange 2011)

They first commercial products were "DictatingMachines" and "Telephone Recorders".

The devices were in the peak of success of their lifecycle when Lemelson changed this paradigm in 1974 (Wire recorder, 2011).

The sound was recorded when a thin wire rolled in a spool traversed over the recording head as referred from. The moving wire was magnetized and when moved again along the reader head, they were played back. The quality of the recorded sound was best of that time and if broke, could be simply knotted back rendering only tiny portion around that knot wasted. The rest of the sections played on perfectly (Wire recorder, 2011).

### Problems with the Prototype

The steel wire was thin and highly prone to tangling and it used to become quite a mess when they were not handled properly. Breakage was another problem. This was mainly due to the fact that wires, which are primarily metals like steel, were prone to corrosion.

Although a big hit in its era, these problems were shouting for a new and improved system of recording (Jerome H. Lemelson et al., 1974).

### Result of Contradiction Analysis

The contradiction posed by the improvements in magnetic wire recording system could be summarized as follows. Usage of a better material of wire should improve the tangling and rusting problem. This is the positive. Having this positive feature in our process poses the negative of the "head finding it difficult to read the data owing to different material property".

The positive, hence could be defined in the language of TRIZ as "Internal damaging factors". The negative clearly could be said as "Complexity of inspection and measurement".

## Inventing Idea

The idea for solving the abovementioned contradiction could be achieved by the navigators offered by the Altschuller matrix when the positives and negatives are looked up in the table. It says to – perform Separation, do a quick Jump, replace expensive long-life object by an inexpensive short-life object and/or perform Segmentation.

Separation helps the invention process by separating out the necessary thin metal strip from a thick inflexible wire and coating this layer on a thin and flexible film. The system could also be invented using the navigator – segmentation. In this case a single magnetic inflexible wire is segmented into a magnetizable thin layer and depositing it on a flexible and thin film base.

## Resultant Artifact

**US Patent No. 3,842,432          Date of Patent: Oct 1, 1974**

**Title: Magnetic Tape Transducing System**

**Inventors: Jerom H. Lemelson et al**

**Abstract:**

Magnetic tape was the answer to the problems in Wire recording. Typically, a tape consisted of three layers. A thin magnetizable layer on a long, narrow strip of plastic covered with a protective coating. The tape moves past the recording head at a continuous rate of speed. The recording head acquires the signal corresponding to the voice to be stored and this in turn magnetizes the tape as it passes below it. As the tape is played back, the magenetized pattern on surface of the tape is read by the head and it converts that into electrical signal which is further converted into sound signal as required (Adapted from Jerome H. Lemelson et al., (1974) & Tape Recorder, 2012).

**fig. 9.46.** An Audio Cassette with tape (Dimensions info 2009)

**Extracting** of "An Audio Cassette with tape" is shown in fig. 9.47.

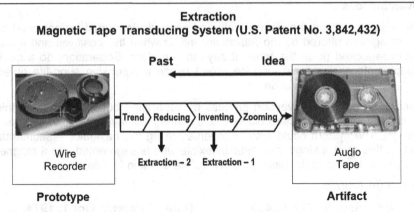

**Extraction**
**Magnetic Tape Transducing System (U.S. Patent No. 3,842,432)**

Past — Idea

Trend > Reducing > Inventing > Zooming

Wire Recorder — Extraction – 2 — Extraction – 1 — Audio Tape

Prototype — Artifact

## Extraction – 1

| LC | No. | Navigator | Substantiation for Extraction |
|----|-----|-----------|-------------------------------|
| ++ | 01 | Change in the aggregate state ... | The relatively rigid wire is replaced with a thin plastic magnetic tape. |
| ++ | 03 | Segmentation | A new magnetic tape is segmented into a magnetic layer and a base layer. |
| ++ | 05 | Separation | The magnetic property of metal is separated out and deposited on thin and flexible base. |
| + | 10 | Copying | A copy of magnetic property of the Wire is placed on the Tape in a layer |
| ++ | 12 | Local property | The local property of the Tape is modified so that only the top surface accommodates the magnetic property |
| + | 17 | Use of composite materials | Homogeneous material is replaced by a composite of a plastic layer and a magnetic layer |
| ++ | 18 | Mediator | The plastic tape acts as a mediator to hold the extremely |
| + | 25 | Use of flex. covers and thin films | Thin flexible film is used to hold the magnetic material |
| + | 35 | Unite | The Plastic Film and Magnetic Film are made to unite together. |

## Extraction - 2

**Standard Contradiction:** A better wire material should improve the rusting and tangling problem but the reading head could find it difficult to read because of the new material property.

**Radical Contradiction:** Recording interface should be flexible to prevent tangling and breaking but it should be thick and inflexible (metallic wire property) because only metals are thought to be magnetized.

**FIM (shortly):** *No breakage, rusting and tangling.*

**fig. 9.47.** An Audio Cassette with tape (Lemelson, 1974).

## Magnetic Tape Transducing System (U.S. Patent No. 3,842,432)

**TREND:** It started off with an idea of recording telephone signals onto a length of steel piano wire. This is when wire recording of music and voice was born. By 1930, advances in electronics allowed the first commercially successful wire recorders to be introduced. The sound was recorded on thin wire by magnetizing it as the wire moved in a spool along the recording head. This is played back by passing the same magnetized spool of wire along the reader (Video Interchange, 2011).

*A Wire Recorder*

**PROBLEM:** The wires used to rust, break and tangle.

### REDUCING

**FIM:** X-resource, together with available or modified resources, and without making the object more complex or introducing any negative properties guarantees attainment of the following IFR (No breakage, rusting and tangling)

Standard (Technical) Contradiction Model

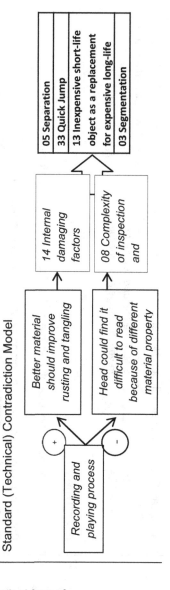

---

Radical (Physical) Contradiction Model

| Recording interface | ⇧ | Should be thin and flexible to prevent breakage and tangling | & | Should be thick and inflexible (metallic wire property) because only metals are thought to be magnetized |

An Audio Tape

### INVENTING

Navigator to model the idea of 'Flexible and durable recording interface' concept:

05: Separation: The magnetic property of metal is separated out and deposited on thin and flexible base.

03: Segmentation: Single magnetic wire is segmented into a magnetic layer and a base layer.

### ZOOMING

Are Contradictions removed? – Yes;   Super Effects: Spool size significantly reduced.

Negative Effects: None

### BRIEF DESCRIPTION

A Magnetic Tape was a thin magnetizable layer on a long, narrow strip of plastic covered with a protective coating. The tape moves past the recording head at a continuous constant rate. The recording head acquires the signal corresponding to the voice to be stored and this in turn magnetizes the tape as it passes over it. As the tape is played back, the magnetized pattern on surface of the tape is read by the head and it converts that into electrical signal which is further converted into sound signal as required.

**fig. 9.48 - end.** Reinventing in standard form of
Magnetic Tape Transducing System (U.S. Patent No. 3,842,432)

## 9.13.  Safety / Risk Mitigation:

### Advanced Reflectors (U.S. Patent No. 4,127,693)

### Prior Art (Engineering Level)

The reflectors with pyramid shaped indentations can be commonly seen today in bicycles as a warning devices. A person with a light source gets alerted or the cyclist's presence with this reflecting device. The reflectors were tradition-ally made of plastic reflecting material with pyramid shaped indentations on their surface. These pyramidal reflectors perform the function of reflecting the light back to its source by the principle of retro reflection, where the light inci-dent on the surface at any angle is reflected back to its source.

**fig. 9.49.** A Pyramidal Reflector (Flickr 2011)

Due to its importance, these reflectors find their use in several places like highway signs, median of roads and warning displays (Lemelson, 1978).

### Problems with the Prototype

Lemelson saw the following problems in the prototype. Since these "Pyramid Reflectors" were located in regions of high dust and dirt like highways, the gaps between the pyramidal indentations used to gather dust with the passage of time.

Although these irregularly shaped reflectors were the best shapes to serve the purpose, their design itself was a contradiction to the requirement as they ac-cumulated dust and dirt and their efficiency got affected over a period of time.

The problems of dust accumulation was big as it required very high mainte-nance but the system itself was hard to dump as the reflectivity was very high. (Lemelson, 1978).

### Result of Contradiction Analysis

The pyramid shaped indentations have very good reflectivity. This could be de-fined as the positive aspect in the contradiction existing within the problem of "Pyramidal reflectors". Hence this could be defined by TRIZ as "Shape".

By having such an aspect of pyramidal reflectors, one can comprehend the "dust accumulates in the system over time". This goes well with the TRIZ lan-guage of "Reliability" as the reliability of the system is affected since it does not anymore reflect properly, the incident light.

The Altshuller matrix yields three navigators for this contradiction: 02.Preliminary Action, 16.Partial / Excess Effect, 17.Composite Materials.

## Inventing Idea

One could develop the invention idea by using the navigator number 17. Composite Materials, that is, Applying "a layer of transparent resin on top of pyramid indentations ensures the same reflectivity level is maintained and eliminates dust accumulation problem owing to the smooth surface of the resin" (Lemelson, 1978).

## Resultant Artifact

### US Patent No. 4,127,693          Date of Patent: Nov 28, 1978

### Title: Reflex reflectors with pyramid-shaped indentations

### Inventors: Jerom H. Lemelson

**Abstract:**

The structures include molded, extruded or embossed plastic or glass which is formed with a plurality of irregular surface formations such as cavities, short or elongated protrusions defining irregular surface formations in or against which dirt or dust may collect and form light blocking material which substantially reduces the efficiency of the reflector or display.

The new and improved structures provided herein include the provision of a coating or covering sheet of suitable transparent material which retains the interface defined by the irregular surface formations in the lens sheet or reflector so that it may properly continue to function substantially as intended.

The coating is mainly a type of resin with high transparency and the final surface created on top of the pyramid indentations in smooth and very easy to maintain as no dust accumulates on this surface anymore' (Lemelson, 1978).

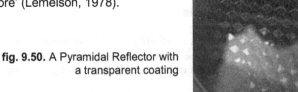

**fig. 9.50.** A Pyramidal Reflector with a transparent coating

**Extracting** of "Advanced Reflectors" is shown in fig. 9.51.

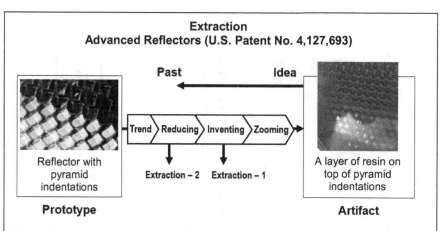

Extraction
**Advanced Reflectors (U.S. Patent No. 4,127,693)**

**Past**          **Idea**

Trend ⟩ Reducing ⟩ Inventing ⟩ Zooming ⟩

Reflector with
pyramid
indentations

Extraction – 2    Extraction – 1

A layer of resin on
top of pyramid
indentations

**Prototype**                              **Artifact**

## Extraction – 1

| LC | No. | Navigator | Substantiation for Extraction |
|----|-----|-----------|-------------------------------|
| ++ | 12 | Local property | The rough surface of the reflector is made smooth by a layer of resin. |
| ++ | 17 | Use of composite materials | The top layer of predominantly plastic reflector comprises transparent resin. |
| + | 18 | Mediator | Another material is used as mediator to prevent dust accumulation. |
| + | 19 | Transition into another dimension | A surface above the actual reflector is created. |
| + | 34 | Matryoshka | The Reflector is nested within the transparent resin layer. |
| + | 35 | Unite | The reflector and epoxy resin layer are made to unite. |
| ++ | 39 | Preliminary counter action | A layer of resin is applied beforehand to counteract the future dust formations. |

## Extraction - 2

**Standard Contradiction:** Having pyramidal shapes on the reflector is good for reflectivity but it accumulates dust over a period time.

**Radical Contradiction:** Pyramid indentations should be present on the reflector's surface as they increase the reflectivity and they should not be present on the surface as they accumulate dust over time.

**FIM:** X-resource, together with available or modified resources, and without making the object more complex or introducing any negative properties guarantees attainment of the following Ideal Final Result: *Prevention of dust accumulation without affecting the reflectivity.*

**fig. 9.51.** Advanced Reflectors (Lemelson, 1978).

## Advanced Reflectors (U.S. Patent No. 4,127,693)

**TREND:** Pyramid shaped structures are provided in reflecting display devices that are used in Highways or in Automobile lights. The structures are primarily irregular shaped such as cavities or protrusions in order to reflect back light falling from several angles towards the observer. But over prolonged time, the dust collects on this surface thereby reducing the efficiency of reflector.

**PROBLEM:** How can we prevent the dust accumulating in the crevice, without affecting the reflectivity?

### REDUCING

**FIM:** X-resource, together with available or modified resources, and without making the object more complex or introducing any negative properties guarantees attainment of the following IFR (Prevention of dust accumulation without affecting the reflectivity)

Standard (Technical) Contradiction Model

*Reflector with pyramid indentations*

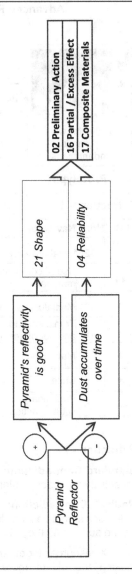

fig. 9.52 - beg. Reinventing in standard form of Advanced Reflectors (U.S. Patent No. 4,127,693)

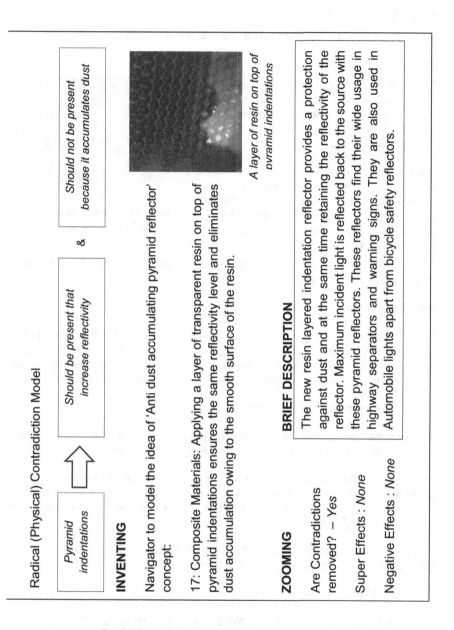

Radical (Physical) Contradiction Model

Pyramid indentations

&

Should be present that increase reflectivity

Should not be present because it accumulates dust

A layer of resin on top of pyramid indentations

## INVENTING

Navigator to model the idea of 'Anti dust accumulating pyramid reflector' concept:

17: Composite Materials: Applying a layer of transparent resin on top of pyramid indentations ensures the same reflectivity level and eliminates dust accumulation owing to the smooth surface of the resin.

## BRIEF DESCRIPTION

The new resin layered indentation reflector provides a protection against dust and at the same time retaining the reflectivity of the reflector. Maximum incident light is reflected back to the source with these pyramid reflectors. These reflectors find their wide usage in highway separators and warning signs. They are also used in Automobile lights apart from bicycle safety reflectors.

## ZOOMING

Are Contradictions removed? – Yes

Super Effects : None

Negative Effects : None

fig. 9.52 - end. Reinventing in standard form of Advanced Reflectors (U.S. Patent No. 4,127,693)

## 9.14.  Sport:

### Improved Roller Skate (U.S. Patent No. 4,273,345)

### Prior Art (Engineering Level)

Roller skating is a form of recreational activity which has evolved into a sport these days. The earliest known skate is from 1860's.Initially made from wood, the rollers advanced into Ball bearing type Rollers. They are thought to have been used initially in Holland, now it has evolved into a full-fledged sport. A more modern version of the sport involves banked tracks on an indoor track. These days In-line skates have taken over the older versions of skates.

**fig. 9.53.** An Inline Skate
(Wikimedia 2012)

They have rollers arranged in a line. This arrangement helps the used with a great deal of balance (Roller Skate, 2012).

### Problems with the Prototype

The challenge in using this skate is mainly in its control. They had in-line wheels and one could achieve great speeds riding it on a flat surface. This system needed further improvisation because slowing down was a real challenge in this easy-to-attain-high-speed device (Jerome H. Lemelson et al.,1981).

### Result of Contradiction Analysis

The positive requirement of the contradiction existing in the current Roller skate system could be defined as "Speed and ease of operation".

The negative that arises by having such a system is that "As the speed of the skate increases, the unreliability of the skate also increases' in terms of stopping or braking. TRIZ defines the positive requirement as "Ease of Use" and the negative "Reliability".

The Contradiction analysis gives the following four navigators – Self-servicing, Change in colour, Equipotentiality or Transition into another dimension.

### Inventing Idea

The Inventing idea could be achieved with the navigator "Transition into another dimension". Which means that some modification done on the other dimension of the skate results in the elimination of contradiction.

The side of the skate contains brake pads reaching close to the ground which is engaged with the ground when the user tilts the legs outwards. Another dimension – Front and back have pivot points to perform static acrobatics.

**Resultant Artifact**

**US Patent No. 4,273,345**          **Date of Patent: Jun 16, 1981**

### Title: Roller Skate

### Inventors: Jerom H. Lemelson, Effraim Ben-Dor

**Abstract:**

Roller skates which utilize a single row of narrow wheels to support each skate include housings in a blade-like support for the skate wheels.

A friction plate or bar is disposed along the side of the support and is dragged along the skating surface to stop movement of the skate.

A plastically deformable pivot is mounted at each end of the blade-like support to permit spinning movement thereon.

This way, the skate is also constructed to permit the user to perform in a manner similar to performance on ice skates including the performance of such acrobatics as reverse skating, turning or twirling on one foot, and a sliding movement in a lateral direction of the blade of the skate for effecting slowing down or stopping the movement of the skater (Lemelson et al, 1981).

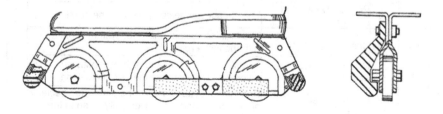

**fig. 9.54.** An Inline Skate with brakes and pivot – side and back views (Lemelson et al.,1981).

**Extracting** of "Improved Roller Skate" is shown in fig. 9.55.

**Extraction**
**Improved Roller Skate (U.S. Patent No.4,273,345)**

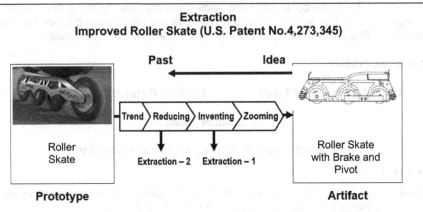

**Past**                                    **Idea**

Trend ⟩ Reducing ⟩ Inventing ⟩ Zooming

Roller
Skate

Extraction – 2     Extraction – 1

Roller Skate
with Brake and
Pivot

**Prototype**                                    **Artifact**

**Extraction – 1**

| LC | No. | Navigator | Substantiation for Extraction |
|----|-----|-----------|-------------------------------|
| + | 05 | Separation | The risk is taken out of the system by placing brakes on the sides of the skate |
| + | 12 | Local property | The free rolling is controlled by the friction of the brakes |
| + | 17 | Use of composite materials | The brakes are made of rubberized material, pivot a composite |
| ++ | 18 | Mediator | The brake disc acts as mediator to control the speed |
| + | 24 | Asymmetry | The brake pad is placed only on one side of the skate |
| + | 28 | Previously installed cushions | The impact can be cushioned by the brakes |
| ++ | 35 | Unite | The braking system and pivot system is united with the inline roller skate |
| + | 39 | Preliminary counter-action | The side brakes and end pivot are installed beforehand to eliminate the risk of crashing |
| + | 40 | Uninterrupted useful function | Continuous action can now be performed because of full control and maneuverability given to the skate |

**Extraction - 2**

**Standard Contradiction:** The inline roller skates should be fast and easy to operate but as the speed increases, the unreliability to stop also increases.

**Radical Contradiction:** The roller skates should move freely at one time but it should not move at another point of time when required.

**FIM (shortly):** *Complete braking and stationary acrobatics possible.*

**fig. 9.55.** Improved Roller Skate (Lemelson, 1981).

## Improved Roller Skate (U.S. Patent No.4,273,345)

**TREND:** Roller skating is a form of recreational activity which has evolved into a sport. Skates mounted on wooden rollers date from the 1860s, and soon wooden wheels replaced the rollers. The ball-bearing skates followed later. The skates had wheels – inline and one could achieve great speeds on a flat surface. The slowing down however was a big challenge as there was no device integrated into the system to enable that.

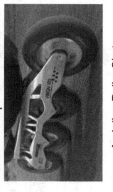

*An Inline Roller Skate*

**PROBLEM:** Hard to brake or perform stationary acrobatics.

### REDUCING

**FIM:** X-resource, together with available or modified resources, and without making the object more complex or introducing any negative properties guarantees attainment of the following IFR (Braking and stationary acrobatics possible)

Standard (Technical) Contradiction Model

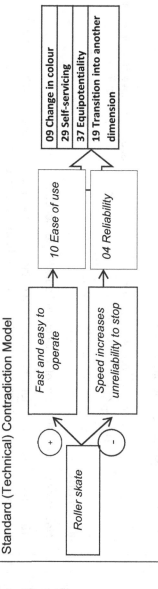

fig. 9.56 - beg. Reinventing in standard form of Improved Roller Skate (U.S. Patent No. 4,273,345)

Radical (Physical) Contradiction Model

| Skate Wheels |
| --- |

⬆

| Should move fast and free at one time | & | Should not move at another point of time |
| --- | --- | --- |

### INVENTING

Navigator to model the idea of 'Fully controllable Skate' concept:

19 : Transition into another dimension: Side of the skate contains brake pads reaching close to the ground. Front and back have pivot points to perform static acrobatics.

### ZOOMING

Are Contradictions removed? – *Yes*; Super Effects : *None*; Negative Effects : *None*

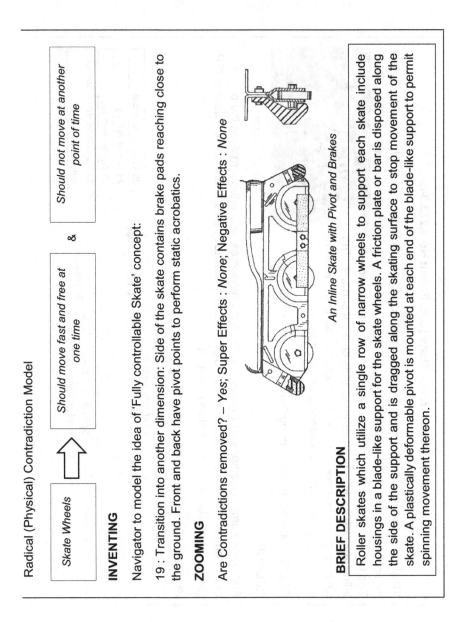

*An Inline Skate with Pivot and Brakes*

### BRIEF DESCRIPTION

Roller skates which utilize a single row of narrow wheels to support each skate include housings in a blade-like support for the skate wheels. A friction plate or bar is disposed along the side of the support and is dragged along the skating surface to stop movement of the skate. A plastically deformable pivot is mounted at each end of the blade-like support to permit spinning movement thereon.

**fig. 9.56 - end.** Reinventing in standard form of Improved Roller Skate (U.S. Patent No. 4,273,345)

## 9.15. Recreation:

### Towed Watercraft Steering Method (U.S. Patent No. 5,462,001)

**Prior Art (Engineering Level)**

A watercraft is a craft that carries a person is normally towed behind a motor boat. These include water skis, a single water ski, a surfboard, and other similar shaped boards. Such devices are controlled by shifting the person's weight that is standing on top of it. The maneuverability of the craft depends of the ability of the person who is riding it and his skills that might involve training (Lemelson, 1995).

**Problems with the Prototype**

For a learner, this system has always been a problem. At first, the person needs to learn to balance himself on the craft and then upon passage of time, he learns the steering mechanism by shifting his bodyweight to steer the craft in the required direction.

**fig. 9.57.** A towed Watercraft (Nellis AF 2012)

This learning curve is normally long and there is also a probability that the user might quit owing to the sophisticated body flexing the system requires. A new system which did not involve such complex movements was required.

**Result of Contradiction Analysis**

The required system of Watercraft must be easily steerable. But such a system could involve complex mechanisms. These positives and negatives could be represented as 'Ease of Use and "Complexity of Construction" respectively in TRIZ. The Altshuller Matrix yields the following results for this contradiction-Change in colour, Self-servicing, Equipotentiality, Transition into another dimension. The following section discusses about the inventing idea using one of these navigators.

**Inventing Idea**

The Idea for inventing the new and improved Watercraft could be obtained using the navigator "Transition into another dimension". The steering mechanism is shifted from "person" on top of the craft to the "rudder" on the bottom of the craft.

That is, the person rather than controlling the craft from the top with body motion, now controls the craft using kinetics of Rudder below the Craft (Lemelson, 1995).

**Resultant Artifact**

**US Patent No. 5,462,001**        **Date of Patent: Oct 31, 1995**

### Title: Roller Skate

### Inventors: Jerom H. Lemelson

**Abstract:**

A pullable, steerable watercraft is operable to be towed behind a power boat or kite.

The watercraft has a steering mechanism within the watercraft body, and has a dome-shaped steering control recessed into the top surface thereof.

The steering mechanism includes a movable mechanical shaft or arm which extends through the interior of the watercraft and which pivots about a vertical axis in order to rotate a rudder (Lemelson, 1995).

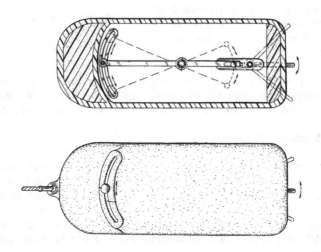

**fig. 9.58.** A steerable towed watercraft – Internal and external views
(Lemelson, 1995)

**Extracting** of "Towed Watercraft Steering Method" is shown in fig. 9.59.

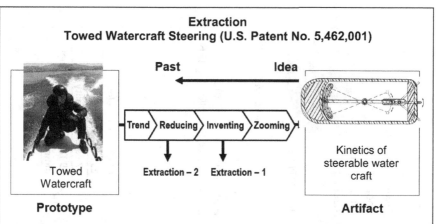

## Extraction
### Towed Watercraft Steering (U.S. Patent No. 5,462,001)

**Past**      **Idea**

Trend ⟩ Reducing ⟩ Inventing ⟩ Zooming

Extraction – 2    Extraction – 1

Towed Watercraft

Kinetics of steerable water craft

**Prototype**        **Artifact**

### Extraction – 1

| LC | No. | Navigator | Substantiation for Extraction |
|---|---|---|---|
| + | 05 | Separation | The physics of steering with body is taken out and a new steering mechanism is installed on this basis. |
| + | 07 | Dynamization | Parts of the craft (Rudder) are made dynamic to steer the craft. |
| ++ | 10 | Copying | A copy of Boat steering mechanism is installed on the craft. |
| ++ | 11 | Inverse action | Rather than controlling the turning of craft from above (motion of user), it is controlled from below (Rudder). |
| + | 18 | Mediator | Rather than the mediator 'person' steering the craft, the mediator 'Rudder' steers it now. |
| ++ | 19 | Transition Into another dimension | The steering mechanism is shifted from 'person' on top of the craft to the 'Rudder' on the bottom of the craft. |
| + | 35 | Unite | The Watercraft is united with a steering Rudder. |

### Extraction - 2

**Standard Contradiction:** The watercraft should be easily steerable but it involves complex mechanism if rider controls it with his body movement.

**Radical Contradiction:** The person should move the body as craft needs to be steered but he should not move the body as steering involves complex moves.

**FIM:** X-resource, together with available or modified resources, and without making the object more complex or introducing any negative properties guarantees attainment of the following Ideal Final Result: *Craft steered without user's body movement.*

**fig. 9.59.** Towed Watercraft Steering Method (Lemelson, 1995).

## Towed Watercraft Steering (U.S. Patent No. 5,462,001)

*Towed Watercraft*

**TREND:** A watercraft is a craft that carries a person is normally towed behind a motor boat. These include water skis, a single water ski, a surfboard, and other similar shaped boards. These devices are controlled by shifting the weight of the user. Such control depends upon the ability of the passenger to stand on the craft and shift weight and varies with the speed of the motor boat.

**PROBLEM:** Watercraft challenging to steer. It is steered with body weight tilting and so only experienced can control the craft.

**REDUCING**

**FIM:** X-resource, together with available or modified resources, and without making the object more complex or introducing any negative properties guarantees attainment of the following IFR (Craft steered without user's body movement)

Standard (Technical) Contradiction Model

Watercraft

(+) Easily steerable craft → 10 Ease of use

(–) Involves complex mechanisms → 07 Complexity of construction

09 Change in colour
29 Self-servicing
37 Equipotentiality
19 Transition into another dimension

**fig. 9.60 - beg.** Reinventing in standard form of
Towed Watercraft Steering Method (U.S. Patent No. 5,462,001)

Radical (Physical) Contradiction Model

The person

↑

Should move the body as craft needs to be steered

&

Should not move the body as steering involves complex moves

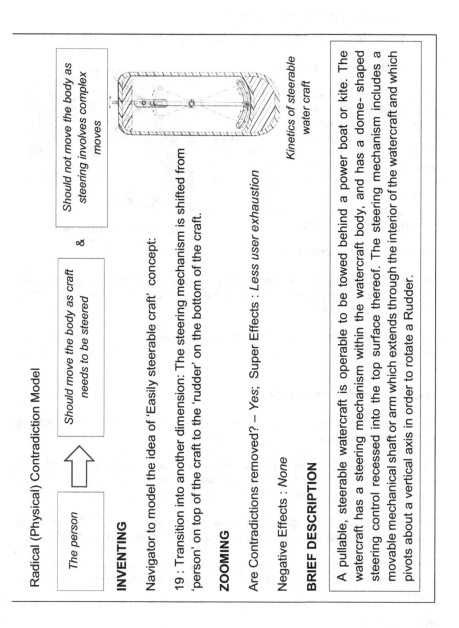

Kinetics of steerable water craft

### INVENTING

Navigator to model the idea of 'Easily steerable craft' concept:

19 : Transition into another dimension: The steering mechanism is shifted from 'person' on top of the craft to the 'rudder' on the bottom of the craft.

### ZOOMING

Are Contradictions removed? – Yes;  Super Effects : Less user exhaustion

Negative Effects : None

### BRIEF DESCRIPTION

A pullable, steerable watercraft is operable to be towed behind a power boat or kite. The watercraft has a steering mechanism within the watercraft body, and has a dome- shaped steering control recessed into the top surface thereof. The steering mechanism includes a movable mechanical shaft or arm which extends through the interior of the watercraft and which pivots about a vertical axis in order to rotate a Rudder.

**fig. 9.60 - end.** Reinventing in standard form of
Towed Watercraft Steering Method (U.S. Patent No. 5,462,001)

## 9.16.  Service and Maintenance:

### Wheel Inspection Apparatus (U.S. Patent No. 3,148,535)

### Prior Art (Engineering Level)

In the years before 1960, the inspection of an automotive Wheel was done manually. Every time a wheel needed to be inspected, the Car had to go to the workshop for the wheel to be removed manually. After 'jacking up' the car and loosening the Bolts, the wheel would come loose and it is then carried from station to station depending on the nature of inspection it needed (Lemelson et al,1964).

fig. 9.61. A manual wheel removal in progress (Honda Tuning 2012)

### Problems with the Prototype

The total time consumed to perform the checks on all the wheels was very high. Apart from being labour intensive, the customers had to wait too long or pick up the car at a later point of time.

Additional wheel balancing and adjustment was needed every time the wheel was fixed back. A new system to eliminate all these challenges was in demand.

### Result of Contradiction Analysis

The contradiction analysis could be structured in this problem of Wheel Inspection System as follows.

The "plus" in the system would be an "An easy to use simple device for inspection without removal of Wheel" but by doing so, we are faced with a "minus" contradiction that the "Simple system could affect proper inspection".

To convert it in the language of TRIZ, the "plus" transforms into 'Complexity of construction' and the 'minus' transforms into "Complexity of inspection".

The Altshuller Matrix yields the navigators to be used as – Dynamization, Preliminary action, Full use of thermal expansion and Replacement of mechanical matter.

### Inventing Idea

The idea for the resultant artifact could be obtained with the navigators Dynamization and Preliminary action.

Dynamization navigator contributes to the system when the Stationary interface holding the Wheels is made dynamic. Navigator – Preliminary Action also contributes to the inventing idea.

An automatic system to inspect the wheels is constructed in advance. This enables inspection without the removal of Wheels.

**Resultant Artifact**

**US Patent No. 3,148,535       Date of Patent: Sep 15, 1964**

**Title: Wheel Inspection Apparatus**

**Inventors: Jerom H. Lemelson et al**

**Abstract:**

The following diagram illustrates the sequence of movements required for prepositioning a vehicle in apparatus for automatically inspecting all four rubber tyres for internal flaws, cracks or tears. The vehicle is driven over a lifting fixture which comprises a hydraulic cylinder rigidly secured in the floor and is adapted when operated by a remote control to raise the ramp which engages the frame, axles or body of the vehicle and lifts it up.

Each wheel can be precisely lowered into its tanks which contain rollers and a plurality of inspection devices. The tanks are positioned in a way as to accommodate many types of car wheels and varying width between its two wheels.

Once the vehicle is lowered into the water, all the four rollers rotated simultaneously, making the wheels rotate into the water. Once lowered, the water level inside the tank is maintained so that the inspection transducers mounted on opposite walls of the tank may transmit ultrasonic waves through the water. This way all the Wheels could be inspected at once, without its removal from the vehicle (Abstract adapted from Lemelson et al,1964).

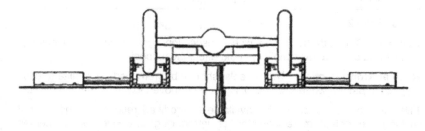

**fig. 9.62.** A simultaneous wheel inspection device (Lemelson et al,1964)

**Extracting** of "Wheel Inspection Device" is shown in fig. 9.63.

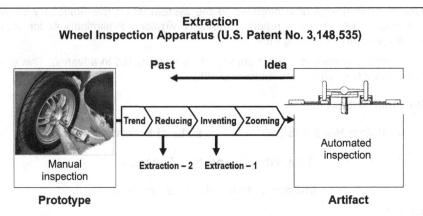

**Extraction**
**Wheel Inspection Apparatus (U.S. Patent No. 3,148,535)**

**Past**                                    **Idea**

Trend > Reducing > Inventing > Zooming

Manual inspection          Extraction – 2   Extraction – 1          Automated inspection

**Prototype**                                    **Artifact**

**Extraction – 1**

| LC | No. | Navigator | Substantiation for Extraction |
|----|-----|-----------|-------------------------------|
| + | 01 | Change in the aggregate state of an | Individual inspection devices are combined together to form on comprehensive testing |
| ++ | 02 | Preliminary action | An automatic system to check the wheels is constructed in advance. This enables inspection without the removal of wheels. |
| ++ | 07 | Dynamization | Stationary interface holding the Wheels is made dynamic (rollers), on which wheels revolve |
| ++ | 11 | Inverse action | The Wheel is left to stay with its system and the Machine is flexible to take the whole Car. This was the other way in prototype artifact. |
| + | 14 | Use of pneumatic or hydraulic constructions | Hydraulic systems help the wheel stay in its system. |
| + | 19 | Transition into another dimension | Rather than inspecting the car at ground level, it is lifted to a higher dimension in space for inspection. |
| + | 35 | Unite | Inspecting of all four Wheels are united and run in parallel. |

**Extraction - 2**

**Standard Contradiction:** The wheel inspection system should be easy to handle and use but this simplicity could affect the inspection.

**Radical Contradiction:** The wheels should be removed to facilitate easy inspection but the wheels should also not be removed to save time and effort.

**FIM:** X-resource, together with available or modified resources, and without making the object more complex or introducing any negative properties guarantees attainment of the following Ideal Final Result: *Inspection completed without the removal of Wheels.*

**fig. 9.63.** Wheel Inspection Device (Lemelson, 1964).

## Wheel Inspection Apparatus (U.S. Patent No. 3,148,535)

*A Manual Wheel Inspection*

**TREND:** Before 1960s, the inspection of an automotive Wheel was done manually. Every time a wheel needed to be inspected, the car had to go to the workshop for the wheel to be removed manually. After 'jacking up' the car and loosening the bolts, the wheel would come loose and it is then carried from station to station depending on the nature of inspection it needed

**PROBLEM:** Automotive Wheels have to be removed every time they need to be checked.

### REDUCING

**FIM:** X-resource, together with available or modified resources, and without making the object more complex or introducing any negative properties guarantees attainment of the following IFR (Inspection completed without the removal of Wheels)

Standard (Technical) Contradiction Model

Wheel inspection system

(+) → An easy to use, simple device for inspection without removal of Wheel → *07 Complexity of construction*

(–) → Simplicity could affect proper inspection → *08 Complexity of inspection*

→ 07 Dynamization
02 Preliminary action
27 Full use of thermal expansion
04 Replacement of mechanical matter

**fig. 9.64 - beg.** Reinventing in standard form of Wheel Inspection Apparatus (U.S. Patent No. 3,148,535)

Radical (Physical) Contradiction Model

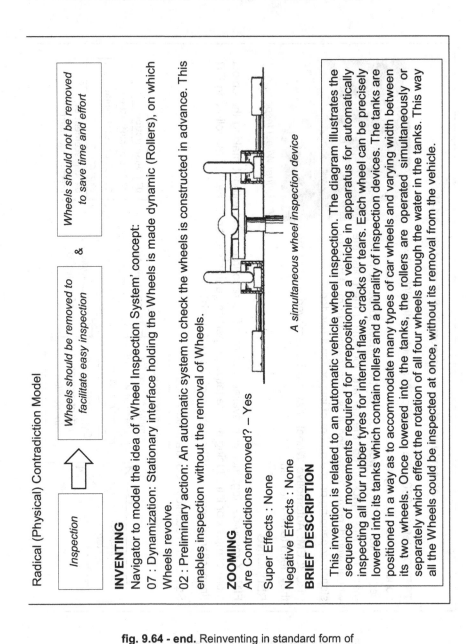

*Inspection*

⟹

*Wheels should be removed to facilitate easy inspection*

&

*Wheels should not be removed to save time and effort*

**INVENTING**

Navigator to model the idea of 'Wheel Inspection System' concept:

07 : Dynamization: Stationary interface holding the Wheels is made dynamic (Rollers), on which Wheels revolve.

02 : Preliminary action: An automatic system to check the wheels is constructed in advance. This enables inspection without the removal of Wheels.

**ZOOMING**

Are Contradictions removed? – Yes

Super Effects : None

Negative Effects : None

*A simultaneous wheel inspection device*

**BRIEF DESCRIPTION**

This invention is related to an automatic vehicle wheel inspection. The diagram illustrates the sequence of movements required for prepositioning a vehicle in apparatus for automatically inspecting all four rubber tyres for internal flaws, cracks or tears. Each wheel can be precisely lowered into its tanks which contain rollers and a plurality of inspection devices. The tanks are positioned in a way as to accommodate many types of car wheels and varying width between its two wheels. Once lowered into the tanks, the rollers are operated simultaneously or separately which effect the rotation of all four wheels through the water in the tanks. This way all the Wheels could be inspected at once, without its removal from the vehicle.

**fig. 9.64 - end.** Reinventing in standard form of
Wheel Inspection Apparatus (U.S. Patent No. 3,148,535)

## 9.17. Apparel:

### Reflective Thread (U.S. Patent No. 3,050,824)

#### Prior Art (Engineering Level)

Nighttime pedestrian safety is a real challenge. The pedestrians, who normally use the streets, are subject to such risk because of the non-reflective clothing they wear. One solution to the problem of nighttime pedestrian safety has been to attach reflectors to the pedestrian's clothing. The light reflected from the reflectors would alert the driver of the person's presence. Another way to serve this purpose was to coat the clothes with glass bead particles that reflect off the light (Lemelson et al., 1962).

**fig. 9.65.** A suit with reflective stickers (Ecouterre 2012)

#### Problems with the Prototype

The outer clothing of the average pedestrian is usually made of a dark material which absorbs most of the light rather than reflecting it. As a result, a motor vehicle operator travelling at only moderate speeds will often not be able to see a pedestrian walking or standing in his path of travel in time to stop or turn his vehicle.

While the above mentioned reflectorized cloth does serve the purpose as a reflecting medium, it has a number of disadvantages. In the first place, it is not suitable for use as wearing apparel due to the stiffness, weight and appearance of the coated sheet. In addition, the small glass spheres are easily dislodged when the garment is cleaned or washed, and as a result, this reflectorized cloth has been limited in use to that of a reflective cape or the like (Lemelson et al., 1962).

#### Result of Contradiction Analysis

The contradiction analysis for this problem could be performed as here below. A dress should be simple and easy to produce. However, that simple normal dress would not be able to hold the light reflecting capability. Hence the "plus" could be defined as avoiding "Ease of Manufacture" and the "minus" as "Brightness of Lighting". The A-matrix gives away the following navigator recommendations – Replacement of mechanical matter, Mediator, Inexpensive short-life object as a replacement for expensive long-life and Segmentation.

#### Inventing Idea

The inventing idea could be obtained by using the navigator – Mediator. Microscopic glass beads with high refractive indices are used as mediator in the threads of clothes to perform the function of reflecting the light in the dark. A dress made out of such a thread is completely reflective and eliminated all the previously described contradictions.

## Resultant Artifact

### US Patent No. 3,050,824          Date of Patent: Aug 28, 1962

### Title: Reflective Thread

### Inventors: Jerom H. Lemelson et al

## Abstract:

The reflex reflecting elements of this invention are preferably small spheres or beads having a diameter of between about 1 and 50 mm, depending primarily upon the diameter of the thread. These spheres or beads must be made of glass, but ordinary glass would not be suitable because of its refractive index of 1.5 or less (Refractive index of 1.5 to 2.0 is considered good). The Cataphot Corporation of Toledo4, Ohio manufactures reflex reflecting beads having a refractive index in excess of 1.85 and within the aforesaid preferred diameter range. The reflex reflective elements are dispersed at random throughout the textile element, particularly if dispersed in sufficiently dense concentrations, but more preferably are arranged in a circular or helical pattern or array about the longitudinal axis of the thread or filament (Lemelson et al., 1962).

## Integrating the Reflex Reflecting Elements with the Textile Element

The reflex reflecting elements may be dispersed or integrated with the textile element in a number of ways. One way of joining these two elements is to form an extrusion mixture comprising the plastic material and the reflective beads. Alternatively, the plastic alone is extruded first and then the reflective beads are made to adhere to the surface. A coating over the beads (such as with polyvinyl chloride, rubber hydrochloride, methyl methacrylate) is provided for its protection (Lemelson et al., 1962).

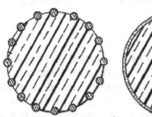

| | |
|---|---|
| **fig. 9.66.** Cross section of thread with embedded glass beads (Lemelson et al, 1962). | **fig. 9.67.** A sweat shirt made from reflective thread (Night Gear 2012) |

**Extracting** of "Reflective Thread" is shown in fig. 9.68.

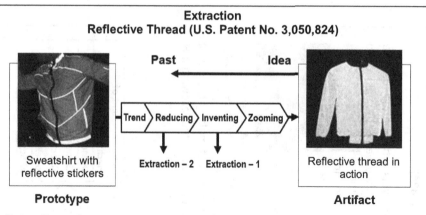

**Prototype**                                                   **Artifact**

### Extraction – 1

| LC | No. | Navigator | Substantiation for Extraction |
|---|---|---|---|
| + | 02 | Preliminary action | Reflective Glass Beads are placed inside the thread rather than placing stickers on the Clothes |
| ++ | 05 | Separation | The reflective property of the Stickers are taken out, miniaturized and located in Thread |
| + | 10 | Copying | Copy of the reflective property is placed inside the Thread. |
| + | 11 | Inverse action | Rather than manufacturing the Clothes and placing the reflector on it, the Thread is embedded with several reflecting elements during manufacture. |
| ++ | 12 | Local property | To achieve the reflective ability, the property of the Thread is changed |
| ++ | 17 | Use of composite mater. | Rather than using homogeneous material on thread, it is combined with reflecting Glass Beads and the glue base. |
| ++ | 18 | Mediator | A layer of microscopic Glass Beads are used in Thread as mediators to reflect light. |
| + | 34 | Matryoshka | The reflective beads are nested within the Thread and base composite. |
| ++ | 35 | Unite | The reflective Glass pieces are combined with the thread. |

### Extraction - 2

**Standard Contradiction:** Non-reflective clothing is simple to produce but its light reflective capability at night is compromised.

**Radical Contradiction:** The cloth should be simple to enable it to be manufactured easily but it should also be sophisticated to possess the light reflective ability.

**FIM (shortly):** *Cloth by itself reflects light.*

**fig. 9.68.** Reflective Thread (Lemelson, 1962).

## Reflective Threads  (U.S. Patent No. 3,050,824)

**TREND:** Nighttime pedestrian safety is a real challenge. The pedestrians, who normally use the streets, are subject to such risk because of the non-reflective clothing they wear. One solution to the problem of nighttime pedestrian safety has been to attach one or more reflectors to the clothes. It is however not always suitable to be used in wearing apparel due to the stiffness, weight and appearance of the coated sheet.

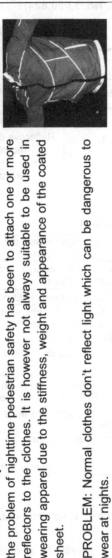

*A sweatshirt with reflective stickers*

**PROBLEM:** Normal clothes don't reflect light which can be dangerous to wear at nights.

**REDUCING**

**FIM:** X-resource, together with available or modified resources, and without making the object more complex or introducing any negative properties guarantees attainment of the following IFR (Cloth by itself reflects light).

Standard (Technical) Contradiction Model

| 04 Replacement of mechanical matter |
| 18 Mediator |
| 13 Inexpensive short-life object as a replacement for expensive long-life |
| 03 Segmentation |

*09 Ease of manufacture*

*35 Brightness of lighting*

*Simple to produce*

*Light reflection capability affected*

*Non reflective clothing*

**fig. 9.69 - beg.** Reinventing in standard form of Reflective Thread (U.S. Patent No. 3,050,824)

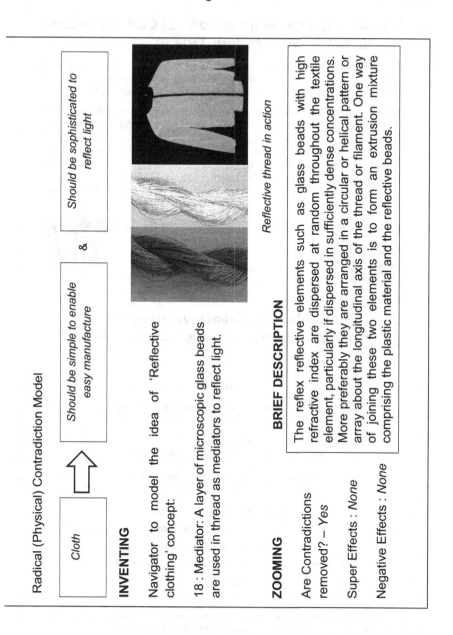

Radical (Physical) Contradiction Model

| Cloth | ⟹ | *Should be simple to enable easy manufacture* | & | *Should be sophisticated to reflect light* |

## INVENTING

Navigator to model the idea of 'Reflective clothing' concept:

18 : Mediator: A layer of microscopic glass beads are used in thread as mediators to reflect light.

*Reflective thread in action*

## BRIEF DESCRIPTION

The reflex reflective elements such as glass beads with high refractive index are dispersed at random throughout the textile element, particularly if dispersed in sufficiently dense concentrations. More preferably they are arranged in a circular or helical pattern or array about the longitudinal axis of the thread or filament. One way of joining these two elements is to form an extrusion mixture comprising the plastic material and the reflective beads.

## ZOOMING

Are Contradictions removed? – *Yes*

Super Effects : *None*

Negative Effects : *None*

**fig. 9.69 - end.** Reinventing in standard form of Reflective Thread (U.S. Patent No. 3,050,824)

# 10 Modeling the Great Inventions of Carl and Werner von Siemens with Modern TRIZ [159]

"For someone who is serious and prepared to act, the phase 'I want' is filled with magical power! Indeed, one should never shy away from obstacles and setbacks and never lose sight of one's goals!" – 1854 [160]

"I have always had more express interest in the present and the future, than in the past. I also think that it may prove more useful and stimulating for the coming generation if I can convincingly demonstrate that a young man without inherited resources or influential guardians, moreover, without proper prior training, relying exclusively on himself, can rise and do something useful in the world." – Harzburg, June 1889 [161]

*Werner von Siemens*

### Technische Universität Berlin

*"Wir haben die Ideen für die Zukunft"*

Global Production Engineering
International Masters Program

## Modeling the Great Inventions of Carl and Werner von Siemens with Modern TRIZ

by

**YANG ZHAO**

A thesis submitted in partial fulfillment of the requirements for the degree of
'Master of Science in Global Production Engineering'

Under the guidance of
**Professor, Dr. Dr. Sc. techn. Michael Orloff**

**Fakultät V – Verkehrs- und Maschinensysteme**
Produktionstechnisches Zentrum

**Technische Universität Berlin**

**Global Production Engineering**

**Dean: Prof. Dr.-Ing. Günther Seliger**

**Department of Assembly Technology and Factory Management**

**Berlin, Germany**

---

[159] Part 10 is based on the edited and shortened fragments of the master thesis by Yang Zhao (China), a certified graduate of MTRIZ course at GPE Program, TUB

[160] excerpts from Special Issue 160 Years of Siemens, October 2007, SIEMENS Internet Archive (www.siemens.com)

[161] quotation with translation from Werner von Siemens (Hrsg. Prof. Wilfried Feldenkirchen) *Lebenserinnerungen.* – Piper Verlag, München, 2008

## 10.1. Overview of the Siemens and the Historical Inventions of Brothers Siemens

In this chapter, the history of the Siemens Company and the founder the company Brothers Siemens will be generally introduced first, and after that the most important inventions by Brothers Siemens will also be explained briefly.

### 10.1.1. The History of Siemens

Siemens AG is a German engineering conglomerate, the largest of its kind in Europe. Siemens has international headquarters located in Berlin, Munich and Erlangen. The company has three main business sectors: Industry, Energy, and Healthcare; with a total of 15 divisions.

Generally speaking, the most important events in over 160 years of the company's development can be summarized as follows:

1847 – The founding of the company: 31-year-old Werner von Siemens and university mechanical engineer Johann Georg Halske establish the Telegraphen-Bauanstalt von Siemens & Halske (Telegraph Construction Company of Siemens & Halske) to manufacture pointer telegraphs. The company begins operations on October 12, 1847 in a back building in Berlin. For decades, the largest share of total sales is generated by the production of electrical telegraphs.

**fig. 10.1 (left).** Johann Georg Halske (b. July 30, 1814, Hamburg – d. March 18, 1890, Berlin)

**fig. 10.1 (right).** Sir William Siemens in 1850

1855 – Founding of a Russian subsidiary: starting in 1853, Siemens & Halske begins expanding the Russian telegraph network. In 1855, Siemens & Halske establishes its first foreign agency in St. Peterburg, headed by Carl von Siemens, the brother of the company founder.

1858 – Founding of an independent English subsidiary: in 1850, 3 years after the company is founded, a sales office is opened in England, agented by Sir William Siemens, a brother of the company's founder. In 1858, the London office is converted into an independent company – Siemens, Halske & Co. headed by Sir William Siemens (see in fig. 10.1).

1863 – First Siemens' cable plant in Woolwich on the Thames: Siemens, Halske & Co. opens its own cable plant in Woolwich near London in early 1863 in order to avoid dependence on the quality and prices of English cable suppliers.

1890 – Werner von Siemens transfers the business to his successors: when Siemens & Halske is transformed into a joint-stock company in 1890, Werner von Siemens retires from the operational management of the family company. Henceforth, his brother Carl and his sons Arnold and Wilhelm will be responsible for company's growing business.

1891 – Introduction of the 8.5-hour workday: after Siemens & Halske in Berlin introduces the 9-hour workday in 1873, daily working hours are reduced again by 30 minutes in 1891.

1897 – Siemens & Halske becomes a publicly listed company: the company is listed on the stock exchange for the first time in 1897 to get a broader financial basis for continuing growing and remaining competitive.

1903 – Founding of Siemens-Schukertwerke GmbH: in March 1903, the heavy-current divisions of Siemens & Halske are merged with Schuckert & Co., a joint stock company in the electrical engineering field, to form Siemens-Schuckertwerke GmbH.

1914 – Berlin Siemensstadt: to secure its expansion at its traditional location, Siemens & Halske acquires a virtually uninhabited piece of land northwest or Berlin in 1897.

**fig. 10.2 (above).** Siemens' first cable plant in Woolwich in 1863

**fig. 10.2 (middle).** Shares of Siemens

**fig. 10.2 (below).** Berlin Siemensstadt in 1914

1919 – Carl Friedrich von Siemens becomes head of the company: Carl Friedrich von Siemens, the third and youngest son of the company's founder, takes the helm.

1919 – Founding of Osram GmbH KG (see in fig. 10.3): German leading light bulb manufactures combine to form Osram GmbH KG, in which Siemens holds a 40 percent stake.

1923 – Establishment of Fusi Denki Seizo in Japan: to systematically tap the Asian market, Siemens-Schuckertwerke establishes a joint venture – Fusi Denki Seizo KK (see in fig. 10.3) – with Furukawa of Japan in August 1923.

**fig. 10.3 (left):** Osram GmbH KG

**fig. 10.3 (right):** Fusi Denke Seizo KK

1924 – Beginning of assembly line production: since 1924, Siemens promotes the standardization and reorganization of processes toward assembly line production.

1928 – Establishment of Siemens-Planiawerke AG and Vereinigte Eisenbahn-Signalwerke GmbH: in 1928, Gebrüder Siemens & Co. is merged with the Planis plant of Rütgerswerke AG to form Siemens-Planiawerke AG für Kohlefabrikate. In the same year, Siemens joins forces with AEG to establish Vereinigte Eisenbahn-Signalwerke GmbH.

1932 – Founding of Siemens-Reiniger-Werke AG: in 1932, the Siemens-Reiniger-Werke AG is formed to bring all Siemens' medical engineering activities under one roof.

1941 – Hermann von Siemens takes charge of the company: Hermann von Siemens, a grandson of Werner von Siemens, becomes chairman of the Supervisory Board.

**fig. 10.4.** Hermann von Siemens

1957 – Establishment of Siemens-Electrogeräte AG: the Siemens management spins off the company's home appliance and entertainment electronics activities to form a new company, Siemens-Electrogeräte AG.

1966 – Founding of Siemens AG: with a view to technological progress and structural change in the global electricity market, Siemens & Halske, Siemens-

Schukertwerke and Siemens-Reiniger-Werke are joined together legally and organizationally. Siemens AG is established on October 1, 1966. This move lays the basis for a successful repositioning of the expanding electrical concern.

1967 – Establishment of Bosch-Siemens Hausgeräte GmbH.

In 1967, in order to increase the competitiveness of the products in and outside Germany, Siemens and Bosch agree to bundle their consumer product activities in Bosch-Siemens Hausgeräte GmbH (BSHG).

1969 – Founding of Kraftwerk Union AG: in 1969, Siemens and AEG establish Kraftwerk Union AG.

1979 – New location in Munich-Perlach: a "think tank for data technology" is set up in Munich-Perlach between 1971 and 1984 to boost the company's international competitiveness.

**fig. 10.5 (above).** Shares of Siemens AG in 1966

**fig. 10.5 (below).** New location in Munich-Perlach

1989 – Reorganization: in 1989, the seven company-wide Groups are reorganized into 15 more manageable operating units, each with its own clear profile. This decentralized structure provides the basis for further market success in the age of globalization.

1994 – Expansion of business in China: with the aim of coordinating all activities in the growth market of China, Siemens establishes Siemens Ltd. China in 1994.

2001 – Launch of the Siemens share on the New York Stock Exchange: the Siemens share, which has been traded on the German stock exchange since March, 8, 1899, is listed of the first time on the New York Stock Exchange (NYSE) on March 12, 2001.

2008 – New corporate structure: the operating business is divided at the start of 2008 into three sectors: Industry, Energy and Healthcare. The Managing Board is adapted to match the new structure and at the same time is reduced from eleven to eight members.

The brief information about Siemens can be seen in Table 10.1.

Table 10.1. Siemens AG

| Type | Aktiengesellschaft * | Industry | Conglomerate |
|------|----------------------|----------|--------------|
| Founded | 1847 in Berlin | Founder | Werner von Siemens |
| Headquarters | Munich, Germany | Area served | Worldwide |
| Products | Communication systems, power generation technology, industrial and buildings automation, lighting, medical technology, railway vehicles, water treatment systems, home appliances, fire alarms, PLM software | Services | Business services, financing, project engineering and construction |
| Revenue | €75.98 billion (2009/2010) | Operating income | €5.916 billion (2009/2010) |
| Net income | €3.899 billion (2009/2010) | Total assets | €102.83 billion (September 2010) |
| Total equity | €29.07 billion (September 2010) | Employees | 405,000 (September 2010) |
| Divisions | Industry Sector, Energy Sector, Healthcare Sector & Megastructure and Cities Sector | Website | www.siemens.com |

* Aktiengesellschaft is a German term that refers to a corporation that is limited by shares, i.e. owned by shareholders, and may be traded on a stock market.

Nowadays, Siemens is a global powerhouse with activities in nearly 190 regions, which enables the company to offer customers fast, local, tailor-made solutions. About 336,000 employees work at 1,640 locations around the globe, including 176 R&D facilities.

### 10.1.2. Brief Introduction of Brothers Siemens

The Brothers Siemens are sons of Christian Ferdinand Siemens (31 July 1787 – 16 January 1840) and wife Eleonore Deichmann (1792 – 8 July 1839).

The family tree of Siemens family can be seen in fig. 10.6.

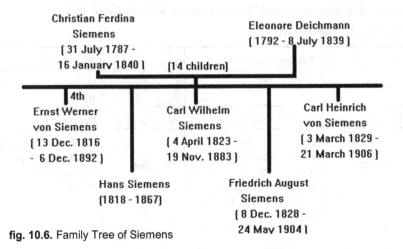

| Christian Ferdina Siemens [ 31 July 1787 - 16 January 1840 ] | | Eleonore Deichmann [ 1792 - 8 July 1839 ] |
|---|---|---|
| | [14 children] | |

4th
**Ernst Werner von Siemens** [ 13 Dec. 1816 - 6 Dec. 1892 ]

**Carl Wilhelm Siemens** [ 4 April 1823 - 19 Nov. 1883 ]

**Carl Heinrich von Siemens** [ 3 March 1829 - 21 March 1906 ]

Hans Siemens [1818 - 1867]

Friedrich August Siemens [ 8 Dec. 1828 - 24 May 1904 ]

**fig. 10.6.** Family Tree of Siemens

Ernst Werner von Siemens & Carl Wilhelm Siemens will be introduced in details individually in following.

### 10.1.2.1. Carl Siemens

Carl Wilhelm Siemens (en: Charles William Siemens, known as Sir William Siemens) (4 April 1823 – 19 November 1883) was an engineer born in Germany, the village Lenthe, today part of Gehrden, near Hanover, and spent most of his life working in Britain.

**fig. 10.7.** Carl Wilhelm Siemens

William began his studies to become an engineer when he was fifteen. He attended a highly respected School of Trade and Commerce, Magdeburg Gewerbe Schule.

After he had completed his course at Magedeburg school, William went on to Goettingen University. There he attended lectures on physical geography and technology, high mathematics, theoretical chemistry and practical chemistry and physics, etc. He also worked with the renowned scientist and inventor, Wilhelm Weber, in his Magnetic Observatory.

William became an apprentice engineer when he left the university at nineteen. His progress in the engineering factory was so rapid that his two years apprenticeship was cut down to one.

Due to the financial worry on educating the younger members of the family, Wilhelm left for London on 10 March, 1843. He was acting as an agent for his brother Werner and hoped to earn enough money by selling a patent in England to help support the education of his many brothers and sisters.

Wilhelm was trained to be a mechanical engineer, and most of his important work at the early stage was non-electrical. The greatest achievement of his life was the regenerative furnace. He discarded the older notions of heat as a substance, but accepted it as a form of energy. Working with this new thought, Wilhelm gained advantage over other inventors at his time and made his first attempt to economize heat. In 1847, at the factory of John Hich, of Bolton, he constructed a four horse-power engine. It had a condenser provided with regenerators and utilized the superheated steam. But the use of superheated steam was attended with many practical difficulties, and the invention was not entirely successful. In 1850, the Society of Arts acknowledged the value of the principle and awarded Wilhelm a gold medal for his regenerative condenser.

Since 1859 Wilhelm devoted a great part of his time on the field of electrical invention and research. In 1872, he became the first President of the Societ of Telegraph Engineers which became the Institution of Electrical Engineers, the forerunner of the Institution of Engineering and Technology. In June 1862, he was elected a Fellow of the Royal Society and in 1871 delivered their Bakerian Lecture.

The electric pyrometer, perhaps the most elegant and original of all Wilhem's inventions, is the link connecting his electrical with his metallurgical researches. He pursued two major themes in his inventive efforts, one based on the science of heat while the other on the science of electricity; and the electric thermometer was a delicate cross-coupling which connected both.

Wilhelm was later knighted – becoming Sir William – a few months before his death. He died on the evening of Monday, 19th November, 1883, and was buried on Monday, 26th November in Kensal Green Cemetery.

### 10.1.2.2. Werner Siemens

Werner Siemens ,the "von" was added in 1888, was born on December 13th, 1816, the forth of 14 children of a tenant farmer in Lenthe near Hanover. His family's tight financial situation precluded the formal schooling appropriate to upper middle-class ambitions, so that Werner Siemens left "Gymnasium" without taking the final exams in 1834 in order to gain access to engineering training by joining the Prussian army.

**fig. 10.8.** Ernst Werner von Siemens in 1880

His three-year training at the artillery and engineering school in Berlin created a solid foundation for his future work in the field of electrical engineering.

**fig. 10.9** (above). Siemens' house in Lenthe where Werner von Siemens was born.

**fig. 10.9** (below). A monument in honour of Werner von Siemens in the village where he was born.

In 1847, Werner Siemens constructed a pointer telegraph, which aimed to provide rapid and reliable communication for the army; it was absolutely more reliable and far superior to previous equipment.

And this invention also laid the foundation for the "Siemens & Halske Telegraph Construction Company" that later founded by Werner Siemens jointly with the master mechanic Johann Georg Halske in Berlin in 1st October, 1847.

The company mainly manufactured and repaired telegraphs, and it built offices in Berlin, London, Paris, St. Peterburg, and other major cities. In 1849, Werner Siemens left the army in order to fully dedicate himself into the company business.

In addition to business activities, Werner Siemens also dedicated himself into scientific research. In 1866 he discovered the dynamo-electric principle, which probably was the most important contribution he made in the field of electrical engineering.

This discovery laid the foundation of access to electricity using as a source of power. On 17th Jan. 1867, the report entitled "On the conversion of mechanical energy into electric current without the use of permanent magnets" was sent to the Berlin Academy of Sciences.

In this report, Werner Siemens predicted that "Technology now has the means to generate electrical current of unlimited strength in an inexpensive and convenient way wherever mechanical energy is available. This fact will become of great importance in several of its sectors. " Heavy-current engineering was then known and developed with a breathtaking pace.

Werner Siemens repeatedly introduced electric current into new applications: in 1879, the first electric railway was presented at the Berlin Trade Fair and the first electric streetlights were installed in Berlin's Kaisergalerie, in 1880 the first electric elevator was built in Mannheim, and the world's first electric streetcar went into service in Berlin-Lichterfelde in 1881.

Even the German word for "electrical engineering", "Elektrotechnik", was coined by Werner Siemens.

The entrepreneur Werner Siemens was successful not only because of his discovery of fundamental technical principles, but also because he, as a businessman, thought the whole process from invention to marketable product and system solutions. Werner Siemens received numerous honors all his life, and in 1888 he was raised to the nobility by Emperor Friedrich III.

On 6th December, 1892, Werner von Siemens died one week before his seventy-sixth birthday, at Charlottenburg (now a part of Berlin), Germany.

**fig. 10.10.** Werner von Siemens a few months before his death

As a German inventor and industrialist of the 19th century, Werner von Siemens was the pioneer of the electro industry and brought a great technological advancement with many of his important discoveries. He earned a prominent position among the multitude of awards for achievements in science and technology.

## 10.2  Modern TRIZ Investigation and Modeling in Case Studies of the Historical Inventions of Brothers Siemens

### 10.2.1. Inventions of Carl Siemens

Carl Siemens had been trained as a mechanical engineer, he pursued two major themes in his inventive efforts, on based upon the science of heat, and the other based upon the science of electricity. In this section, the case studies will focus on the inventions by Carl Siemens regarding the two aspects. The Indo-European telegraph line installed by Siemens represents his highest achievement in electrical field; the regenerative furnace is the greatest single invention and achievement of his life; while the electric thermometer was a delicate cross-coupling which connected both.

#### 10.2.1.1. TELEGRAPH LINE

With the rise of industrialization in the 19th century, people needed not only rapid modes of transport for themselves and their products but also modern means of transmitting all kinds of news. Reliable communication could be decisive for accessing colonies as well as linking markets and trade partners.

#### The Prototype – Ancient Message

Types of ways to communication and send messages existed in ancient times, and some of them are even still used today.

*Smoke Signals:* smoke signals were first created and used by the Native North Americans and the Chinese. The North American natives used smoke signals between camps, while the Chinese used smoke signals along the Great Wall of China to send military warning and messages. Smoke signal is a form of optical telegraph, which means these messages can be sent over distances only within one's eyeshot. That means a long-distance message sent by smoke signal needs several or even more the intermediate stations, and that increase the cost to send a message a lot. Also, what is very important is that everyone sending and receiving the smoke signals must know that they mean. A code needs to be worked out so that everyone can understand the messages being sent.

*Carrier Pigeons:* pigeons have been used to carry messages for thousands of years. Pigeons can fly very fast and very far (80 kilometers per hour and in two days they can cover 1,600 kilometers to reach their loft). About 2,800 years ago, Greeks used pigeons to deliver news of the winners from Olympic Games. They were also used to deliver messages for the military from some of the earliest recorded battles in history through World War II. But the disadvantages of carrier pigeons is obvious, the reliability is the shortage of this kind of communication. For example, the messages can be interrupted due to the lost of the pigeons or even the death of the pigeons. Especially in a war, the carrier pigeons delivering military messages can be shot by the enemies. Besides, it also costs a lot of time to send a message if compared with modern technology to send a message.

## Contradiction Analysis

The "plus-state" of the "Ancient message" can be determined as in As-Matrix as number 10 (ease of use), because the ways to send message by smoke signal and pigeons are easy to learn and do not need the users to master any special knowledge. The "minus-state" of the "Ancient message" can be determined as in As-Matrix as number 25 (loss of time), because normally it costs lots of time to deliver a message as abovementioned.

## Inventing Idea

Combing the plus and minus state factors above, the results will show several points in As-navigators. The target is to shorten the time to deliver a message. After combing the As-Matrix plus-10 and minus-25, the idea navigator resulted:

• Navigator 04: replacement of mechanical matter – according to this navigator, application of electrical technologies can be used in order to reduce the delivering time by improving the delivering speed.

• Navigator 02: preliminary action – according to this navigator, devices are necessary for applying electrical technologies and tubes for passing through the message are also needed.

## Result Artifact (Solution) – Telegraph Line

Initially focused on telecommunications by improving the pointer telegraph, Siemens installed in 1848 the first long-distance telegraph line in Europe (500km route from Berlin to Frankfurt) and a little over two decades later it became a renowned international company when, in 1870, installed the Indo-European Telegraph line (11,000km) making it possible to transmit a message from London to Calcutta in 28 minutes as opposed to 30 days.

The plan to construct the telegraph line from England to India had been repeatedly discussed since about the 1850s. As a result, individual section had been built along the route at first until the reliable, uninterrupted link was completed in 1870.

The first dispatch by this line was sent on 12 April, 1870, after the construction phase of almost 20 years. Sir William invited prominent guests to the telegraph station in London, establishing a connection to Teheran, the intermediate station on the way to India, before successfully demonstrating a direct link to Calcutta. Since this initial dispatch, the Indo-European telegraph line has been in operation for more than 60 years until 1931. It was not technical deficiencies that finally caused the decline, but the rise of wireless radio connections after World War I. Individual sections of the line remained in operation, and according to Corporate Archives sources, original iron telegraph masts were still in service in Iran in 1965.

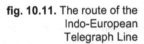

**fig. 10.11.** The route of the Indo-European Telegraph Line

Extracting of invention "Telegraph Line" is shown in fig. 10.12.

## Extracting – Telegraph Line (C.S.)

### Extracting-1

| LC | No. | Navigator | Substantiation for the Extracting |
|----|-----|-----------|-----------------------------------|
| ++ | 02 | **Preliminary action** | Telegraph lines and stations need to be set up before the messages are delivered. |
| ++ | 04 | **Replacement of mechanical matter** | Traditional matters are replaced by electrical matter. |
| + | 10 | **Copying** | Messages are copied by translating into telegraph codes. |
| + | 17 | **Use of composite materials** | Special composite materials are required to produce telegraph cables. |
| + | 18 | **Mediator** | Electrical matters are used as a mediator to transfer the signal. |

### Extracting-2

**Standard Contradiction:** The ancient ways to deliver messages makes the usage easy, **but** the delivering time would be too long.

**Radical Contradiction:** The delivering time should be long so that it can be easily mastered by users, but it should also be short, so that the waiting time would not be too long and the message can be more timely.

**FIM:** X -resource, together with avai   lable or modified resources, and without making the object more complex or introducing any negative properties, guarantees attainment of the following **IFR:** [messages delivering in shorter time].

**fig. 10.12.** Extracting of invention "Telegraph Line"

**TREND**

The traditional methods (like smoke signals, carrier pigeons) to deliver messages cost too much time, and some of them are even not at a high level of reliability.

PROBLEM: How to deliver a message in a shorter time?

**REDUCING**

**FIM (shortly)**: [messages delivering in shorter time].

**Standard (Technical) Contradiction**

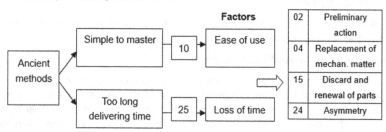

| | | |
|---|---|---|
| **Navigators** | | |
| 02 | Preliminary action | |
| 04 | Replacement of mechan. matter | |
| 15 | Discard and renewal of parts | |
| 24 | Asymmetry | |

**Radical (Physical) Contradiction**

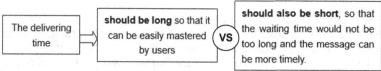

**INVENTING**

Group of navigators to model the idea for the "messages fast delivering" concept:

**02. Preliminary action**: Telegraph lines and stations need to be set up before the messages are delivered.

**04. Replacement of mechanical matter**: Traditional matters are replaced by electrical matter.

**ZOOMING**

Have the contradictions been removed? – Yes. – ~~No~~.

Super-effects: Yes. It does not only improve delivering speed but also improves the reliability of message delivering. Negative effects: Large scale of telegraph system construction are prerequisite.

**BRIEF DESCRIPTION**

> The electric telegraph is a now outdated communication system that transmitted electric signals over wires from location to location that translated into a message. The 'Telegraph Line': The telegraph line from London to Calcutta was a great challenge, it stretches more than 9,000miles. After the line went into service in 1870, the Siemens company became famous as a general contractor for worldwide telecommunications.

**fig. 10.13.** Reinventing of invention "Telegraph Line"

### 10.2.1.2. REGENERATIVE FURNACE

The regenerative furnace is one of the greatest inventions by Carl Siemens, using the process known as the Siemens-Martin process.

Generally speaking, it absorbs heat from the outgoing waste gases and returns it later to the incoming cold gases, saving 70-80% of the fuel.

Sir Carl Wilhelm Siemens developed the Siemens regenerative furnace in the 1850s and claimed it in 1857.

This furnace operates at a high temperature by using regenerative preheating of fuel and air for combustion.

#### The Prototype – Conventional Furnace

Before the idea of regenerative furnace was introduced by Carl Siemens, the furnaces are designed without thermal regenerative devices, which means in the conventional furnace a very large part of heat of combustion is lost by being carried off in the hot gases which pass up the chimney. The high temperature smoke was exhausted into air directly after one-off combustion.

#### Contradiction Analysis

The "plus-state" of the "furnace without thermal regenerative devices" can be determined as in A-Matrix as number 07 (complexity of construction), because the ordinary furnace exhausts gases into air directly that it does not require a complex structure. The "minus-state" of the "furnace without thermal regenerative devices" can be determined as in A-Matrix as number 39 (loss of energy), because the heat within the high temperature gases was lost during the process of gas exhausting without thermal regenerative.

#### Inventing Idea

Combing the plus and minus state factors above, the results will show several points in As-navigators. The target is to keep the simplicity of the furnace structure and to reduce energy consumption which will be significantly economically and environmentally important. After combing the A-Matrix plus-07 and minus-39, the idea navigator resulted:

• Navigator 34: matryoshka (nested doll) – according to this navigator, the furnace can be reconstructed into structure with hollow space, so the structure can be not too complex to fulfill the requirement.

• Navigator 36: feedback – according to this navigator, specific thermal regenerative devices were introduced that the heat within the high temperature gases influenced the new coming originally cold gases.

#### The Result Artifact (Solution) – Regenerative Furnace

In the regenerative furnace the hot gases pass through a regenerator, or chamber stacked with loose bricks, which absorb the heat. When the bricks are well heated the hot gases are diverted so to pass through another similar chamber, while the air necessary for combustion, before it enters the furnace, is made to traverse the heated chamber, taking up as it goes the heat which has been stored in the bricks.

After a suitable interval the air currents are again reversed. The process is repeated periodically, with the result that the products of combustion escape only after being cooled, the heat which they take from the furnace being in great part carried back in the heated air.

The regenerators are the distinctive feature of the furnace and consist of firebrick flues filled with bricks set on edge and arranged in such a way as to have a great number of small passages between them. The bricks absorb most of the heat from the outgoing waste gases and return it later to the incoming cold gases for combustion.

Through this method, an open-hearth furnace can reach temperatures high enough to melt steel, but Siemens did not initially use it for that.

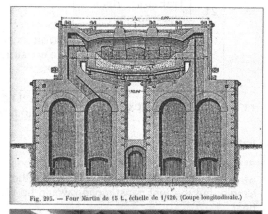

Fig. 295. — Four Martin de 15 t., échelle de 1/120. (Coupe longitudinale.)

**fig. 10.14** (above). Blueprint of Siemens regenerative furnace

**fig. 10.14** (below). Regenerative furnace produced in Guangdong, China

Extracting of invention "Regenerative furnace" is shown in fig. 10.15.

**Regenerative Furnace (C.S.)**

**Extracting-1**

| LC | No. | Navigator | Substantiation for the Extracting |
|---|---|---|---|
| + | 08 | **Periodic action** | The whole process is repeated periodically. |
| + | 11 | **Inverse action** | After a certain interval the air currents are reversed. |
| + | 21 | **Transform damage into use** | The thermal regenerative process is energy saving process and environment friendly. |
| ++ | 34 | **Matryoshka (nested doll)** | Similar chambers are designed in the construction to fully absorb the heat within the gases to be exhausted. |
| ++ | 36 | **Feedback** | The heat absorbed from the gases return to new incoming gases. |

**Extracting-2**

**Standard Contradiction:** The conventional furnace exhausts the waste gases directly into air would make the structure of the furnace simple, but the fuel consumption would be high due to the waste of heat.

**Radical Contradiction:** The structure of furnace should be simple so that it could be easily designed and constructed, but it should also be complex, so that the heat can be recollected from the waste gases and it can be more energy saving oriented.

**FIM**: X-resource, together with available or modified resources, and without making the object more complex or introducing any negative properties, guarantees attainment of the following **IFR**: [furnaces with

**fig. 10.15.** Extracting of invention "Regenerative furnace"

**fig. 10.16.** Reinventing of invention "Regenerative furnace" – next page!

## TREND

In conventional furnace, fuel consumption is high mainly due to the direct exhaust of high temperature waste gases. This design is neither economically advantageous, nor environmental friendly.

PROBLEM: How can we avoid or reduce the lost of heat of combustion?

## REDUCING

**FIM**: X-resource, together with available or modified resources, and without making the object more complex or introducing any negative properties, guarantees attainment of the following **IFR**: [furnaces with high efficiency].

**Standard Contradiction**

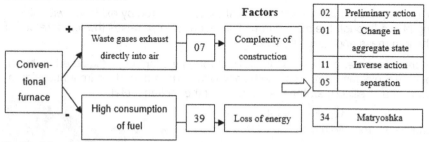

**Radical Contradiction**

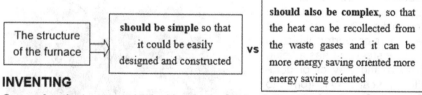

## INVENTING

Group of navigators to model the idea for the "high efficiency furnace" concept:

**34. Matryoshka (nested doll)**: Similar chambers are designed in the construction to fully absorb the heat within the gases to be exhausted.

**36. Feedback**: The heat absorbed from the gases return to new incoming gases.

## ZOOMING

Have the contradictions been removed? – Yes. –No.

Super-effects: Yes. It makes the steel melting in an open-hearth furnace environment available. Negative effects: Improved structure of furnace is required.

## BRIEF DESCRIPTION

The 'Regenerative furnace': This was one of the most important innovations by Carl Siemens. The idea saves energy consumption (70-80% of the fuel) and has economical signification. Besides, it also makes the steel melting in an open-hearth furnace environment available.

### 10.2.1.3. OPEN HEARTH

Open-hearth process, also called Siemens-martin process, steelmaking technique that for most of the 20th century accounted for the major part of all steel made in the world.

#### The Prototype – Bessemer process converter

The Bessemer process was the first inexpensive industrial process for the mass-production of steel from molten pig iron. The process is named after its inventor, Henry Bessemer, who took out a patent on the process in 1855.

The process was independently discovered in 1851 by William Kelly. The process had also been used outside of Europe for hundreds of years, but not on an industrial scale.

The key principle is removal of impurities from the iron by oxidation with air being blown through the molten iron. The oxidation also raises the temperature of the iron mass and keeps it molten.

In the Bessemer process, air is blown through molten metal (liquefied by heat), because oxygen combines chemically with carbon and other impurities (manganese and silicon) and burns them out of the molten metal.

Then the purified molten metal can be poured into molds.

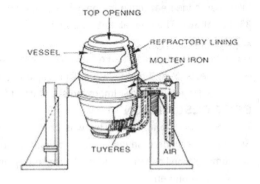

The process, completed in 15 to 20 min, is carried on in a large container, egg-shaped (see in fig. 10.17), called the Bessemer converter, which is made of steel and has a lining of silica and clay or of dolomite. The capacity is from 8 to 30 tons of molten iron; the usual charge is 15 or 18 tons.

**fig. 10.17** (above). Bessemer converter, Kelham Island Museum, Sheffield, England

**fig. 10.17** (below). Bessemer converter component

## Contradiction Analysis

The "plus-state" of the "Bessemer process" can be determined as in As-Matrix as number 25 (loss of time), because only 15 to 20 minutes are taken to complete the process, which makes the usage of steel instead of wrought iron in bridges making or framework for buildings available. The "minus-state" of the "Bessemer process" can be determined as in As-Matrix as number 01 (productivity), because the volume of Bessemer process is only from 8 to 30 tons per time which is not sufficient for large scale construction requirement at that era.

## Inventing Idea

Combing the plus and minus state factors above, the results will show several points in As-navigators. The target is to increase the productivity of the process. After combing the As-Matrix plus-25 and minus-01, the idea navigator resulted:

• Navigator 02: preliminary action – according to this navigator, the air can be heated up before processing, this can help save a lot of energy and time.

• Navigator 18: mediator – according to this navigator, the waste energy can be transferred to the air but special device are required to complete this transferring process.

## The Result Artifact (Solution) – Open Hearth

The first open hearth might have been used in the $8^{th}$ century, but the first commercial use for steelmaking was done in the $19^{th}$ century. A well known and widely used process Siemens Regenerative Furnace was developed around 1850 by Sir Wilhelm Siemens. Operating at high temperature this process can save nearly 80% of the fuel by using regenerative preheating for fuel. However, the process failed because the lining of the furnace also melted.

Later in 1865, the French engineer Pierre Martin took out a license from Siemens and first applied his furnace for making steel. Their process was known as the Siemens-Martin process, and the furnace as an "open-hearth" furnace.

Martins developed a furnace with wall lining that were heat resistant. The process which is known as Siemens-martins process began in 1864. For 100 years, Siemens-martins process and open hearth furnace ruled the steel making industry.

As early as 1895 in the UK it was being noted that the heyday of the Bessemer process was over and that the open hearth method predominated. The Iron and Coal Trades Review said that it was "in a semi-moribund condition. Year after year, it has not only ceased to make progress, but it has absolutely declined."

The Siemens-Martin open-hearth furnace consists of upper and lower furnace. The top furnace is to melt the metal and the lower furnace is the regenerative chamber where the combustion air and the generator gas are preheated. In this process, exhaust gases from the furnace are pumped into a chamber containing bricks, where heat is transferred from the gases to the bricks. The flow of the furnace is then reversed so that fuel and air pass through the chamber and are heated by the bricks. Through this process the temperature reaches up to 1800°C, which is high enough to melt steel.

In order to remove the elements or impurities like carbons, manganese, silicon, phosphorus and other elements, either the substances are oxidized and escape as gas or float on the liquid steel.

This process is economical because by using regenerative firing heat is produced with less fuel consumption. Unlike conventional furnace, here temperature of the furnace can be increased up to 1800°C and nearly 600 tons of steel can be produced.

The most appealing characteristic of the Siemens regenerative furnace is the rapid production of large quantities of basic steel, used for example to construct high-rise buildings. The usual size of furnaces is 50 to 100 tons, but for some special processes they may have a capacity of 250 or even 500 tons.

The Siemens-Martin process complemented rather than replaced the Bessemer process. It is slower and thus easier to control. It also permits the melting and refining of large amounts of scrap steel, further lowering steel production costs and recycling an otherwise troublesome waste material. Its worst drawback is the fact that melting and refining a charge takes several hours.

This was an advantage in the early 20th century, as it gave plant chemists time to analyze the steel and decide how much longer to refine it. But by about 1975, electronic instruments such as atomic absorption spectrophotometers had made analysis of the steel much easier and faster.

One of the chief benefits of using an open hearth furnace is the ability to extract the impurities from the pig iron as it is subjected to the extreme temperatures. The end result is steel that is more durable and able to withstand greater levels of stress.

Over time, the open hearth furnace has lost ground to new technologies that made it possible to remove the impurities and produce higher grades of steel, while also reducing the cost of production.

Much of the reduction in production costs came about due to the development of alternative methods that were more energy-efficient, such as the electric arc furnace or the oxygen furnace.

While no longer in common use around the world, the open hearth furnace is still utilized in some countries, although the production is normally on a much smaller scale than a few decades ago.

**Open Hearth (C.S.)**
**Extracting-1**

| LC | No. | Navigator | Substantiation for the Extracting |
|----|-----|-----------|-----------------------------------|
| ++ | 02 | **Preliminary action** | The air can be heated up before processing; this can help save a lot of energy and time. |
| + | 03 | **Segmentation** | Two chambers are designed into individuals. |
| + | 10 | **Copying** | Usage and application of regenerative furnace design theory. |
| + | 12 | **Local property** | Two chambers have different functions. |
| ++ | 18 | **Mediator** | The waste energy can be transferred to the air but special device are required to complete this transferring process. |
| ++ | 21 | **Transform damage into use** | The thermal regenerative process is energy saving process and environment friendly. |
| + | 29 | **Self-servicing** | Reuse of waste energy. |

**Extracting-2**

**Standard Contradiction:** The Bessemer process takes short time to complete each batch reduce the loss of time, but each batch produces limited volume, which is insufficient to satisfy the requirement of large scale construction.

**Radical Contradiction:** The volume per batch should be large so that it can satisfy the requirement of large scale construction, but it should also be small so that it has advantage in time consuming.

**FIM**: X-resource, together with available or modified resources, and without making the object more complex or introducing any negative properties, guarantees attainment of the following **IFR**: [to increase the productivity of the process].

**fig. 10.18.** Extracting of invention "Open hearth"

**fig. 10.19.** Reinventing of invention "Open hearth" – next page!

**TREND**

In Bessemer process, the volume per batch is only about 8 to 30 tons which can not satisfy the requirement of the large scale construction.

PROBLEM: How to increase the productivity of the steel making process?

**REDUCING**

**FIM**: X-resource, together with available or modified resources, and without making the object more complex or introducing any negative properties, guarantees attainment of the following **IFR**: [to increase the productivity].

**Standard Contradiction**

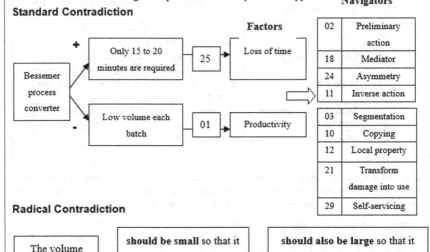

Navigators

| 02 | Preliminary action |
| 18 | Mediator |
| 24 | Asymmetry |
| 11 | Inverse action |
| 03 | Segmentation |
| 10 | Copying |
| 12 | Local property |
| 21 | Transform damage into use |
| 29 | Self-servicing |

**Radical Contradiction**

| The volume per batch | ⟹ | **should be small** so that it has advantage in time consuming | **vs** | **should also be large** so that it can satisfy the requirement of large scale construction |

**INVENTING**

Group of navigators to model the idea for the "high productivity steel making" concept:

**02. Preliminary action:** The air can be heated up before processing; this can help save a lot of energy and time.

**18. Mediator:** The waste energy can be transferred to the air but special device are required to complete this transferring process.

**ZOOMING**

Have the contradictions been removed? – Yes. – ~~No~~.

Super-effects: Yes. It is slower and thus easier to control.

Negative effects: It takes several hours to complete each batch.

**BRIEF DESCRIPTION**

Though the open-hearth process has been almost completely replaced in the most industrialized countries by the basic oxygen process and the electric arc furnace, it nevertheless accounts for about one-sixth of all steel produced worldwide.

## 10.2.1.4. AN ELECTRIC THERMOMETER

A thermometer is a device using variety of different principles to measure temperature or temperature gradient. A thermometer has two important elements: the temperature sensor (such as the bulb on a mercury thermometer) in which some physical change occurs with temperature, and some means of converting this physical change into a value (such as the scale on a mercury thermometer) (see in fig. 10.20). Today, thermometer is widely used in variety of different fields, medical treatment, industrial monitoring and also your daily life.

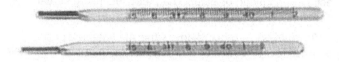

**fig. 10.20:** Oral clinical mercury thermometer

### The Prototype – Mercury-in-glass Thermometer

According to many authors recorded, the first thermometer was developed by a Muslim scientist Abū Alī ibn Sīnā (known as Avicenna in the West) in the early eleventh century, and by several European scientists in the sixteenth and seventeenth centuries, notably Galileo Galilei. Basically speaking, they used instruments to measure level of water in a closed tube to scale the temperature, because the position of water/air interface moves in the closed tube due to the expansion and contraction of the air caused by the change of the temperature outside. Later, many other thermometers are introduced by scientists around the world. In 1724, for the first time, Daniel Gabriel Fahrenheit developed a thermometers using mercury (which has a high coefficient of expansion), one of the major types of thermometer still used today, which has a finer scale and greater reproducibility. Compared with the previous types of thermometers, the new type (mercury-in-glass thermometer) has several advantages. For example, the water would expand its volume extremely upon solidification if the temperature went below 0°C, which means it would be not suitable to measure the temperature below 0°C. But mercury solidifies (freezes) at -38.83°C, and unlike water, mercury does not expand upon solidification, which makes the measurement of temperatures below 0°C available.

### Contradiction Analysis

The "plus-state" of the "mercury-in-glass thermometer" can be determined as in As-Matrix as number 08 (complexity of inspection and measurement), because no preparation work are necessary before measurement and results can be got directly by reading from the scale. The "minus-state" of the "mercury-in-glass thermometer" can be determined as in As-Matrix as number 06 (precision of measurement), because more and more, the mercury-in-glass thermometers are not sufficient for high precision measurement.

## Inventing Idea

Combing the plus and minus state factors above, the results will show several points in As-navigators. The target is to improve the precision of measurement but keep the facility of usage of thermometers. After combing the As-Matrix plus-08 and minus-06, the idea navigator resulted:

• Navigator 04: replacement of mechanical matter – according to this navigator, electrical, magnetic, or electromagnetic application can be considered in order to improve the precision. Generally speaking, compared with traditional methods, electrical, magnetic or electromagnetic technologies can help to get more precise results in measurement.

• Navigator 18: mediator – according to this navigator, special devices or methods are used to make transfers between temperatures and electrical, magnetic or electromagnetic signals.

### The Result Artifact (Solution) – Thermocouple & Electrical Thermometer

In year 1821, German-Estonian physicist Thomas Johann Seebeck discovered the thermoelectric effect (also called Seebeck effect) as known today, which laid the foundation of a new type of thermometer, thermocouple, a widely used type of temperature sensor. This increased the accuracy and the operating ranges extremely. Although the thermocouple improves the properties of thermometers in precision and the operating ranges a lot, it still has many disadvantages and limitations. For example, special extension cables and cold junction compensations are required, because thermocouple was designed based on the thermoelectric effect abovementioned, two junctions are always necessary.

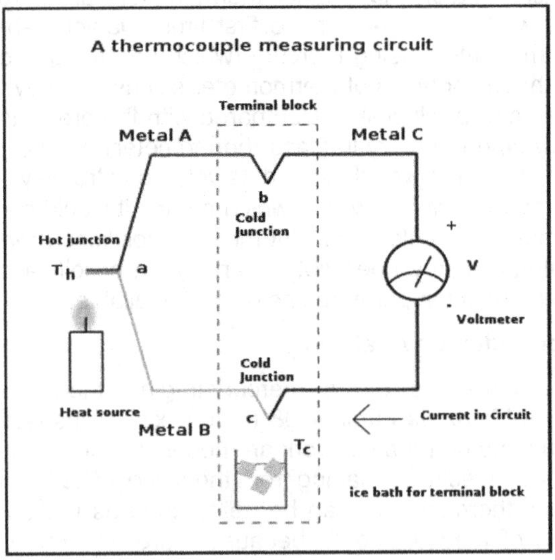

**fig. 10.21.:**
A thermocouple
measuring circuit

Later in 1871, the application of the tendency of electrical conductors to increase their electrical resistance with rising temperature was first described by Sir William Siemens, and the necessary methods of construction were established by Callendar, Griffiths, Holborn and Wein between 1885 and 1900.

Resistance thermometers, which are also called electrical resistance thermometers, are constructed in a number of forms, and they offer greater stability, accuracy and repeatability in some cases than thermocouples. While thermocouples use the Seebeck effect to generate a voltage; resistance thermometers use electrical resistance and require a power source when to operate.

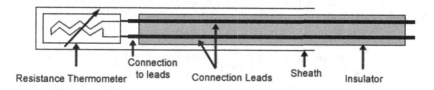

Resistance Thermometer   Connection to leads   Connection Leads   Sheath   Insulator

**fig. 10.22.** Construction of Resistance Thermometers

---

### Extracting-1

| LC | No. | Navigator | Substantiation for the Extracting |
|---|---|---|---|
| + | 02 | **Preliminary action** | Prepared work are required when use electric thermometers to measure the temperature. |
| ++ | 04 | **Replacement of mechanical matter** | Electrical material and methods are applied and used instead of the traditional methods. |
| ++ | 18 | **Mediator** | Devices are used to make transfers between temperatures and electrical signals according to the thermoelectric effect. |

### Extracting-2

**Standard Contradiction:** The mercury-in-glass thermometer is easy to operate, but the results gained from this type of thermometers are not sufficient for some of the high precision measurements.

**Radical Contradiction:** The measurement should be simple so that the operation costs less time and steps, but it should also be complex, so that more accurate results to fulfill higher precision measurement are attainable.

**FIM:** X-resource, together with available or modified resources, and without making the object more complex or introducing any negative properties, guarantees attainment of the following **IFR:** [temperature measurement with higher accuracy].

---

**fig. 10.23.** Extracting of invention "Resistance Thermometers"

**fig. 10.24.** Reinventing of invention "Resistance Thermometers" – next page!

## TREND

In traditional temperature measurement (mercury-in-glass thermometer), the operation is simple. But due to the theory of the traditional measurement is based on the thermal expansion which is greatly influenced by other factors, it normally provides results with great error which is not tolerable in high precision measurement.

PROBLEM: How to measure temperature more accurately?

## REDUCING

FIM: X-resource, together with available or modified resources, and without making the object more complex or introducing any negative properties, guarantees attainment of the following **IFR**: [temperature measurement with higher accuracy].

**Standard (Technical) Contradiction**

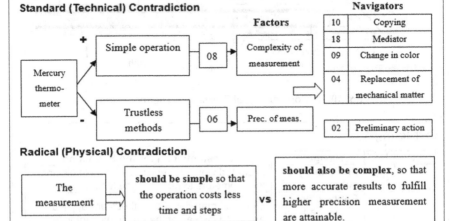

**Radical (Physical) Contradiction**

| The measurement | **should be simple** so that the operation costs less time and steps | vs | **should also be complex**, so that more accurate results to fulfill higher precision measurement are attainable. |

## INVENTING

Group of navigators to model the idea for the "high accuracy thermometer" concept:

**04. Replacement of mechanical matter**: Electrical material and methods are applied and used instead of the traditional methods.

**18. Mediator**: Devices are used to make transfers between temperatures and electrical signals according to the thermoelectric effect.

## ZOOMING

Have the contradictions been removed? – Yes. – ~~No~~.

Super-effects: Yes. It does not only improve the accuracy, but also extend the working range. Negative effects: No.

## BRIEF DESCRIPTION

The 'Electric thermometer': also called electrical resistance thermometer, first described by Sir William Siemens in 1871, and later established by Callendar, Griffiths, Holborn and Wein between 1885 and 1900, offers greater stability, accuracy and repeatability than traditional thermometers. It is widely used today in many application fields.

## 10.2.2. Inventions of Werner Siemens

As a German inventor and industrialist, Werner Siemens was the pioneer of the electro industry and brought about a great technological advancement with many of his important discoveries. In this section, the case studies will focus on the most important and representative inventions by Werner Siemens. His invention of the telegraph that used a needle to point to the right letter, instead of using Morse code led to the foundation of the electrical and telecommunications company Siemens as well known today. Besides Werner Siemens made outstanding contributions to the expansion of electrical engineering and is therefore known as the founding father of the discipline in Germany that he designed the world's first electric elevator.

### 10.2.2.1. POINTER TELEGRAPH

Before the telegraph, there was the optical telegraph, a chain of towers topped by large pivoting cross members, and spaced as far apart as the eye could see. Developed by the Frenchman Claude Chappe at the end of the 18th century, optical telegraph lines once stretched from Paris out to Dunkirk and Strasbourg, and were in service for more than half a century.

### The Prototype – Optical telegraph

Email leaves all other communication systems far behind in terms of speed. But the principle of the technology – forwarding coded messages over long distances – is nothing new. It has its origins in the use of plumes of smoke, fire signals and drums, thousands of years before the start of our era. Coded long distance communication also formed the basis of a remarkable but largely forgotten communications network that prepared the arrival of the internet: the optical telegraph.

Already in antiquity, post systems were designed that made use of the changing of postmen. In these stations, the message was transferred to another runner or rider, or the horseman could change his horse. These organised systems greatly increased the speed of the postal services. The average speed of a galloping horse is 21 kilometres an hour, which means that the distance in time between Paris and Antwerp could be shortened to a few days. A carrier pigeon was twice as fast, but less reliable. Intercontinental communication was limited to the speed of shipping.

Centuries of slow long-distance communications came to an end with the arrival of the telegraph. Most history books start this chapter with the appearance of the electrical telegraph, midway the nineteenth century. However, they skip an important intermediate step. Fifty years earlier (in 1791) the Frenchman Claude Chappe developed the optical telegraph. Thanks to this technology, messages could be transferred very quickly over long distances, without the need for postmen, horses, wires or electricity.

The optical telegraph network consisted of a chain of towers, each placed 5 to 20 kilometres apart from each other. On each of these towers a wooden semaphore and two telescopes were mounted (the telescope was invented in 1600).

The semaphore had two signalling arms which each could be placed in seven positions. The wooden post itself could also be turned in 4 positions, so that 196 different positions were possible. Every one of these arrangements corresponded with a code for a letter, a number, a word or (a part of) a sentence.

Every tower had a telegrapher, looking through the telescope at the previous tower in the chain. If the semaphore on that tower was put into a certain position, the telegrapher copied that symbol on his own tower.

Next he used the telescope to look at the succeeding tower in the chain, to control if the next telegrapher had copied the symbol correctly. In this way, messages were signed through symbol by symbol from tower to tower.

The semaphore was operated by two levers. A telegrapher could reach a speed of 1 to 3 symbols per minute.

The technology today may sound a bit absurd, but in those times the optical telegraph was a genuine revolution. In a few decades, continental networks were built both in Europe and the United States. The first line was built between Paris and Lille during the French revolution, close to the frontline. It was 230 kilometers long and consisted of 15 semaphores. The very first message – a military victory over the Austrians – was transmitted in less than half an hour. The transmission of 1 symbol from Paris to Lille could happen in ten minutes, which comes down to a speed of 1,380 kilometers an hour.

**fig. 10.25.** Optical Telegraph, France

Faster than a modern passenger plane – this was invented only one and a half centuries later.

**Contradiction Analysis**

The "plus-state" of the "optical telegraph" can be determined as in As-Matrix as number 25 (loss of time), because it the contemporary condition, the optical telegraph saves a lot of time for communication.

The "minus-state" of the "optical telegraph" can be determined as in As-Matrix as number 10 (ease of use), because lots of towers are requisite before the functioning of optical telegraph.

### Inventing Idea

Combing the plus and minus state factors above, the results will show several points in As-navigators. The target is to ease the usage of telegraphing. After combing the As-Matrix plus-25 and minus-10, the idea navigator resulted:

• **Navigator 04:** replacement of mechanical matter – according to this navigator, electrical, magnetic, or electromagnetic technology can be considered to apply. Generally speaking, compared with traditional methods, electrical, magnetic or electromagnetic technologies have advantages in many aspects.

• **Navigator 02:** preliminary action – according to this navigator, infrastructions for electric telegraph are requisite for applying electrical technology before functioning.

### The Result Artifact (Solution) – Pointer Telegraph

Werner von Siemens' first great invention was the pointer telegraph. In his day, people communicated over long distances by optical telegraphy, a technology that was not very user-friendly. It worked, for example, only in daylight, only over short distances, only in clear weather. Compared with the optical telegraphy, the pointer telegraph had great advantages over optical telegraphy and even over traditional Morse code, which it eventually made obsolete. Morse Code is an early form of digital communication and all the Characters are encoded as Dots or Dashes. The pointer telegraph worked day or night, in any kind of weather, using real letters, not coded signs. In 1847, Werner von Siemens needed only a cigar box, some imagination and a few lengths of copper wire to construct an operational pointer telegraph of simple design (see in fig. 10.26). His achievement solved the problem of transmitting messages reliably over longer distances. It also enabled him to lay the foundation stone for the Siemens & Halske Telegraph Construction Company.

The Pointer Telegraph uses the technology of a Step-by-step Sequential Circuit. The Pointer Telegraph is much easier to use than the Morse Telegraph, but much slower.

**fig. 10.26**. Early letter pointer telegraph instrument at Hamar museum

In 1856, Werner von Siemens invents the "H armature". This development is first applied in a magneto-electric pointer telegraph (see in fig. 10.27) with a crank inductor produced for the Bavarian railway company. Unlike its predecessors, this "railway telegraph" works without batteries since the electricity required for operation is generated during telegraphy by turning a crank.

**fig. 10.27** - (left). Pointer Telegraph in 1847
**fig. 10.27** - (righ)t. Battery-free telegraph

| Extracting-1 | | | |
|---|---|---|---|
| **LC** | **No.** | **Navigator** | **Substantiation for the Extracting** |
| ++ | 04 | **Replacement of mechanical matter** | Electrical technologies and matters are used in pointer telegraph. |
| ++ | 02 | **Preliminary action** | Infrastructions for electric telegraph are requisite for applying electrical technology before functioning. |
| + | 15 | **Discard and renewal of parts** | The pointer turns back to the original position after send out every single code for the next time. |

**Extracting-2**

**Standard Contradiction:** the optical telegraph saves time for code communication, but the optical telegraph makes the usage of telegraphing difficult.

**Radical Contradiction:** the usage of telegraphing should be simple so that the complexity of telegraphing can be low, but it should also be complex so that the cost of time can be low.

**FIM:** X-resource, together with available or modified resources, and without making the object more complex or introducing any negative properties, guarantees attainment of the following **IFR**: [ease the usage of telegraphing].

**fig. 10.28.** Extracting of invention "Pointer Telegraph"

**fig. 10.29.** Reinventing of invention "Pointer Telegraph" – next page!

## TREND

Optical telegraph is widely used in the early 19ᵗʰ. It has advantages in saving time, but it also has many disadvantages in reliability, complexity of usage and other aspects.
PROBLEM: How to ease the usage of telegraphing?

## REDUCING

**FIM**: X-resource, together with available or modified resources, and without making the object more complex or introducing any negative properties, guarantees attainment of the following **IFR**: [ease the usage of telegraphing].

**Standard (Technical) Contradiction**

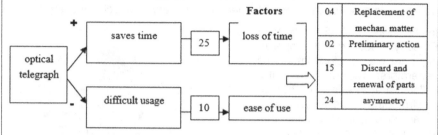

**Radical (Physical) Contradiction**

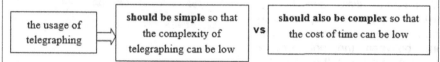

## INVENTING

Group of navigators to model the idea for the "easy-using telegraph" concept:
**02. Preliminary action**: Infrastructions for electric telegraph are requisite for applying electrical technology before functioning.
**04. Replacement of mechanical matter**: Electrical technologies and matters are used in pointer telegraph.

## ZOOMING

Have the contradictions been removed? – ~~Yes~~. – No.
Super-effects: Yes. The new method of telegraphing can work in any weather and becomes more reliable. Negative effects: No.

## BRIEF DESCRIPTION

In 1847, Werner von Siemens improves the pointer telegraph invented by the Englishman Charles Wheatstone by electrically synchronizing the transmitter and receiver. Thanks to this innovation, the apparatus – which is made of cigar boxes, tin plate, pieces of iron and insulated copper wire – is vastly superior to previous equipment. It has a range of 50 km.

### 10.2.2.2. DYNAMO (ELECTRIC GENERATOR)

In 1866, Werner von Siemens discovered the dynamo-electric principle and thus enabled electricity to be put to practical use. The dynamo can convert mechanical energy into electrical energy in an economical way. Its invention laid the foundation for today's world of electrical engineering.

### The Prototype – Faraday disk

A dynamo (from the Greek word "dynamis": meaning power), originally another name for an electrical generator, generally means a generator that produces direct current with the use of a commutator. In electricity generation, an electric generator is a device that converts mechanical energy to electrical energy.

In the year 1831-1832, Michael Faraday discovered the operating principle of electromagnetic generators. The principle, later called Faraday's law, is that an electromotive force is generated in an electrical conductor that encircles a varying magnetic flux. He also built the first electromagnetic generator, called the Faraday disk (see in fig. 10.30), a type of homopolar generator, using a copper disc rotating between the poles of a horseshoe magnet. It produced a small DC voltage.

**fig. 10.30.** Faraday disk

This design was inefficient due to self-cancelling counter-flows of current in regions not under the influence of the magnetic field. While current was induced directly underneath the magnet, the current would circulate backwards in regions outside the influence of the magnetic field.

This counter-flow limits the power output to the pickup wires and induces waste heating of the copper disc. Later homopolar generators would solve this problem by using an array of magnets arranged around the disc perimeter to maintain a steady field effect in one current-flow direction.

Another disadvantage was that the output voltage was very low, due to the single current path through the magnetic flux. Experimenters found that using multiple turns of wire in a coil could produce higher, more useful voltages.

Since the output voltage is proportional to the number of turns, generators could be easily designed to produce any desired voltage by varying the number of turns. Wire windings became a basic feature of all subsequent generator designs.

## Contradiction Analysis

The "plus-state" of the "Faraday disk" can be determined as in As-Matrix as number 07 (complexity of construction), because it has a simple structure. The "minus-state" of the "Farady disk" can be determined as in As-Matrix as number 36 (power), because the current created by Faraday disk is insufficient.

## Inventing Idea

Combing the plus and minus state factors above, the results will show several points in As-navigators. The target is to improve the output current of the dynamo. After combing the As-Matrix plus-07 and minus-36, the idea navigator resulted:

• Navigator 40: uninterrupted useful function – according to this navigator, the structure of the disk can be designed into a style with continuous magnetic filed covering every side of the disk.

## The Result Artifact (Solution) – Siemens' dynamo

In 1827, Hungarian Anyos Jedlik started experimenting with electromagnetic rotating devices which he called electromagnetic self-rotors. In the prototype of the single-pole electric starter, both the stationary and the revolving parts were electromagnetic.

He formulated the concept of the dynamo about six years before Siemens and Wheatstone but did not patent it as he thought he was not the first to realize this. His dynamo used, instead of permanent magnets, two electromagnets opposite to each other to induce the magnetic field around the rotor. It was also the discovery of the principle of dynamo self-excitation.

The first practical designs for a dynamo were announced independently and simultaneously by Dr. Werner Siemens and Charles Wheatstone. On January 17, 1867, Siemens announced to the Berlin academy a "dynamo-electric machine" (first use of the term) which employed self-powering electromagnetic field coils rather than permanent magnets to create the stator field.

On the same day that this invention was announced to the Royal Society Charles Wheatstone read a paper describing a similar design with the difference that in the Siemens design the stator electromagnets were in series with the rotor, but in Wheatstone's design they were in parallel.

The use of electromagnets rather than permanent magnets greatly increases the power output of a dynamo and enabled high power generation for the first time.

This invention led directly to the first major industrial uses of electricity.

For example, in the 1870s Siemens used electromagnetic dynamos to power electric arc furnaces for the production of metals and other materials.

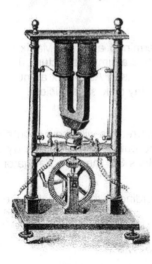

fig. 10.31 - (left). Pixxi's dynamo
fig. 10.31 - (right). Siemens' dynamo

## Siemens' dynamo (W.S.)
### Extracting-1

| LC | No. | Navigator | Substantiation for the Extracting |
|----|-----|-----------|-----------------------------------|
| ++ | 29 | **Self-servicing** | Stationary and revolving parts are previously installed for auxiliary use. |
| + | 08 | **Periodic action** | The stationary and revolving parts rotate with fixed period. |
| + | 10 | **Copying** | Application of Faraday's law. |
| ++ | 40 | **Uninterrupted useful function** | Elimination of the self-cancelling counter-flows of current. |
| + | 07 | **Dynamization** | The rotor and stator both rotate into different directions to create higher current. |
| + | 35 | **Unite** | Multi-turns of wire combined to increase the output current. |

### Extracting-2

**Standard Contradiction:** The Faraday disk has a simple structure that reduces the complexity of construction, but the output current is not sufficient.

**Radical Contradiction:** The structure should be simple to limit the complexity of construction, but it should also be complicated to satisfy the requirement of sufficient output current.

**FIM**: X-resource, together with available or modified resources, and without making the object more complex or introducing any negative properties, guarantees attainment of the following **IFR**: [dynamo with high output current].

fig. 10.32. Extracting of invention "Siemens' dynamo"

fig. 10.33. Reinventing of invention "Siemens' dynamo" – next page!

## TREND

The Faraday disk is the first electromagnetic generator, but it can not provide sufficient output current and has limitation in many aspects.

PROBLEM: How to increase the current created?

## REDUCING

**FIM**: X-resource, together with available or modified resources, and without making the object more complex or introducing any negative properties, guarantees attainment of the following **IFR**: [dynamo with high output current].

## Standard (Technical) Contradiction

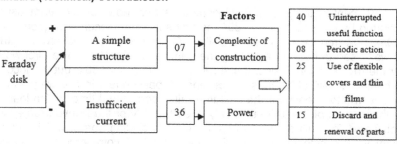

**Navigators**

| | |
|---|---|
| 40 | Uninterrupted useful function |
| 08 | Periodic action |
| 25 | Use of flexible covers and thin films |
| 15 | Discard and renewal of parts |

| | |
|---|---|
| 10 | Copying |
| 29 | Self-servicing |
| 35 | Unite |
| 07 | Dynamization |

## Radical (Physical) Contradiction

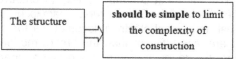

| The structure | **should be simple** to limit the complexity of construction | **vs** | **should also be complicated** to satisfy the requirement of sufficient output current |

## INVENTING

Group of navigators to model the idea for the "high output current dynamo" concept:

**40. Uninterrupted useful function:** Elimination of the self-cancelling counter-flows of current.

**29. Self-servicing:** Stationary and revolving parts are previously installed for auxiliary use.

## ZOOMING

Have the contradictions been removed? – Yes. –No.

Super-effects: No.

Negative effects: Complicated structure are necessary.

## BRIEF DESCRIPTION

In 1866, Werner von Siemens discovered the dynamo-electric principle and thus enabled electricity to be put to practical use. It also laid the foundation for today's world of electrical engineering.

### 10.2.2.3. ELECTRIC RAILWAY

#### The Prototype – Steam locomotive

A locomotive is a railway vehicle that provides the motive power for a train. Generally speaking, the locomotives can be classified according to the motive power into several types that they may generate their power from fuel (wood, coal, petroleum or natural gas), or they may take power from an outside source of electricity. It can commonly be classified into: steam locomotive, gasoline locomotive, diesel locomotive, electrical locomotive, hybrid locomotive and so on.

In the 19th century the first railway locomotives were powered by steam, usually generated by burning coal. In 1804 the first successful locomotive (see in Fig. 5.2.3a), powered by steam, was built by Cornish inventor Richard Trevithick and was used to haul a train along the tramway of the Penydarren ironworks, near Merthyr Tydfil in Wales. It first ran on 21 February 1804, but the locomotive hauled a train of 10 tons of iron and 70 passengers in five wagons over 14km, it was too heavy for the cast iron rails used at that time, which leads to the abandoning of the train after only three trips. After the Penydarren experiment, Trevithick built a series of locomotives, including one of which ran at a colliery in Tyneside in northern England. The first commercially successful steam locomotive Salamanca was built by Matthew Murray for the narrow gauge Middleton Railway in 1812. And after that in 1813 Puffing Billy was built by Christopher Blackett and William Hedley for the Wylam Colliery Railway, which is now on display in the Science Museum in London as the oldest locomotive in existence.

By far, the steam locomotive remained the most common type of locomotive until after World War 2. But since the middle of the 20th century, steam locomotives are being replaced by electrical and diesel-electrical locomotives for its disadvantages. Steam locomotives are less efficient than their more modern diesel and electrical counterparts and require much greater manpower to operate and service. According to the figures from British Rail, the cost of crewing and fuelling a steam locomotive was some two and a half times that of diesel power.

As a result, the steam locomotives create steam and the steam has negative influence to the environment especially in cities. And besides, steam locomotives require much higher labor costs than the new counterparts especially after the Second World War. As a result, non-steam technologies became much more cost-efficient and by the end of the 1970s, most western countries had completely replaced steam locomotives in passenger service.

But in other parts of the world, steam locomotives are still at service. Steam locomotives remained in commercial use in parts of Mexico into the late 1970s; be in regular use until 2004 in People's Republic of China; be at service in some mountainous and high altitude rail lines even today. And as of 2006 DLM AG (Switzerland) continues to manufacture new steam locomotives.

**fig. 10.34** - (left). Trevithick's locomotive, 1804
**fig. 10.34** - (right). A steam locomotive at the Gare du Nord, Paris, 1930

## Contradiction Analysis

The "plus-state" of the "steam locomotive" can be determined as in As-Matrix as number 01 (productivity), because the locomotives significantly increase the productivity of transportation both in goods and passengers. The "minus-state" of the "steam locomotive" can be determined as in As-Matrix as number 39 (loss of energy), because it cost much energy consumption and the steam has negative influence to the environment especially in cities.

## Inventing Idea

Combing the plus and minus state factors above, the results will show several points in As-navigators. The target is to improve the efficiency of the energy consumption. After combing the As-Matrix plus-01 and minus-39, the idea navigator resulted:

- Navigator 04: replacement of mechanical matter – according to this navigator, electrical, magnetic, or electromagnetic technology can be considered to apply. Generally speaking, compared with traditional methods, electrical, magnetic or electromagnetic technologies have advantages in many aspects.

- Navigator 02: preliminary action – according to this navigator, railway electrification (like the power source, and new railway to adapt the electric locomotive) are requisite for applying electrical technology in locomotive field.

## The Result Artifact (Solution) – Electric Locomotives

An electric locomotive is a locomotive powered by electricity from overhead lines, a third rail or an on-board energy storage device (such as a chemical battery or fuel cell).

The first electric passenger train (see in fig. 10.35) was presented by Werner von Siemens at Berlin in 1879. The locomotive was driven by a 2.2 kW series wound motor and the train, consisting of the locomotive and three cars, reached a maximum speed of 13 km/h. During four months, the train carried 90,000 passengers on a 300 meter long circular track. The electricity (150 V DC) was supplied through a third, insulated rail situated between the tracks.

A contact roller was used to collect the electricity from the third rail. The world's first electric tram line opened in Lichterfelde near Berlin, Germany, in 1881.

Electricity is used to eliminate smoke and take advantage of the high efficiency of electric motors; however, the cost of railway electrification means that usually only heavily used lines can be electrified.

**fig. 10.35.** Experimental train von Siemens, 1879

| Extracting-1 | | | |
|---|---|---|---|
| **LC** | **No.** | **Navigator** | **Substantiation for the Extracting** |
| ++ | 02 | **Preliminary action** | Railway electrification (like the power source, and new railway to adapt the electric locomotive) are requisite for applying electrical technology in locomotive field |
| ++ | 04 | **Replacement of mechanical matter** | Electrical technologies are used within the field of locomotive industry instead of the diesel locomotives. |
| + | 12 | **Local property** | The structure of the locomotive is changed to adapt the new working principle. |
| + | 37 | **Equipotentiality** | The working principle is changed so that much labor costs are saved. |

**Extracting-2**

**Standard Contradiction:** The steam locomotives provide higher transportation capability, but it creates steam which has negative influence in cities.

**Radical Contradiction:** The steam locomotive should create steam as power resource to haul the train, but it also should not create steam to eliminate the negative influence to environment especially in cities.

**FIM** (shortly): [locomotives without steam].

**fig. 10.36.** Extracting of invention "Electric Locomotive"

**fig. 10.37.** Reinventing of invention "Electric Locomotive" – next page!

**TREND**

Steam locomotives have been widely used before the Second World War. But the steam locomotives create steam which has negative influence to the environment especially in cities.

PROBLEM: How to eliminate the steam created by locomotives?

**REDUCING**

**FIM**: X-resource, together with available or modified resources, and without making the object more complex or introducing any negative properties, guarantees attainment of the following **IFR**: [locomotives without steam].

**Standard (Technical) Contradiction**

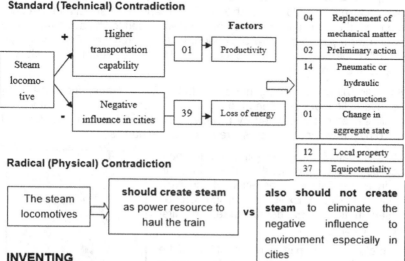

| Navigators | |
|---|---|
| 04 | Replacement of mechanical matter |
| 02 | Preliminary action |
| 14 | Pneumatic or hydraulic constructions |
| 01 | Change in aggregate state |
| 12 | Local property |
| 37 | Equipotentiality |

**Radical (Physical) Contradiction**

| The steam locomotives | ⟹ | should create steam as power resource to haul the train | **vs** | also should not create steam to eliminate the negative influence to environment especially in cities |
|---|---|---|---|---|

**INVENTING**

Group of navigators to model the idea for the "Non-steam locomotive" concept:

**02. Preliminary action**: Railway electrification (like the power source, and new railway to adapt the electric locomotive) are requisite for applying electrical technology in locomotive field.

**04. Replacement of mechanical matter**: Electrical technologies are used within the field of locomotive industry instead of the diesel locomotives.

**ZOOMING**

Have the contradictions been removed? – ~~Yes~~. – No.

Super-effects: Yes. With the new technology, the labor costs are also reduced.

Negative effects: Large scale of railway electrification is requisite and the cost is high.

**BRIEF DESCRIPTION**

An electric locomotive is a locomotive powered by electricity from overhead lines, a third rail or an on-board energy storage device. Electricity is used to eliminate smoke and take advantage of the high efficiency of electric motors, but the cost of railway electrification is also high.

### 10.2.2.4. ELECTRIC ELEVATOR

An elevator (or lift in the Commonwealth) is a type of vertical transport equipment that efficiently moves people or goods between floors (levels, decks) of a building, vessel or other structures. Elevators are generally powered by electric motors that either drive traction cables or counterweight systems like a hoist, or pump hydraulic fluid to raise a cylindrical piston like a jack.

### The Prototype – Steam & hydraulic Elevator

From about the middle of the 19th century, elevators were powered, often steam-operated, and were used for conveying materials in factories, mines, and warehouses. In 1823, two architects Burton and Hormer built an "ascending room" as they called it, this crude elevator was used to lift paying tourists to a platform for a panorama view of London. In 1835, architects Frost and Stutt built the "Teagle", a belt-driven, counter-weighted, and steam-driven lift was developed in England.

In 1846, Sir William Armstrong introduced the hydraulic crane, and in the early 1870s, hydraulic machines began to replace the steam-powered elevator. The hydraulic elevator is supported by a heavy piston, moving in a cylinder, and operated by the water (or oil) pressure produced by pumps.

The first elevators designed expressly for passenger use were introduced in the 1850s. In 1854, in a dramatic demonstration at the New York Crystal Palace Exhibition, Elisha Graves Otis (see in fig. 10.38) demonstrated the first "safety elevator."

With the elevator set up in a prominent part of the exhibition hall, he stood on the elevator platform as it was raised four stories. He then had the suspension rope cut.

The audience gasped, but the platform did not hurtle to the ground. Instead it stood locked and safely suspended above the ground. Four years later in 1857, Otis installed the first passenger elevator in E.V. Haughwout & Co., a store located on Broadway in New York, NY.

**fig. 10.38.** Otis demonstrating safety elevator 1854 NYC

Powered by a steam engine, the elevator at Haughwout was the talk of the city, as thousands of curious visitors flocked to the store.

### Contradiction Analysis

The "plus-state" of the "steam & hydraulic elevator" can be determined as in As-Matrix as number 22 (speed), because the steam & hydraulic elevator extremely improves the speed to deliver passengers and goods to altitude. The "minus-state" of the "steam & hydraulic elevator" can be determined as in As-

Matrix as number 04 (reliability), because it is limited in the height to which it could rise.

**Inventing Idea**

Combing the plus and minus state factors above, the results will show several points in As-navigators. The target is to improve the reliability of steam & hydraulic elevator. After combing the As-Matrix plus-22 and minus-04, the idea navigator resulted:

● Navigator 04: replacement of mechanical matter – according to this navigator, electrical, magnetic, or electromagnetic technology can be considered to apply. Generally speaking, compared with traditional methods, electrical, magnetic or electromagnetic technologies have advantages in many aspects. Compared with the steam & hydraulic power, electricity provides more stable working condition.

● Navigator 28: previously installed cushions – according to this navigator, special devices should be set to ensure the safety of using electric elevator.

**The Result Artifact (Solution) – Electric Elevator**

In 1880, Werner von Siemens demonstrated the first electric powered elevator at the Mannheim Pfalzgau exhibition.

After he discovered the dynamo-electric principle in 1866, he concentrated on finding practical applications for this new technology. One of his most famous constructions was the first electric railway, which ran round a Berlin industrial exhibition in 1879 and was very popular with the visitors. What is not so well-known is that the idea of the electric elevator was to some extent a by-product of this. The pioneer of electricity was soon to have an opportunity to put this idea into practice.

In April 1880, the organizers of the "Mannheim Pfalzgao Trade & Agricultural Exhibition" asked whether Siemens could build an "electric elevator" for their exhibition. Von Siemens accepted the commission and work commenced shortly afterwards.

With a motor mounted on the bottom of the cab, Siemens' electric elevator (see in fig. 10.39) used a gearing system to climb wall shafts fitted with racks.

**fig. 10.39.** Werner von Siemens, Electric elevator, 1880 Mannheim

Although novel, this electric elevator was still too crude to compete with the existing steam-driven, hydraulic elevator technology. Intended simply as an illustration broad applicability of his pioneering work in D.C. traction motors, Siemens had little interest in pursuing the electric elevator further.

Siemens instead focused on large projects, such as electric trains and electric power systems. Electric-powered elevators offered two significant advantages.

First, electric power was clearly becoming universally available, and any building likely to be equipped with an elevator would also have electric power.

Second, hydraulic elevators were severely limited in the height to which they could rise, while electric elevators, using a simple cable and pulley system, had virtually no height limit. For many years, electric elevators used either direct current (DC) motors or alternating current (AC) motors.

Today, almost all elevators use one of two types of AC motors: the most common are geared motors for elevators moving at speeds up to 500 feet per minute (153 m per minute), while direct-drive motors are used for elevators moving at higher speeds. Some modern high-speed elevators move at up to 2,000 feet per minute (610 m per minute).

---

### Extracting-1

| LC | No. | Navigator | Substantiation for the Extracting |
|---|---|---|---|
| ++ | 04 | **Replacement of mechanical matter** | Electrical technology is applied within this new development. |
| ++ | 02 | **Preliminary action** | Electricity supply is requisite for apply electric elevator. |
| + | 28 | **Previously installed cushions** | Safety devices are built to prevent injury in urgency. |

### Extracting-2

**Standard Contradiction:** the steam & hydraulic elevator provides convenience for climbing to high position, but it has limitation in height it could rise and has low reliability.

**Radical Contradiction:** the height the steam & hydraulic elevator could rise should be high so that it could service more requirement, but it should also be low so that it can maintain its reliability and safety.

**FIM**: X-resource, together with available or modified resources, and without making the object more complex or introducing any negative properties, guarantees attainment of the following **IFR**: [elevator to reach higher position with more reliability].

---

**fig. 10.40.** Extracting of invention "Electric Locomotive"

**fig. 10.41.** Reinventing of invention "Electric Locomotive" – next page!

## TREND

The steam & hydraulic elevator provides much convenience for climbing, but it has limitation in the height it could rise due to the reliability and safety issues.

PROBLEM: How to increase the height the elevator could rise?

## REDUCING

**FIM** (shortly): [elevator to reach higher position with more reliability].

**Standard Contradiction**

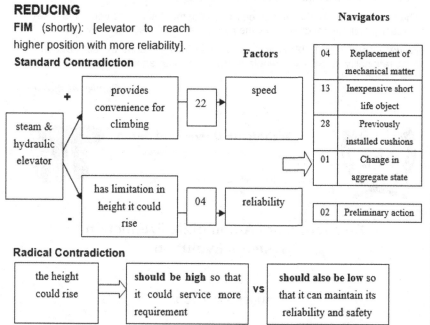

**Radical Contradiction**

| the height could rise | ⟹ | **should be high** so that it could service more requirement | **vs** | **should also be low** so that it can maintain its reliability and safety |

## INVENTING

Group of navigators to model the idea for the "advanced elevator" concept:

**04. Replacement of mechanical matter**: The electric elevator is equipped with electric powered engine instead of the old time steam or hydraulic power.

**28. Previously installed cushions**: Safety devices are built to prevent injury in urgency.

## ZOOMING

Have the contradictions been removed? – ~~Yes~~. – No.

Super-effects: Yes. With the development of electricity supply, it becomes more universally available. Negative effects: Electricity supply should be requisite, but it's not a big issue today.

## BRIEF DESCRIPTION

In 1880, Werner von Siemens demonstrated the first electric powered elevator at the Mannheim Pfalzgau exhibition. It provides much convenience to people and changes the ways of people to live even in recent days.

# 11 Reinventing of Automobile Production Systems Evolution [162]

The projection of technical systems was still an art 100 years ago. Today it's become an exact science through systems development.

The appearance of TRIZ and its quick development is no accident. It is a necessary development dictated by the modern scientific-technical revolution.

Work "according to TRIZ" continues to displace work "done by feel". But, human rationality is not idle: people will always think about more complicated assignments.[163]

*Genrikh Altshuller*

**Technische Universität Berlin**

*"Wir haben die ideen für die zukunft"*

Global Production Engineering
International Masters Program

## Reinventing of Automobile Production Systems Evolution

by

**SUGENG WAHYUDI**

A thesis submitted in partial fulfillment of the requirements for the degree of
'Master of Science in Global Production Engineering'

Under the guidance of
**Professor, Dr. Dr. Sc. techn. Michael Orloff**

**Fakultät V – Verkehrs- und Maschinensysteme**
Produktionstechnisches Zentrum

**Technische Universität Berlin**

**Global Production Engineering**

**Dean: Prof. Dr.-Ing. Günther Seliger**

**Department of Assembly Technology and Factory Management**

**Berlin, Germany**

---

[162] Part 11 is based on the edited fragments of the master thesis by Sugeng Wahyudi (Indonesia / Singapore), a certified graduate of MTRIZ course at GPE Program, TUB
[163] compiled and translated by M.Orloff from different works of G.Altshuller

## 11.1 From Craft Production to Mass Production System

In this section, the case studies will focus on the innovations occurred in the evolution of Craft Production to Mass Production System.

### 11.1.1 Interchangeability of Parts

For centuries, after the Industrial Revolution in England in the late 18th century, economic production was based on the notion of craftsmen. Everything was crafted by skilled workers who had the requisite materials, tools, and most important, skills.

Although in the late of 1800s – when the first manufacturers of automobiles flourished – machinery and mechanization had been used as primary instruments of production, the skills of the craftsmen still played an important role. The skills and methods was the source of pride that made the craftsman was exclusive to each other.

**The Prototype – Independent manufacturing stations with individual gauging system**

In a Craft Production System, no two automobiles were built exactly the same. This was due to the parts manufactured had high inconsistency of dimensions. This made the assembly process could only be performed by skilled craftsmen, who still needed long time to fix and fit the parts each other. Abovementioned problems were due to the parts were manufactured in several machine shops that have their own gauging systems.

The first company formed exclusively to build automobiles was Panhard et Levassor (P&L) in France[164]. It was a very famous manufacturer at that time and its Système Panhard, a breakthrough design with front-mounted engine and rear wheel drive, became standard for all other automobiles.

Although using the same blueprint, the company could not produce even two identical cars. It was because the parts were supplied by different machine shops, and each of them used their own gauging system and specifications. This condition was exacerbated by limitation of the technology in processing the raw material.

When these parts came to the automobile company for fitting, the parts specifications could be described as "approximate". The first things need to do by the workers were try to fit the parts to each other until the whole vehicle was complete. The job required a lot of skill, time and effort since each part did not suit to each other. This caused a "dimensional creep" phenomenon, since by the time the workers reached the last part, the whole vehicle could differ significantly in dimension from the other car that was built by using the same blueprints[165].

---

[164] Gorgano, G. N. (1985) *Cars: Early and Vintage, 1886-1930.* – Grange-Universal, London

[165] Womack, J. P., Jones, D. T. and Roos, D. (1990) *The Machine that Changed the World.* – Rawson Associates, New York

Those were the reasons why automobile company at that time could not produced cars in large quantity with the same dimension and quality.

**fig. 11.1.** Fixing and fitting of car body in a craft work-shop

## The Result of Contradiction Analysis

The "plus-state" of the "independent manufacturing stations with individual gauging system" can be determined as in A-Matrix as **number 22 (speed)**, because the parts could be produced in several parallel machine shops that increase rate of availability of parts. The "minus-state" of the "independent manufacturing stations with individual gauging system" can be determined as in A-Matrix as **number 05 (precision of manufacture)**, because those parts became not fit to each other due to different gauging systems.

**fig. 11.2.** Machine shop of Auto Hacket Motorcar Co., Michigan, in early 1900s

## Inventing Idea

Combining the plus and minus state factors above, the results will show several points in As-navigators. The target is to keep separate manufacturing stations so that the rate of parts production will still be maintained while they are able to produce parts with consistent dimensions that are fit to each other. After combining the A-Matrix 22 and 05, the idea navigator resulted :

• **Navigator 02 (preliminary action)** – according to this navigator, the manufacturing machine shops can prepare the parts in advanced by using standard gauging system so that the assembly process can be processed without loss of time for fixing and fitting the parts.

• **Navigator 18 (mediator)** – according to this navigator, specific gauging devices were used as mediators between the company and the suppliers in order to have standardized measurement of parts.

## The Product-artifact (Solution) – Interchangeability of Parts

> The way to make automobiles is to make one automobile just like
> another automobile.[166]
>
> *Henry Ford*

Started in 1903, Ford Motor Company adopted the Craft Production during its early years as the common production system available at that time. When it introduced its famous Model T in 1908, the production volume rate was very small, while the demand was increasing. When Ford discovered the disadvantages of this production system, he began to improve it one by one.

Many people think that the key to Mass Production was the continuous assembly line. This opinion is not quite correct because the continuous assembly line could not work if the parts could not be attached to each other in a fast and simple way. To achieve this, the assembly line should be supplied by standard parts so that all of the parts could be fitted without any difficulties. This was the concept that called "interchangeability of parts".

To achieve this condition, Ford insisted that all the machine shops that supplied parts to his company needed to use the same gauging system in their manufacturing process. This was a simple idea, but a breakthrough at that time, since there were no companies had thought about that.

By the standardization of the gauging system, all the parts delivered to Ford's assembly line could be easily fitted each other, and reduce the assembly process time.

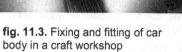
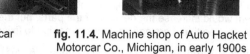

**fig. 11.3.** Fixing and fitting of car body in a craft workshop

**fig. 11.4.** Machine shop of Auto Hacket Motorcar Co., Michigan, in early 1900s

With this condition, Ford could also reduce the number of highly-skilled fitters in his company and saved labor cost. In the end, the company could produce more cars in lower price. Simultaneously, the quality of the products could be relatively uniform.

---

[166] ibid.

| Unidentical parts produced by independent machine shops | Identical parts by Interchangeability concept |
|---|---|
| WAS | IS |

| LC | No. | Naviga-tor | Substantiation for the Extracting – 1 |
|---|---|---|---|
| ++ | 02 | Prelimi-nary ac-tion | Prepare parts in advanced by using standard gauging system so that the assembly process can be done without loss of time. |
| + | 10 | Copying | Standard gauging system are copied and used in every machine shop that supplies the parts to the company. |
| + | 11 | Inverse action | Instead of letting the machine shops use their own gauging system, a standard gauging system needs to be enforced to them. |
| ++ | 18 | Mediator | A specific gauging device is used as mediator between the company and the suppliers in order to have standardized parts measurement. |
| + | 38 | Homoge-neity | The gauging devices that used by machine shops shall have similar structure and made of similar material in order to give standard measurement result. |

**Extracting-2**

**Standard Contradiction:** Independent manufacturing stations with their own gauging system would increase the delivery of the parts, but the parts produced would not fit to each other so that it prolong assembly time due to fixing and fitting activities.

**Radical Contradiction:** The manufacturing stations should be unified so that they could produced parts that fit to each other, but they should also be separated, so that the manufacturing speed could increase.

**FIM** (shortly): [ *automobiles with consistent similarity* ].

**fig. 11.5.** Extracting: Interchangeability of Parts

## TREND

In a Craft Production System, no two cars were built exactly alike. This product variation issue was due to the parts manufactured had high inconsistency of dimensions. This also made the assembly process could only be performed by skilled craftsmen, who needed long time to fix and fit the parts each other. These problems were caused due to the parts were manufactured in several independent machine shops that have their own gauging systems.

*Fixing and fitting activities in Craft Production*

PROBLEM: How can we build automobiles with consistent similarity ?

### REDUCING

**FIM:** X-resource, together with available or modified resources, and without making the object more complex or introducing any negative properties, guarantees attainment of the following **IFR**: [ *automobiles with consistent similarity* ].

### Standard (Technical) Contradiction

**Factors**

| | |
|---|---|
| 22 | Speed |
| 05 | Precision of manufacture |

Parallel process that increased delivery rate of parts

Parts were not fit to each other due to different gauging system

Independent manufacturing stations with individual gauging systems

(+)  (−)

**Navigators**

| 02 | Preliminary action |
|----|--------------------|
| 04 | Replacement of mechanical matter |
| 09 | Change in color |
| 29 | Self-servicing |

| 18 | Mediator |
|----|----------|

**fig. 11.6 - beg.** Reinventing: Interchangeability of Parts

## Radical (Physical) Contradiction

The manufacturing stations → **should be unified**, *so that they could produce parts that were fit to each other* & **should be separated**, *so that manufacturing speed could increase*

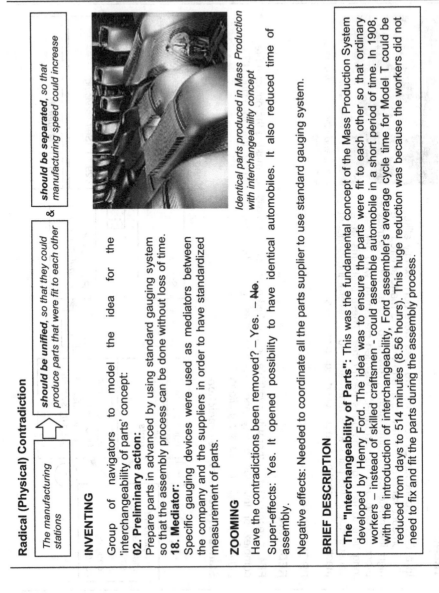

*Identical parts produced in Mass Production with interchangeability concept*

### INVENTING

Group of navigators to model the idea for the 'interchangeability of parts' concept:

**02. Preliminary action:**
Prepare parts in advanced by using standard gauging system so that the assembly process can be done without loss of time.

**18. Mediator:**
Specific gauging devices were used as mediators between the company and the suppliers in order to have standardized measurement of parts.

### ZOOMING

Have the contradictions been removed? – Yes. – ~~No~~.

Super-effects: Yes. It opened possibility to have identical automobiles. It also reduced time of assembly.

Negative effects: Needed to coordinate all the parts supplier to use standard gauging system.

### BRIEF DESCRIPTION

**The "Interchangeability of Parts":** This was the fundamental concept of the Mass Production System developed by Henry Ford. The idea was to ensure the parts were fit to each other so that ordinary workers – instead of skilled craftsmen - could assemble automobile in a short period of time. In 1908, with the introduction of interchangeability, Ford assembler's average cycle time for Model T could be reduced from days to 514 minutes (8.56 hours). This huge reduction was because the workers did not need to fix and fit the parts during the assembly process.

**fig. 11.6 - end.** Reinventing: Interchangeability of Parts

### 11.1.2 Dedicated Machine Tools

In the era of Craft Production System, general purpose machines or tools were used to all sorts of processes. Most of the manufacturing activities by using these machines were still performed manually by craftsmen.

### The Prototype – General Purpose Tools

In a craft automobile industry, the company just acted as designer and assembler. Major parts were made and supplied by many independent machine shops run by individual craftsmen. These craftsmen had their own machineries and tools that used for making all kind of parts ordered by the automobile company. Most of the tools were general ones, with some of them – by their own innovations – have been adapted to some specific functions.

In 1890s, Panhard et Levassor workforce was mainly composed of skilled craftsmen who hand-built only several hundred automobiles per years. The craftsmen were using general machines or tools that were used to all sort of processes: drilling, grinding, cutting, stamping, forming, and other operations on metal and wood.[167]

These general machines have flexibility in processing many kinds of parts. But they required set-up and adjustment by highly skilled machinist. The product result also had poor accuracy that required a lot of fixing.

With the raising demand of cars in the early of 1900s, it was impossible to rely on those ordinary machines, since the parts were not produced in correct and consistent dimension and shape. This obstructed productivity because the fitters needed long time to fit the parts during assembly. The production cost became high because it required highly skilled workers to do the jobs.

There were also not many skillful workers available at that time that caused high competition among the car manufacturers to employ them. This competition drove up wages and put the craftsmen in higher bargain position to get best paid job. In the end, it would increase the labor cost and the companies need to sell the products in a higher price.

**fig. 11.7.** Car body assembly with general tools

**fig. 11.8.** Finishing activity in craft manufacturer

---

[167] ibid.

## The Result of Contradiction Analysis

The "plus-state" of the "general purpose tools" can be determined as in $A_S$-Matrix as **number 02 (universality, adaptability)**, because the craftsman needed to provide and use only few tools to perform many kind of jobs. The "minus-state" of the "general purpose tools" can be determined as in $A_S$-Matrix as **number 05 (precision of manufacture)**, the tools had low precision in producing the parts that caused the parts did not have correct and consistent dimension and shape.

## Inventing Idea

Combining the plus and minus state factors above, the results will show several points in $A_S$-navigators. The target is to increase the precision characteristic of the tools so that in the end the workers do not require long time to fit the parts each other during assembly process. After combining the $A_S$-Matrix 02 and 05, the idea navigators resulted :

- **Navigator 12 (local property)** – according to this navigator, the transition from a similar to a different structure of machine tools is required so that every tool can complete its function under the best condition.

- **Navigator 02 (preliminary action)** – according to this navigator, it is necessary to make partial or complete change of the tools so that they can be put to work from the best position and are available without loss of time.

- **Navigator 05 (separation)** – according to this navigator, the incompatible property of the tools shall be separated so that each tool can have the best property to complete its work.

## The Product-artifact (Solution) – Dedicated Machine Tools

Therefore in 1909, I announced one morning without any previous warning, that in the future we were going to build only one model, that the model was going to be 'Model T', and that the chassis would be exactly the same for all cars, and I remarked: 'Any customer can have a car painted in any colour that he wants, so long as it is black. [168]

*Henry Ford*

To achieve standardization with high quality of precision of parts, high volume of products, and low cost, dedicated machines for special purpose of tasks were required.

As has been mentioned earlier, the fundamental key of the Mass Production System was the interchangeability of parts that laid in standardization of gauging system. But this was not enough. Ford required this interchangeable parts need to be produced in lower cost. In a Craft Production System – where a single machine could do many tasks – many set-ups and adjustment needed to

---

[168] McNairn, W. and McNairn, M. (1978) *Quotations from the Unusual Henry Ford.* – Quotamus Press, Redono Beach, CA

be performed by skilled machinist that require long time and high labor cost. So, the key of producing inexpensive interchangeable parts laid in machines that could perform jobs in high volume with low set-up costs between pieces.

At his factory in Highland Park, Ford provided machines dedicated to produce a single item only and also nearly automated, so they could produce high volume of highly accurate parts with tight tolerances required to eliminate hand fitting.

For example, Ford introduced two types of dedicated machines, fifteen machines for milling blocks and thirty machines just for milling heads.

Ford also purchased stamping presses, used to make sheet-steel parts, with die spaces large enough to handle only a specific part.

By making machines that could only do single task, Ford could dramatically reduced set-up time. His engineers also perfected simple jigs and fixtures for holding the work piece in this dedicated machine that allowed unskilled workers – with a short period of training – to operate them.

This meant the company could reduce its dependency on special machinist and in the end it could reduce the labor cost.

This dedicated machines strategy was running well because Ford only produced one model of car in a long period of time[169] (Model T was produced from 1908 – 1927, followed by Model A from 1927 – 1931).

This was made possible by the homogeneity of the customer demand at that time that could be satisfied only by low variants of products.

fig. 11.9. Car body assembly with
dedicated press machine

fig. 11.10. Dedicated machine tools
in a mass production manufacturer

---

[169] HFHA – Henry Ford Heritage Association (2009): The Ford Story.
hfha.org/HenryFord.htm#Ford-Motor-Co

| *Craft production with general purpose tools* | *Dedicated machine tools in Mass Production* |

WAS                    IS

| LC | No. | Navigator | Substantiation for the Extracting – 1 |
|----|-----|-----------|----------------------------------------|
| ++ | 02 | **Preliminary action** | Make necessary (partial or complete) change of the tools so that they can be put to work from the best position and are available without loss of time. |
| + | 03 | **Segmentation** | "Disassemble" the universal tool into "individual" tools with their own specific task. |
| ++ | 05 | **Separation** | Separate the incompatible property from the tools so that each tool can have the best property to complete its work. |
| + | 11 | **Inverse action** | Instead of using universal tool for performing many tasks, dedicated tools can be used for each specific task. |
| ++ | 12 | **Local property** | Transition from a similar to a different structure of machine tools so that every tool can complete its function under the best condition. |

**Extracting-2**

**Standard Contradiction:** Universal tools could be used in making many parts, but they had low precision that could not produced parts with correct and consistent dimension and shape.

**Radical Contradiction:** The machine and tools should be common so that the worker could easily do the job, but they also should be unique so that the parts produced had high precision.

**FIM** (shortly): [ *high precision of parts* ].

**fig. 11.11.** Extracting: Dedicated Machine Tools

## TREND

In an automobile company that use Craft Production System, the craftsmen used general purpose machine & tools to perform drilling, grinding, cutting, stamping, forming and other operations on metal and wood. This obstructed productivity because the parts were not produced in the correct and consistent dimension and shape, so in the end the craftsmen needed long time to fit the parts during assembly.

*Craft production with universal tools*

PROBLEM : How could we increase the precision of the parts ?

## REDUCING

**FIM:** X-resource, together with available or modified resources, and without making the object more complex or introducing any negative properties, guarantees attainment of the following **IFR:** [ *high precision of parts* ].

## Standard (Technical) Contradiction

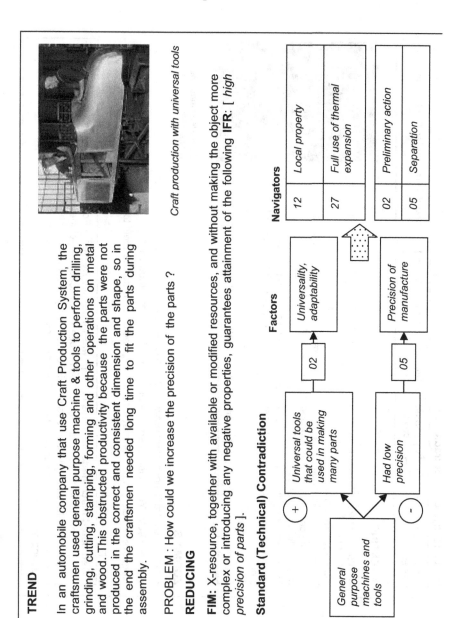

**Factors**

- Universality, adaptability
- Precision of manufacture

(+) Universal tools that could be used in making many parts — 02

(−) Had low precision — 05

General purpose machines and tools

**Navigators**

| 12 | Local property |
|----|----------------|
| 27 | Full use of thermal expansion |

| 02 | Preliminary action |
|----|---------------------|
| 05 | Separation |

**fig. 11.12 - beg.** Reinventing: Dedicated Machine Tools

## Radical (Physical) Contradiction

| The machines and tools | **should be common,** so the worker could easily do the job | & | **should be unique,** so that the parts produced had high precision |

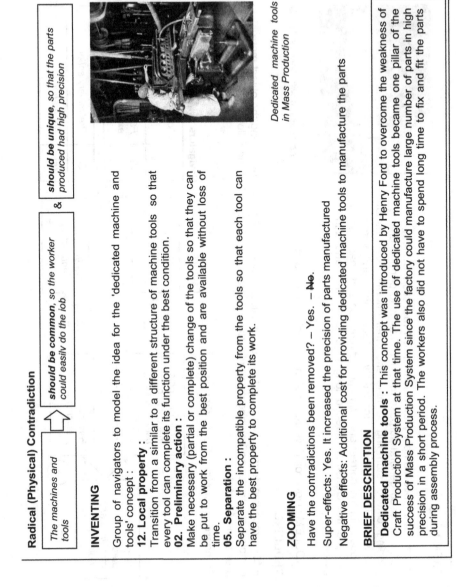

*Dedicated machine tools in Mass Production*

### INVENTING

Group of navigators to model the idea for the 'dedicated machine and tools' concept :

**12. Local property :**
Transition from a similar to a different structure of machine tools  so that every tool can complete its function under the best condition .

**02. Preliminary action :**
Make necessary (partial or complete) change of the tools so that they can be put to work from the best position and are available without loss of time.

**05. Separation :**
Separate the incompatible property from the tools so that each tool can have the best property to complete its work.

### ZOOMING

Have the contradictions been removed? – Yes. – N̶o̶.

Super-effects: Yes. It increased the precision of parts manufactured

Negative effects: Additional cost for providing dedicated machine tools to manufacture the parts

### BRIEF DESCRIPTION

**Dedicated machine tools :** This concept was introduced by Henry Ford to overcome the weakness of Craft Production System at that time. The use of dedicated machine tools became one pillar of the success of Mass Production System since the factory could manufacture large number of parts in high precision in a short period. The workers also did not have to spend long time to fix and fit the parts during assembly process.

**fig. 11.12 - end.** Reinventing: Dedicated Machine Tools

### 11.1.3 Design Simplification

### The Prototype – Engine with Fragmented Parts

The four-stroke internal combustion engine is the type that commonly used for automobile and industrial purposes today (cars, trucks, generators, etc.). This concept was developed in 1877 by Nikolaus Otto[170]. Later on it was leveraged by Wilhelm Maybach, who built the first four-cylinder, four-stroke internal combustion engine for Daimler Motoren Gesellschaft in 1890. The engine[171] has an output of five hp (3.7 kW) at 620 rpm. This was an epoch-making invention that replaced the two-stroke engine concept and later on used widely by many automobile companies.

The four-stroke cycle engine is more efficient, last longer and less polluting than the two-stroke cycle one. The disadvantages of this concept are it requires more moving parts and manufacturing expertise.

In the early 1900s, for the automobile industries that still adapted craft production, the production of four-stroke cycle engine involved many parts need to be manufactured. The cylinders need to be casted separately and the assembly required several highly skilled fitters. This contributed to the long production time and high labor cost in producing cars.

The design of the automobile at that time was also very complex that an owner would need to employ mechanic as his/her personal staff. This caused the vehicle could only be owned by wealthy people, and became one prohibiting factor to make automobiles as common consumer goods.

**fig. 11.13.** The first 4-stroke, 4-cylinder automobile engine invented by Wilhelm Maybach in 1890

**fig. 11.14.** Robert Peugeot in a Peugeot car in 1897. Only wealthy person could afford this kind of vehicle

---

170 Nikolaus August Otto (1832 – 1891) – a German inventor who made the first internal-combustion engine with four-stroke concept
171 Hodzic, M. (2008) Gottlieb Daimler, Wilhelm Maybach and the "Grandfather Clock"; www.benzinsider.com/2008/06/gottlieb-daimler-wilhelm-maybach-and-the-grandfather-clock

## The Result of Contradiction Analysis

The "plus-state" of the "engine with fragmented parts" can be determined as in $A_S$-Matrix as **number 09 (ease of manufacture)**, because it was easier for the manufacturer to produce the parts since they had simple shapes. The "minus-state" of the "engine with fragmented parts" can be determined as in $A_S$-Matrix as **number 10 (ease of use)**, because it required long time and skillful workers to assemble the moveable parts into final product.

## Inventing Idea

Combining the plus and minus state factors above, the results will show several points in $A_S$-navigators. The target is to reduce the number of the parts need to be assembled so that ordinary workers can perform the activity in short period of time. After combining the $A_S$-Matrix 09 and 10, the idea navigators resulted :

- **Navigator 11 (inverse action)** – according to this navigator, it is advised to make the "moveable" parts of the engine fixed, or the opposite.

- **Navigator 35 (unite)** – according to this navigator, similar parts or parts for neighboring operations can be united.

## The Product-artifact (Solution) – Design Simplification

I will build a car for the great multitude. It will be large enough for the family, but small enough for the individual to run and care for. It will be constructed of the best materials, by the best men to be hired, after the simplest designs that modern engineering can devise.[172]

*Henry Ford*

Since Model T were using the 4-cylinder, 4-stroke engine, Ford also found similar problem faced by other automobile companies: long assembly time and high labor cost. This obstructed Ford's ambition to have a cheap, massive production of automobiles. He believed that one of the important keys to achieve that goal was high production rate supported by simplicity of design and ease of attachment.

To overcome those problems, Ford's idea was to develop an engine design that reduced the number of parts needed and made these parts easy to attach. It was C. Harold Wills[173], one of Ford's senior engineer who finally made the new four-cylinder engine block casted in one piece, with a detachable cylinder head, which was very daring at that time (Tebo, 2010).

This design decreased the production time since it reduced the number of moving parts in engines and simplified the assembly process. It also omitted the requirement for skilled fitters that in the end could save the labor cost.

---

[172] Ford, H. and Crowther, S. (1922) *My Life and Work*. – Nevins and Hill, Ford TMC

[173] Childe Harold Wills (1878 – 1940) – an American engineer, one of the first employees of the Ford Motor Co. and contributed to the design of the Model T. Founder of Wills Sainte Claire automobile company

To attract consumers from the average level, Ford had designed cars with unprecedented ease of operations and maintainability. He wanted that ordinary users with low ability in mechanical skills would be able to fix the cars by themselves with modest tool kits.

For example, the owners could remove carbon deposits from the engine valves with the Ford Valve Grinding Tool, which came with the vehicle. Owners could also buy spare parts at Ford dealer and replaced the broken parts by themselves. The procedures of solving 140 common problems of the car were provided in a simple question-and-answer form of Model T's owner's manual.

The simplicity of design and ease of attachment gave Ford tremendous advantages over his competitors that catapulted Ford to the head of the world's automobile industry.

**fig. 11.15.** Ford's 4-cylinder, 4-stroke engine
that casted in one piece

**fig. 11.16.** A Ford Model T that had been modified as farm truck. Simplicity of design made this model of vehicle favored by farmers for modifications

*A complex, 4-cylinder engine designed by Maybach (1890)*

*Design simplification on Ford Model-T engine, a four-cylinder engine block cast in one piece*

WAS                                          IS

| LC | No. | Navigator | Substantiation for the Extracting |
|----|-----|-----------|-----------------------------------|
| + | 02 | **Preliminary action** | Partial change on the design of the machine need to be done. |
| + | 05 | **Separation** | Separate the parts only if necessary. |
| ++ | 11 | **Inverse action** | Try to make the moveable parts of the engine fixed, or the opposite. |
| + | 12 | **Local property** | Every part should exist under condition that it should be able to fit easily during assembly process. |
| + | 16 | **Partial or excess effect** | If it is difficult to assemble the engine from completely separated parts, then some parts of the engine can be unified first during manufacturing. |
| ++ | 35 | **Unite** | Unite similar parts or parts for neighboring operation. |

**Extracting-2**

**Standard Contradiction:** Engine with fragmented parts will give simpler process during manufacturing process, but the company need to hire specialists to assemble the engine in a long time.

**Radical Contradiction:** The engine construction should be complex so that it could carry out its function, but it also should be simple so that ordinary workers could assemble it fast.

**FIM:** X-resource, together with available or modified resources, and without making the object more complex or introducing any negative properties, guarantees attainment of the following **IFR**:

[ *fast assembly process of engine* ].

**fig. 11.17.** Extracting: Design Simplification

The first 4-cylinder engine designed by Wilhelm Maybach (1890)

## TREND

In engine type used during the Craft Production System was comprised of large number of individual parts and cylinders that need to be joined and bolted together by skilled specialist. This caused the assembly process for the engine required a long time. This also caused high cost due to the process could only be performed by specialists.

PROBLEM : How could we reduce the time required to assemble the engine ?

## REDUCING

**FIM:** X-resource, together with available or modified resources, and without making the object more complex or introducing any negative properties, guarantees attainment of the following **IFR:** [ *fast assembly process of engine* ].

## Standard (Technical) Contradiction

**Factors**

Ease of manufacture

09

Simpler process to manufacture the individual parts

(+)

Engine with fragmented parts

Ease of use

10

Needed a specialist to assemble the engine parts in a long time

(–)

**Navigators**

| | |
|----|------------------------|
| 05 | Separation |
| 11 | Inverse action |
| 16 | Partial or excess effect |
| 35 | Unite |

**fig. 11.18 - beg.** Reinventing: Design Simplification

## Radical (Physical) Contradiction

| The engine construction | ⇨ | *should be complex, so that it could carry out its function* | & | *should be simple, so that ordinary workers could assemble it fast* |

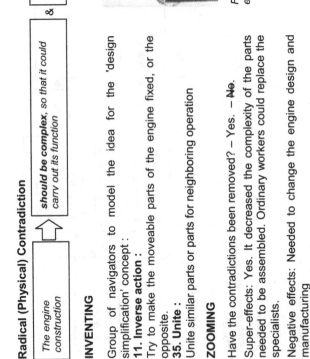

*Ford Model-T Engine, a four-cylinder engine block cast in one piece*

### INVENTING

Group of navigators to model the idea for the 'design simplification' concept :

**11. Inverse action :**
Try to make the moveable parts of the engine fixed, or the opposite.

**35. Unite :**
Unite similar parts or parts for neighboring operation

### ZOOMING

Have the contradictions been removed? – Yes. – ~~No~~.

Super-effects: Yes. It decreased the complexity of the parts needed to be assembled. Ordinary workers could replace the specialists.

Negative effects: Needed to change the engine design and manufacturing

### BRIEF DESCRIPTION

**The "Design Simplification":** This was an important concept that contributed to rise of the Mass Production System. With this concept, less time were required to assemble automobile's engine. Ford reduced the number of moving parts in the engine, i.e. the engine casting comprised of single complex block. By contrast, competitors casted each cylinder individually and bolted them together during assembly. With this innovation, Ford company could reduce the labour cost due to less specialists were required to do the work.

**fig. 11.18 - end.** Reinventing: Design Simplification

### 11.1.4 Delivery of Parts

The static workstation that had been adopted by Craft Production System in automobile company brought consequence that the vehicles sat in a place while the craftsmen need to bring all the tools and materials into the location.

### The Prototype – Self-picking Parts

Since a lot of parts were required to build a vehicle, the job of assembling the car usually could be considered as a simpler task compared with handling the materials that had to be brought in. This was complicated by the location of parts that stored quite far from the assembly stations. The craftsmen also needed to find them first in the stock piles before could get what they were looking for.

In the early years of Ford's Model T production, a craftsman was given a full responsibility in making the whole vehicle, including picking-up all the parts and semi-finished products required.

Since hundreds of parts were required, it would take a long time for the craftsman to go to each material store, picked the parts, and put it in his workstation before he could continue his work.

These repetitive activities were not only time-consuming – it also decreased the productivity of the craftsman and obstructed the production output.

**fig. 11.19, a.** Ford's early assembly line. All required parts were brought near the workstation

### The Result of Contradiction Analysis

The "plus-state" of the "self-picking parts" can be determined as in $A_S$-Matrix as **number 07 (complexity of construction)**, because it did not require complex layout arrangement to introduce the parts storage into the production line. The "minus-state" of the "self-picking parts" can be determined as in $A_S$-Matrix as **number 03 (level of automation)**, because the workers needed to leave their workstation to pick the required parts. This activity also required long time and delay the assembly process.

### Inventing Idea

Combining the plus and minus state factors above, the results will show several points in $A_S$-navigators. The target is to reduce the waiting and transportation time due to the workers need to pick the required parts by themselves. The new alternative will make the workers to be easier to get the parts but without putting all the parts in the production line that will make the line crowded. After combining the $A_S$-Matrix 07 and 03, the idea navigators resulted :

• **Navigator 07 (dynamization)** – it is advised to make the required parts 'moveable' to the location of the workers.

• **Navigator 18 (mediator)** – it is possible to use another object to transfer all the required parts from the store to the workstations.

### The Product-artifact: Delivery of Parts

> Use work slides or some other form of carrier so that when a workman completes his operation, he drops the part always in the same place – which place must always be the most convenient place to his hand – and if possible have gravity carry the part to the next workman for his operation … Use sliding assembly lines by which the parts to be assembled are delivered at convenient distances. [174]
>
> *Henry Ford*

fig. 11.19, b. Ford's assembly line with its improved parts delivery system in 1910s. Some of the parts were put in a sloped-rack that supplied the parts by gravity force

As has been discussed earlier, the problem of Craft Production was that the craftsmen wasted their time to pick-up the parts every time they needed them. The step Ford took to make his plant more efficient was to deliver the parts to each workstation so that the assemblers could remain at the same spot all day. The improvement project was handled by Charles Sorensen[175], Ford's engineer, and Charlie Lewis, Ford's foreman. They arranged storage rooms for light-handling stocks and bulky parts, like engines and axles, closer to the assembly stations. They then arranged with the stock department to bring up batch of materials and parts required and delivered them to each workstation regularly. With this arrangement, the craftsmen just needed to wait in their workstations and could focus with their activities.

fig. 11.20. Parts delivery system in Austin automobile company in 1950s. The workers had lists of major items needed for each car, and they had to ensure correct items (engine, gearbox, etc.) were put in the correct boxes before sending them to the workstations by conveyor

No more time wasted for craftsmen to pick-up the parts and this means higher productivity and output. This innovation would become an important supporting strategy for the implementation of Ford's Mass Production – the parts that came to the worker, and no the opposite concept like had been implemented for many years in Craft Production.

---

[174] McNairn, W. and McNairn, M. (1978) *Quotations from the Unusual Henry Ford.* – Quotamus Press, Redono Beach, CA

[175] Sorensen, S. E. (1956) *My Forty Years with Ford.* – Norton, New York

*Self-picking parts by workers*

*Parts delivery system*

WAS                                          IS

| LC | No. | Navigator | Substantiation for the Extracting – 1 |
|----|-----|-----------|----------------------------------------|
| + | 02 | **Preliminary action** | Prepare the parts in advanced so that the workers do not need to pick them. |
| ++ | 07 | **Dynamization** | Make the parts "moveable" to the location of the workers. |
| + | 11 | **Inverse action** | Instead of having the workers to pick the parts by themselves, the parts are delivered to the workers. |
| ++ | 18 | **Mediator** | Use another object to transfer all the required parts to the location of the car assembled. |
| + | 29 | **Self-servicing** | The parts are "coming by itself" to the location of the workers. |

### Extracting-2

**Standard Contradiction:** Having parts storage separated from the work station requires simple construction of work station area, but the workers had to pick all the parts required all the time.

**Radical Contradiction:** The parts storage should be exist in the work station so that the workers could easily pick the required parts, but the storage should also not be exist so it would not make the work station area to be crowded.

**FIM:** X-resource, together with available or modified resources, and without making the object more complex or introducing any negative properties, guarantees attainment of the following **IFR**:

[ *low assembly time* ].

**fig. 11.21.** Extracting: Delivery of Parts

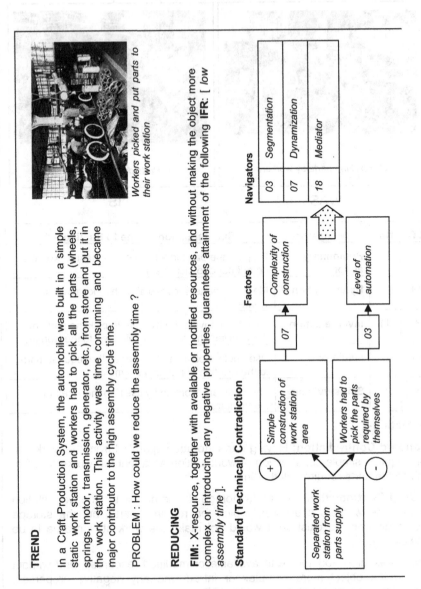

**TREND**

In a Craft Production System, the automobile was built in a simple static work station and workers had to pick all the parts (wheels, springs, motor, transmission, generator, etc.) from store and put it in the work station. This activity was time consuming and became major contributor to the high assembly cycle time.

PROBLEM : How could we reduce the assembly time ?

**REDUCING**

**FIM:** X-resource, together with available or modified resources, and without making the object more complex or introducing any negative properties, guarantees attainment of the following **IFR:** [ *low assembly time* ].

**Standard (Technical) Contradiction**

*Workers picked and put parts to their work station*

**Factors**

**Navigators**

| 03 | Segmentation |
| 07 | Dynamization |
| 18 | Mediator |

fig. 11.22 - beg. Reinventing: Delivery of Parts

## Radical (Physical) Contradiction

*The parts storage* ⟱ **should be exist in work station**, *so that the workers could easily pick the required parts*

&

**Should not be exist**, *so it would not make the work station area to be crowded*

*Parts delivery system*

### INVENTING

Group of navigators to model the idea for the 'delivery of parts' concept :

**07. Dynamization :**
Make the parts 'moveable' to the location of the workers.

**18. Mediator :**
Use another object to transfer all the required parts to the location of the car assembled.

### ZOOMING

Have the contradictions been removed? – Yes. – No.

Super-effects: Yes. It eliminated the transportation time required for workers to pick the parts that would reduce the assembly process time.

Negative effects: Needed a good coordination between people in the store and people in the work stations in terms of number, types and time of parts need to be delivered. Also it required material handlers and system to deliver the parts.

### BRIEF DESCRIPTION

The **"Delivery of Parts"**: This was an important concept that introduced by Henry Ford that inspired the modern supply chain management. With this concept, Ford could reduce the assembly cycle time from hours to minutes. Material handlers were employed to deliver parts from store to the working station, so the workers could stay and focus to his work. Modern automobile companies currently use conveyors to move the parts to the location of the vehicles, in correct type, correct quantity and correct time.

**fig. 11.22 - end.** Reinventing: Delivery of Parts

### 11.1.5 Division of Assembly Line

The craft production of automobile was depended on the skill of the workers. The skilled workers or craftsmen were required to perform major activities in the manufacturing and assembly processes from raw materials until finished vehicles.

### The Prototype – Single-craftsman Operation

The early automobile design had already comprised of thousands of parts and hundreds of activity steps. At that time, only skillful craftsman were able to per-formed most of the complex tasks. However, it took a long time to finish a vehi-cle since he needed to change his work type and adapt with changes of tools from time to time.

In the beginning of Model T production in 1908, the vehicle body was still sup-plied by a craft manufacturer like Fisher Body Company.

The parts of the vehicle such as motor, transmission, generator, springs and wheels would also be assembled on a chassis by crafts-man. This activity took a whole day to complete. The assembly line cy-cle time at that time was 514 minutes, or 8.56 hours (s. Wom-ack). This kind of performance did not satisfied Ford's expectation for producing cars in massive output rate to achieve a state of econo-mies.

**fig. 11.23.** Car frame assembly process performed by single

### The Result of Contradiction Analysis

The "plus-state" of the "single-craftsman operation" can be determined as in A$_S$-Matrix as **number 02 (universality, adaptability)**, because it only required one worker to perform all the activities from the beginning to the end of the as-sembly process. The "minus-state" of the "single-craftsman operation" can be determined as in A$_S$-Matrix as **number 09 (ease of manufacture)**, because the worker required time to change his working pattern and to adapt with changes of tools.

### Inventing Idea

Combining the plus and minus state factors above, the results will show sever-al points in A$_S$-navigators. The target is to reduce the long assembly time due to the worker need some time every time he changes his type of work and tools. The new alternative will make the worker need not to change his work and tools but all the activities to assemble the cars are still carried out. After combining the A$_S$-Matrix 02 and 09, the idea navigators resulted :

- **Navigator 03 (segmentation)** – according to this navigator, it is required to disassemble the assembly process into several steps.

- **Navigator 11 (inverse action)** – according to this navigator, instead of assemble the car alone, the process can be performed by several workers.

- **Navigator 12 (local property)** – according to this navigator, it is necessary to change the structure of the assembly process so that it has different steps with different functions.

- **Navigator 31 (unite)** – according to this navigator, the segmentation of the assembly process can be supported by uniting several process steps that are quite similar or working on the same parts.

### The Product-artifact (Solution) – Division of Assembly Line

> Place the tools and the men in sequence of the operation so that each component part shall travel the least possible distance while in the process of finishing.[176]
>
> *Henry Ford*

The history of managing complexity in the organization begins with Adam Smith[177], who published The Wealth of Nations in 1776, where he introduced concept of "division of labor" – the specialization of labor as a way of workers to achieve greater productivity by breaking large jobs into small tasks.

Another development in increasing productivity and efficiency was developed by Frederick Winslow Taylor. His principle in scientific management was known as *Taylorism* and suggest the importance of task discretion and providing training, detailed instruction and supervision to increase the performance of the workers. [178]

Ford followed above principles by applying standardization of work content and job fragmentation. Thus, not only parts were interchangeable, but labor had become replaceable too because they became merely an appendage of the production apparatus.

The impact of Taylorism at Ford replaced the previously tacit knowledge owned by craftsmen by a more complex but explicit system of clearly defined specific tasks as standards.

By segmenting tasks into smaller units, job contents were reduced and could be more controlled.

---

176 McNairn, W. and McNairn, M. (1978) *Quotations from the Unusual Henry Ford.* – Quotamus Press, Redono Beach, CA

177 Adam Smith (1723 – 1790) – a Scottish economist and social philosopher. Cited as the father of modern economics and capitalism. Author of *The Theory of Moral Sentiments and An Inquiry into the Nature and Causes of the Wealth of Nations;* http://www.encyclopedia.com/topic/Adam_Smith.aspx#4

178 Montgomery, D. (1989) *The Fall of the House of Labour:* The Workplace, the State, and American Labour Activism, 1865-1925. – Cambridge University Press, Cambridge, UK

After implementing the practice of delivery of parts into the workstation and reaching perfect part interchangeability, Ford decided to divide the assembly line into several subassembly divisions so that each worker would only perform a single task and move from vehicle to vehicle around the assembly hall.

There were divisions of workers who performed activities of fitting and assembling, parts supplying, repairing and checking. There were also foremen whose tasks were to check the activities of the workers and informed the line condition to the managers.

By August 1913, after the implementation of division of assembly line, the production cycle time for the average Ford assembler had been incredibly reduced from 514 minutes to 2.3 minutes (99.55% improvement) (s. Womack).

By 1915[179], when the assembly lines at Highland Park had already run in its full capacity, there were more than 7,000 assembly workers there. The workers spoke in more than fifty languages and many of them could barely speak English. This was a proof that division of assembly line implementation in Ford's company was very extreme that the workers need not to communicate each other in order to finished the products.

With dividing the assembly line, the workers could concentrated on a specific task and tools that improved their working efficiency. The production cycle time would decrease and the productivity increased.

On the other hand, craftsmen did not like this concept since it reduced their role and responsibility in the production process. The company then could hire ordinary workers, that were cheaper, to do most of the assembly activities since the tasks became simpler. It also required only few minutes of training to prepare the workers in performing their tasks.

**fig. 11.24.** Workers on a flywheel assembly line at the Ford Motor Company's Highland Park plant, Detroit, in 1913. Each worker completed a different part of the component

**fig. 11.25.** Workers on a Ford assembly line working on the top and bottom ends of the engines (1946)

---

179 Raff, D (1987) *Wage Determination Theory and the Five-Dollar Day at Ford.* – Ph.D. Dissertation, Massachusetts Institute of Technology

*Assembly process by single craftsman*

WAS

*Division of Assembly Line in Ford Motor Company*

IS

| LC | No. | Navigator | Substantiation for the Extracting – 1 |
|----|-----|-----------|----------------------------------------|
| + | 02 | **Preliminary action** | The assembly process needs to be prepared in advanced so that many workers can assemble the product in parallel and in series. |
| ++ | 03 | **Segmentation** | Disassemble the process into several steps. |
| + | 08 | **Periodic action** | Transition from a continuous assembly process (by one worker) to several periodic ones (by many workers). |
| ++ | 11 | **Inverse action** | Instead of assemble the car alone, the process can be performed by several workers. |
| ++ | 12 | **Local property** | Change the structure of the assembly process so that it has different steps with different functions. |
| + | 13 | **Inexpensive short-life object ...** | Low wage-easily replaceable workers as a replacement for high wage-difficult to find craftsmen. |
| ++ | 31 | **Unite** | The segmentation of the assembly process need to be supported by uniting several steps that are quite similar or working on the same parts. |

**Extracting-2**

**Standard Contradiction:** Assembly process by using single worker who built the vehicle from start to end was simple since it only requires one worker to perform all the activities, but it required long assembly time since the worker needed to adapt continuously with different type of tasks and tools.

**Radical Contradiction:** The assembly process steps should be few so that the worker could focus on less activities, but they should also be many since it was the nature of automobile assembly process to have many processing steps.

**FIM** (shortly) [ *low assembly time* ].

**fig. 11.26.** Extracting: Division of Assembly Line

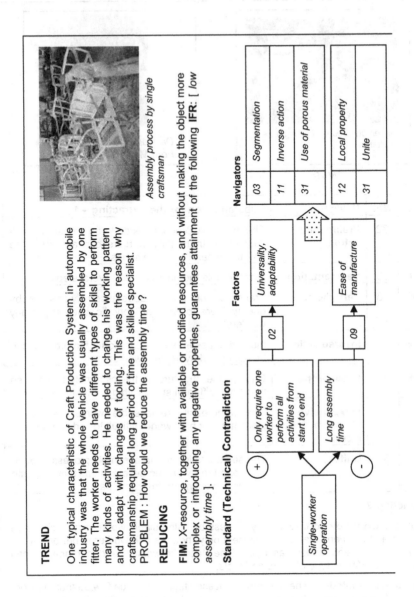

**TREND**

One typical characteristic of Craft Production System in automobile industry was that the whole vehicle was usually assembled by one fitter. The worker needs to have different types of skilsl to perform many kinds of activities. He needed to change his working pattern and to adapt with changes of tooling. This was the reason why craftsmanship required long period of time and skilled specialist.
PROBLEM : How could we reduce the assembly time ?

**REDUCING**

**FIM:** X-resource, together with available or modified resources, and without making the object more complex or introducing any negative properties, guarantees attainment of the following **IFR:** [ *low assembly time* ].

**Standard (Technical) Contradiction**

**fig. 11.27 - beg.** Reinventing: Division of Assembly Line

## Radical (Physical) Contradiction

| The assembly processing steps | ⇨ | *should be few*, so that the worker could focus on less activities | & | *should be many*, because it was the nature of automobile assembly process |

### INVENTING

Group of navigators to model the idea for the 'division of assembly line' concept :

**03. Segmentation :**
Disassemble the assembly process into several steps.

**11. Inverse action :**
Instead of assemble the car alone, the process can be performed by several workers.

**12. Local property :**
Change the structure of the assembly process so that it has different steps with different functions.

**31. Unite :**
The segmentation of the assembly process need to be supported by uniting several process steps that are quite similar/working on the same parts.

### ZOOMING

Have the contradictions been removed? – Yes. – ~~No~~.

Super-effects: Yes. It reduced the time required to assemble an automobile due to each worker could focus on one specific task. Skilled craftsmen could also be replaced by ordinary workers.

Negative effects: Needed a good management in distributing the tasks.

### BRIEF DESCRIPTION

The **"Division of Assembly Line"**: The concept was rooted from Adam Smith's concept of 'division of labour'. Henry Ford revolutionized this concept to support Mass Production System by dividing the assembly line into several stations. This would increase the productivity, because complete familiarity with a single task meant the worker could perform it faster. This basic concept was still used by every manufacturing and assembly companies.

*Division of Assembly Line in Ford Motor Company*

**fig. 11.27 - end.** Reinventing: Division of Assembly Line

### 11.1.6 Moving Assembly Line

In a Craft Production System, the automobile company provided segmented areas for craftsmen to build the vehicle from scratch until finish. Single craftsman perform his job in one area from making frame, installing engines, tires, and fit all the parts and the body of the car.

### The Prototype – Static Work station

The static workstation concept was a common sense in early automobile industry since the vehicle was mainly built by one worker only. The vehicle was built from scratch in one place, and it was the duty of the craftsman to put all the parts. Until 1913, Ford's assembly line still followed the traditional Craft Production, where the vehicle was built in one stationary area. Although he had successfully implemented practices of supplying the parts into the workstations and fragmenting the assembly line, the workers still need to move from one vehicle to another in order to perform his tasks. It was a common sense of practice at that time because people could move easier and faster than the material.

However, Ford noticed weakness from the existing concept: walking activity, even for a small distance, took time. The workers also need to bring his tooling with him and jam-ups as faster workers overtook the slower workers in front of them. This problem need to be resolved since the time wasted actually reduced the productivity of the workers.

fig. 11.28. Ford's Body assembly room at Highland Park Plant, Detroit, in 1910-1911. The vehicle bodies were put on a static support while the components were brought and assembled into them

fig. 11.29. Another typical of static workstation in a craft company of automobiles

### The Result of Contradiction Analysis

The "plus-state" of the "stationary workstation" can be determined as in As-Matrix as **number 07 (complexity of construction)**, because in a static workstation it required only a simple construction to support the car that was being assembled. The "minus-state" of the "stationary workstation" can be determined as in As-Matrix as **number 03 (ease of manufacture)**, because the worker required time to move from one station to another.

## Inventing Idea

Combining the plus and minus state factors above, the results will show several points in A$_S$-navigators. The target is to reduce the long assembly time due to the motion time required by the workers to move between stations. The new alternative will make the assembled cars are still in fixed position to enable the workers to do necessary activities. After combining the A$_S$-Matrix 07 and 03, the idea navigators resulted: **Navigator 07 (dynamization)** – according to this navigator, it is required to make the object to be able to move, instead of having the workers to move between stations.

## Product-artifact (Solution) – Moving Assembly Line

> Along about April 1, 1913, we first tried the experiment of an assembly line...I believe that this was the first moving line ever installed. The idea came in a general way from the overhead trolley that the Chicago packers used in dressing beef.[180]
>
> *Henry Ford*

The moving assembly line was adopted from standard production process techniques which were already established in both the flour milling and meatpackers industries.[181] As early as 1790[182], Oliver Evans[183] applied the principle of process flow in the flour milling industry. He recognized that the input of the material (grain) to the mill was one of the most important ways to control the speed of production. This early practice was then perfected by meatpacking industry started in 1850 by installing moving overhead trolley in slaughter houses to eliminate slow human handling of carcasses. It was Ransome Eli Olds, owner of the Olds Motor Works in Detroit, who first introduced the basic principle of the moving material into the automotive industry in 1901.[184] Always on the hunt for more efficient and lower cost, in 1913 Ford started experimentation of the moving assembly principle in his plants.

Ford's aim of using a moveable assembly line was to reduce the amount of travel workers had to undertake to get from one workstation to the next: "If the workers were going to work like machines, Ford engineers concluded that the entire factory had to work like machine, that the success of assembly line production depended on efficient supply of materials and parts to workstations" (s. Biggs). The development towards the moving assembly line at Ford could divided into three phases. In 1906, Ford had experimented with work slides at the Bellevue plant.

---

[180] McNairn, W. and McNairn, M. (1978) *Quotations from the Unusual Henry Ford.* – Quotamus Press, Redono Beach, CA

[181] Clarke, C. (2005) Automotive Production System and Standardisation: From Ford to the Case of Mercedes Benz. – Physica-Verlag, Heidelberg

[182] Biggs, L. (1996) The Rational Factory, Architecture Technology and Work in America's Age of Mass Production. – The John Hopkins University Press, Baltimore, London

[183] Oliver Evans (1755 – 1819) – an American inventor

[184] Gartman, D. (1986) Auto Slavery: The Labour Process in the American Automobile Industry, 1897 – 1950. – Rutgers University Press, New Brunswick and London

In spring 1913, belt conveyor was installed in the flywheel magneto operation at Ford's Highland Park plant. In August 1913, after completing the implementation of division of assembly line, Ford started to implement this concept into the final assembly line (s. Clarke).

Ford's moving assembly line consisted of two strips of metal plates mounted on a belt rolled along the factory. Thus, with this conveyor, instead of having the workers moved the vehicle from station to station, the conveyor brought the it past the stationary workers. This innovation could reduce cycle time of car assembly process from 2.3 minutes to 1.19 minutes. The overall assembly time could be reduce from 12 hours and 8 minutes to 2 hours and 35 minutes by October 1913, and to 1.5 hours only by April 1914. The difference laid due to reduction of walking activity of the workers. The production output also soared, from 12,000 units in the first year of production to 734,811 units in 1916 – contributed nearly half of the U.S. automobile production. The moving assembly line allowed for a constant flow of production through the plant in which each station was manned by a worker placing standardized parts in a standardized manner onto the moving standardized chassis. Three years after its implementation, the moving assembly line had become standard practice in all U.S. automobile manufacturers.

**fig. 11.30.** Workers assembled automobiles in Ford's moving assembly line

**fig. 11.31.** Moving assembly line at the final assembly station (Henry Ford's automobile factory in Highland Park, Detroit, 1913)

**fig. 11.32.** Workers working on Model T body in a moving assembly line

**fig. 11.33.** Moving assembly line increased Ford Model T output

*Static work station*

*Moving assembly line*

WAS

IS

| LC | No. | Navigator | Substantiation for the Extracting – 1 |
|---|---|---|---|
| + | 01 | **Change in the aggregate state ...** | Change the "degree of flexibility" of the assembly line. |
| + | 02 | **Preliminary action** | The assembly line need to be prepared to reduce workers movement. |
| ++ | 07 | **Dynamization** | Make the object to move so that the workers need only to stand-by in their position. |
| + | 08 | **Periodic action** | Multiple usage |
| + | 11 | **Inverse action** | Instead of having the workers to move between stations, the object should be moved. |
| + | 12 | **Local property** | Change the structure of the work station so that they can deliver the vehicle being built from station to station. |
| + | 18 | **Mediator** | Use some equipment or tools to deliver the vehicle being built between stations. |
| + | 19 | **Transition into another dimension** | The work station is shaped so that it can move. |

**Extracting-2**

**Standard Contradiction:** Static work station required only a simple construction, but it consumed a long time for workers to move from station to station.

**Radical Contradiction:** The position of the assembled cars should be stationary so the worker could do the job on it, but it should also be moveable so that the worker needed not to move between stations.

**FIM** (shortly): [ *low assembly time* ].

**fig. 11.34.** Extracting: Moving Assembly Line

## TREND

In a Craft Production System, the assembly process of automobile was performed in a static work station. When this concept was combined with 'division of assembly line' (different workers performed different tasks), it had a weakness. The assembly time was not in line with the expectation since the workers needed time to move from one station to another.

PROBLEM: How could we reduce the assembly time due to the motion time of workers to move between stations?

## REDUCING

FIM: X-resource, together with available or modified resources; and without making the object more complex or introducing any negative properties, guarantees attainment of the following **IFR**:

[ *low assembly time* ].

## Standard (Technical) Contradiction

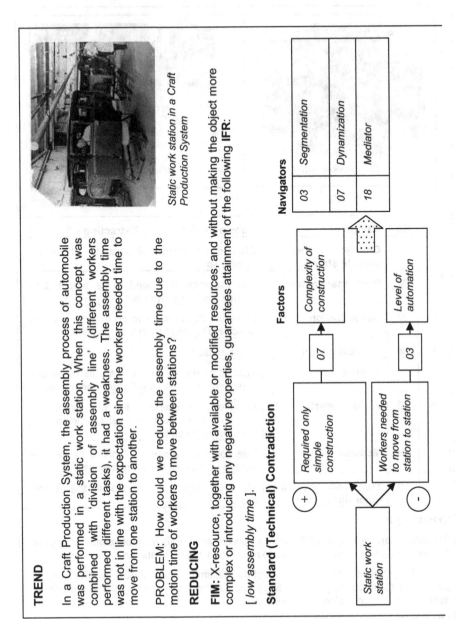

*Static work station in a Craft Production System*

**Factors**

Complexity of construction

Level of automation

Required only simple construction

Workers needed to move from station to station

Static work station

07

03

**Navigators**

| 03 | Segmentation |
| 07 | Dynamization |
| 18 | Mediator |

fig. 11.35 - beg. Reinventing: Moving Assembly Line

## Radical (Physical) Contradiction

| The position of the assembled cars | **should be stationary,** so the worker could do the job in it | & | **should be moveable,** so that the worker needed not to move between stations |
| --- | --- | --- | --- |

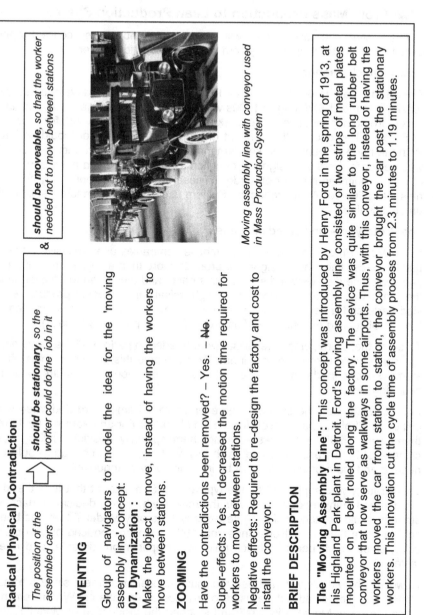

*Moving assembly line with conveyor used in Mass Production System*

### INVENTING

Group of navigators to model the idea for the 'moving assembly line' concept:

**07. Dynamization :**

Make the object to move, instead of having the workers to move between stations.

### ZOOMING

Have the contradictions been removed? – Yes. – ~~No~~.

Super-effects: Yes. It decreased the motion time required for workers to move between stations.

Negative effects: Required to re-design the factory and cost to install the conveyor.

### BRIEF DESCRIPTION

**The "Moving Assembly Line":** This concept was introduced by Henry Ford in the spring of 1913, at his Highland Park plant in Detroit. Ford's moving assembly line consisted of two strips of metal plates mounted on a belt rolled along the factory. The device was quite similar to the long rubber belt conveyor that now serve as walkways in some airports. Thus, with this conveyor, instead of having the workers moved the car from station to station, the conveyor brought the car past the stationary workers. This innovation cut the cycle time of assembly process from 2.3 minutes to 1.19 minutes.

**fig. 11.35 - end.** Reinventing: Moving Assembly Line

## 11.2 From Mass Production to Lean Production System

In this section, the case studies will focus on the innovations occurred in the evolution of Mass Production System to Lean Production System.

### 11.2.1 Quick Changeover

One of the main reasons that Mass Production was able to developed so extensively in the U.S. during in 1900s was that the American market at that time was more homogenous than the market in other parts of the world, for example in Europe. This caused the concept of product standardization could endure. With the huge amount of market demand and low claim of product varieties, it was easier for Ford to design a production system that served huge quantities of cars in a limited number of models. These are the background why companies that adopted Mass Production System used dedicated machines for their production lines.

**The Prototype – Dedicated Machines**

In the Mass Production era, the automobile companies used massive and expensive press lines that designed to produce more than a million units of parts in a year. The dies weighed several tons each, and they need to be aligned in the press with absolute precision. A slight misalignment would produce wrinkled parts and a serious misalignment could make the sheet metal melted in the die, obligating expensive and time-consuming repairs (s. Womack).

To avoid these problems, specialists were needed to performed the die changes, which could take a whole full day. To overcome this, the companies often stamp specific parts in a dedicated presses for months or years, without changing dies.

In Ford Motor Co., the strategy was even more extreme. Ford dedicated machines to all of the parts, even for the simple ones. For an example, by 1914, there were 15,000 machines—among them specially designed milling machines—at Ford's assembly line. With this strategy, no die changes would be needed and the production will not be interrupted by changeover.

This strategy was supported by the typical of U.S. market at that time that did not demand many models. That was the reason why Ford's dedicated machine strategy could be successful for decades with its only Model T that has been sold to more than 15 millions units until 1927 – which produced only in black color, since 1914, because black paint is cheaper, more durable, and dried faster.[185]

However, this strategy demanded too much capital investment since the machines and dies could not be used to produce other variants. When Ford released the new Model A in 1927 and stopped the production of Model T, the

---

[185] McCalley, B. W. (1994) *Model T Ford: The Car That Changed the World.* – Krause Publications, Iola, WI

previous Model T's machines and dies were obsolete and need to be scrapped.

When Toyota tried to adopt the Mass Production to their plant in Japan after the World War II, they realized that they could not implement the principle of dedicated machines.

This was due to Japanese car market was smaller, and the customer demanded various type of cars – that forced the company to do frequent die changes.

The problem was, with existing machines and die construction at that time, it would take a long time for specialists to perform die changes and it would not be feasible to produce a variant in small quantities.

**fig. 11.36.** Dedicated stamping press machine for car fender in Ford's River Rouge Plant, 1927

Therefore, U.S. production equipments were not appropriate for Japan because they developed for the Mass Production that handle fewer type of products, larger lots, and enormous sales.

Moreover, with dedicated machines and dies, it would require too much capital – something that Toyota could not afford. Hence, a breakthrough needed to be done to solve this problem.

### The Result of Contradiction Analysis

The "plus-state" of the "dedicated machines" can be determined as in As-Matrix as **number 10 (ease of use)**, because the machines were ready to use and did not need specialist that required long time to do die–changes in every set-up. The "minus-state" of the "dedicated machines" can be determined as in As-Matrix as **number 02 (universality, adaptability)**, because each machine could not be used for producing other type of parts. This would increase the investment cost.

**fig. 11.37.** A stamping press machine weighed 225 tons that produced 2,700 fenders per day, one fender at each stroke. It was such an accurate machine that the fenders required no further finishing touches before enameling

## Inventing Idea

Combining the plus and minus state factors above, the results will show several points in As-navigators. The target is to avoid high investment cost required to provide dedicated machines to produce many parts.

The new alternative will give possibility to have fewer machines that can produce many type of parts in the same speed and quality as previous concept. After combining the A-Matrix 10 and 02, the idea navigators resulted :

- **Navigator 03 (segmentation)** – according to this navigator, it can be considered to raise the degree of disassembly of the machines.

- **Navigator 07 (dynamization)** – according to this navigator, it is advised to disassemble the machine with its dies so they can be moveable among each other.

## The Result-artifact (Solution) – Machines with Quick Changeover

With regards to the problem faced by Toyota, Taiichi Ohno had an idea to develop die construction with simple die-change techniques so that the changeover activity could be done frequently in a short time. He introduced roller principle to move die in and out of machine and simple adjustment to avoid long time and skilled workers to perform the die change.

Using a few used American presses, Ohno perfected his technique for quick changeover/setup. By the late 1950s, he had reduced the time required to change dies from one day to three minutes and eliminated the need for die-change specialists (s. Womack). The setup on a bolt-forming machine at Toyota Motors could be reduced from eight hours to 58 seconds.[186] These examples were typical kinds of improvements achieved using Single Minute Exchange of Die (SMED), or Quick Changeover principle.

There are two types of setup operations:

- Internal Setup (IED – Internal Exchange of Die): setup activities that can be performed only when the machine is stopped, such as removing old dies and mounting new dies.

- External Setup (OED – Outer Exchange of Die): setup activities that can be completed without stopping the machine, such as transporting dies to or from storage.

The key principle of SMED method to decrease the setup time will be:

**1.** distinguish internal and external setup activities;

**2.** convert internal setup to external setup;

**3.** improve activities in both categories.

---

[186] Shingo, S. (1989) A study of the Toyota Production System from an Industrial Engineering Viewpoint. – Productivity Press, Cambridge, MA

**fig. 11.38.** Toyota's stamping press machines with quick die change concept. Right side is Danly machine and left side is Komatsu machine

By shortening setup time with adopting SMED in the production line, some further benefits can be expected:

- increase machine operating time – increase productivity of the machine;

- can process material in small lots that can reduce finished goods inventories and inventory between processes;

- flexibility of production in responding to fluctuating demand.

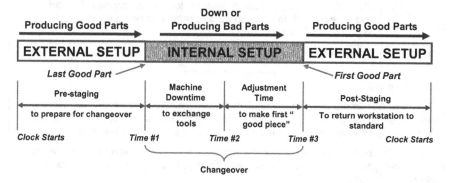

**fig. 11.39.** Principles of setup in a Single Minute Exchange of Die (SMED) concept

Dedicated machine
without changeovers

Machines with quick changeover

WAS                                          IS

| LC | No. | Navigator | Substantiation for the Extracting |
|---|---|---|---|
| + | 02 | Preliminary action | The machines and dies shall be prepared in advanced so that the machine can handle several dies with quick changeover. |
| ++ | 03 | Segmentation | Raise the degree of disassembly of the machines / dies. |
| + | 05 | Separation | Separate the unnecessary activities from the changeover process. |
| ++ | 07 | Dynamization | Disassemble the machine with its dies so that they can be moveable each other. |
| + | 08 | Periodic action | Transition of the machine from a continuous function to a periodic one. |
| + | 11 | Inverse action | Instead of having one machine to produce one type of parts, the machine can be modified to be able to produce many types of parts with simple changeover. |
| + | 15 | Discard and renew of parts | Use the same machine to serve many dies. The out-of-date dies can be discarded and replaced with new dies. |
| + | 20 | Universality | Use universal machine to serve many dies. |

**Extracting-2**

**Standard Contradiction:** The use of dedicated massive and complex machines had an advantage of readiness in use since they did not need any set-up and specialist for die-change, but they could not be used for producing other parts.

**Radical Contradiction:** The press die machine should be common to accommodate many types of parts, but it should also be unique so that it could produce parts with high precision.

**FIM** (shortly): [ universal & simple stamping press machine ].

**fig. 11.40.** Extracting: Machines with Quick Changeover

## TREND

In a Mass Production System, the construction of the stamping press dies were massive, complex and expensive. In order to make another type of part, specialists require a long time (one full day) for set-up (die-change) to ensure absolute precision. To overcome this issue, companies usually used dedicated press sets for every type of part. This required high investment cost that could only be covered by high production rate. This solution became not feasible when the market volume was low or demanded high variation of car types.

PROBLEM : How could we avoid the high-cost-dedicated press dies ?

## REDUCING

**FIM:** X-resource, together with available or modified resources, and without making the object more complex or introducing any negative properties, guarantees attainment of the following **IFR:**

[ *universal & simple stamping press machine* ].

## Standard (Technical) Contradiction

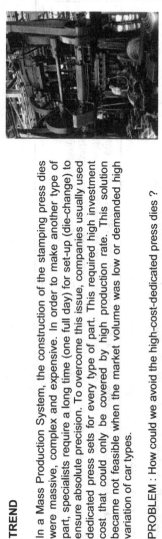

Stamping press in Ford plant, River Rough, 1927

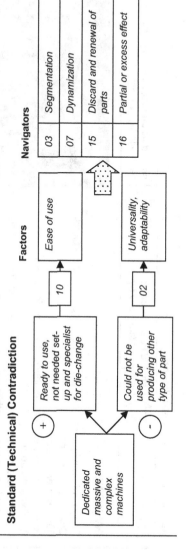

| Navigators | |
|---|---|
| 03 | Segmentation |
| 07 | Dynamization |
| 15 | Discard and renewal of parts |
| 16 | Partial or excess effect |

**fig. 11.41 - beg.** Reinventing: Machines with Quick Changeover

## Radical (Physical) Contradiction

| *The press die* | ⇧ | ***should be common,*** *to accommodate many types of parts* | & | ***should be unique,*** *so it could produce parts with high precision* |

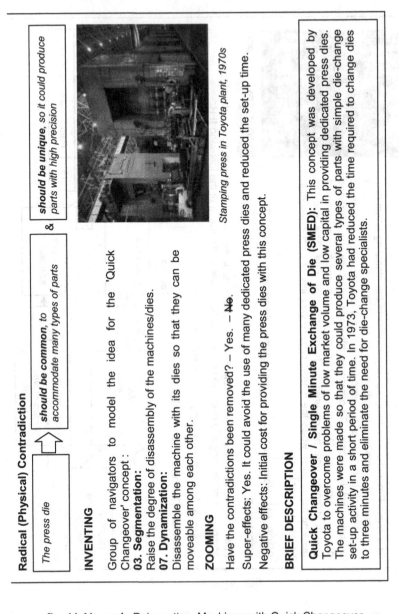

*Stamping press in Toyota plant, 1970s*

### INVENTING

Group of navigators to model the idea for the 'Quick Changeover' concept :

**03. Segmentation:**
Raise the degree of disassembly of the machines/dies.

**07. Dynamization:**
Disassemble the machine with its dies so that they can be moveable among each other.

### ZOOMING

Have the contradictions been removed? – Yes. – ~~No.~~

Super-effects: Yes. It could avoid the use of many dedicated press dies and reduced the set-up time.

Negative effects: Initial cost for providing the press dies with this concept.

### BRIEF DESCRIPTION

**Quick Changeover / Single Minute Exchange of Die (SMED):** This concept was developed by Toyota to overcome problems of low market volume and low capital in providing dedicated press dies. The machines were made so that they could produce several types of parts with simple die-change set-up activity in a short period of time. In 1973, Toyota had reduced the time required to change dies to three minutes and eliminate the need for die-change specialists.

**fig. 11.41 - end.** Reinventing: Machines with Quick Changeover

### 11.2.2 Continuous Flow

Mass Production System focuses on providing affordable products for the "masses". In order to achieve this, it must be supported by another principle: economies of scale – lower cost and price that were achieved with greater output. As prices were lowered, more people could afford to buy the products, resulting in greater sales and therefore greater production, even lower costs and prices, and so on.

### The Prototype – Batch-and Queue Process

In Mass Production, it is necessary to produce high amount of products and also keep high amount of inventory. Machineries were made larger in pursuit of scale economies. The massive expense of the machinery encourages batch production and buildup of huge work-in-process and finished goods inventories (even if there is no customer to buy it). These inventories appeared as assets on company balance sheet, despite the enormous amounts of cash they absorbed.

According to Mass Production System, batch production is necessary to reduce the changeover frequency as minimum as possible, since the more frequent the changeover, the lower the available production time and the higher the labor cost.

As a matter of fact, batch production brings several disadvantages[187]:

**a.** It creates batch delays.

Whenever parts are processed in a batch, the entire batch, except for the one piece being processed, is delayed "in storage" – in either an unprocessed or a processed state until all pieces in the batch have been processed. This is called as "waiting-in-process". Little attention has been paid to such delays because they are typically concealed within processing times. The only reason for increasing batch size is the assumption that this will compensate for delays caused by long changeover. The truth is that batch delays prolong the production cycle time considerably.

**b.** It creates waste of transportation.

Batch production brings consequence of transportation of pile of parts from one station to another. Usually, improving transportation means replace manual handling with tools such as trolleys, forklifts, etc., which actually only improves the work of transport. Transport only increases costs and never adds value. Typically, labor cost in production activity consist of 45 percent of processing, 5 percent each of inspection and delays, and 45 percent of transport. Even when the manual transport is mechanized, transportation costs are just shifted from manual to mechanical – an investment without return. With this in mind, the elimination of transport through layout improvement is required.

---

[187] Shingo, S. (1989) A study of the Toyota Production System from an Industrial Engineering Viewpoint. – Productivity Press, Cambridge, MA

c.   It creates quality problem.

Batch production also creates quality problem because similar defect are usually replicated throughout the batch before it can be caught – usually at the end of line.

In 1970s, Bumper Works, located near Danville, Illinois, U.S., was a typical batch-and-queue factory. Bumper Works made chrome and painted steel bumpers in a variety of styles for pickup trucks. It made large batches of each type of bumper – typically a month's worth – before shifting production to next model and delivered the products to new-car dealers and repair body shops.

Because large batches were considered normal at that time, it was not important that it took sixteen hours to changeover Bumper Works' stamping presses. Batches of raw materials also came in large amount and they were considered unavoidable. Thus, Bumper Works had a warehouse in its plant to store tons of sheets of steel as supplies.

And because the chroming company – performing the key step in the middle of the production process – also worked in a batch mode, Bumper Works piled up semi-finished bumpers in its intermediate goods warehouse until there was an enormous batch and then shipped them to the chroming company all at once.

When the chroming company shipped the materials back in a batch, they were run through a final assembly operation (to install inner reinforcing bars, attachment brackets, and cosmetic coverings), stored once more in a finished goods warehouse, and sent in a batch to the customer according to a predetermined schedule.

**fig. 11.42.** Example of mass production with large batch of inventory. These were jeeps produced for WWII allies

**fig. 11.43.** Batch production and transport of SsangYong New Actyon in Sollers-Far East plant, Korea. Sollers is a Russian automobile company headquartered in Moscow

In 1985, Bumper Works was signed on as a supplier for a small volume of Toyota business in North America, and in 1987 won a sole-source contract for the bumpers on the new version of Toyota's small pickup. The owner of Bumper Works, Shahid Khan – a successful immigrant from Pakistan who started his career as production worker at Bumper Works – wanted to place his company as Toyota's sole bumper supplier for North America.

But there was one problem: Bumper Works' production system was still a classic Mass Production with batch-and-queue. This concept would cause Toyota need to wait for fulfillment of each product type.

Bumper Works would also deliver a huge finished goods products to Toyota, that will force Toyota to increase its inventory area and cost. Internally, the batch-and-queue process also gave burden to Bumper Works since it needed to provide huge area for storing inventories.

The inventories would also consume huge amount of inventory cost and capital.

Toyota then took Khan and his senior managers for a trip to Japan in 1989 and walked them through showcase lean suppliers. There, they learned for the first time the lean concept of Toyota to overcome the disadvantages of the batch-and-queue process.

**The Result of Contradiction Analysis**

The "plus-state" of the "batch-and-queue process" can be determined as in $A_S$-Matrix as **number 36 (power)**, because less power was required due to less changeover and transporting frequency between stations. The "minus-state" of the "batch-and-queue process" can be determined as in $A_S$-Matrix as **number 25 (loss of time)**, because the parts needed to wait until the batch was completed before it could go to next station for the sequential process.

**Inventing Idea**

Combining the plus and minus state factors above, the results will show several points in $A_S$-navigators. The target is to avoid long waiting time of the whole batch to be finished. The new alternative will give possibility to have shorter period to finish and deliver parts to next (internal and external) stations / customers. After combining the $A_S$-Matrix 10 and 02, the idea navigators resulted:

- **Navigator 02 (preliminary action)** – according to this navigator, it is needed to prepare the production line in advanced so that each part can be transferred to next process individually.

- **Navigator 40 (uninterrupted useful function)** – according to this navigator, it is necessary to eliminate idle running and interruption of the parts processing activity.

**The Result-artifact (Solution) – Continuous Flow Process**

After World War II, Taiichi Ohno and his technical collaborators, including Shigeo Shingo[188], concluded that the real solution for the disadvantages of batch-

---

[188] Shigeo Shingo (1909 – 1990) – a Japanese industrial engineer, one of the world's leading experts on manufacturing practices and the Toyota Production System

and-queue process was to create continuous flow in small-lot production when only few units of products were needed, not millions.

Ford actually had introduced a piece of continuous flow in his assembly line, especially with the moving assembly line concept. But the parts supplied to the assembly line are produced entirely in large lots. At Toyota, the manufacturing and assembly processing both were performed as one-piece flow operations.

Toyota created a comprehensive system in which various parts, whether they processed within the plant or supplied from outside, were produced in small lots and flew directly into final assembly. This became one of the fundamental principles of the Toyota system and a significant difference between Ford and Toyota.

Production cycles could be reduced significantly by eliminating batch delays by using one-piece flow concept. Yet, one-piece lot transport involved increase of transportation from one process to the next, a problem solved by improving layout, such as placing succeeding processes next to one another. After layout was improved, more efficient transport means needed to be provided, such as conveyors. This allows materials to flow smoothly between processes that reduced production cycles and cut transport time dramatically.

In May 1990, Toyota sent several lean *sensei* (master in the TPS system) from its Operation Management Consulting Division to Bumper Works as personal tutors. The group was established in 1969 by Taiichi Ohno to promote lean thinking within Toyota and its supplier companies.

The first thing the lean *sensei* noted at Bumper Works was the massive inventories and batches. The production line did not "flow" at all. Since immediate right-sizing the massive stamping presses to permit continuous single-piece flow was not possible, the first solution was to reduce their changeover times and shrink batch sizes. Changeover times were then down from sixteen hours in the mid-1980s to around two hours.

fig. 11.44. A view of the continuous assembly line at the new Toyota plant outside St. Petersburg, Russia

The next important activity was re-layout of the production area. Transportation between stations were minimized in order to have a continuous flow of materials, from the input of raw material until the output of finished goods. The concept of one-piece flow was implemented as maximum as possible. By this concept, metal sheets as raw material flowed directly from the receiving dock to the blanking machine, which cut the steel into required size. The blank sheets then went immediately to the adjacent cell of stamping presses for giving shape. Next, they were shipped at frequent intervals in small batches to the outside chroming company and returned to the welding shop adjacent to the stamping presses. There, several parts of the bumpers were welded together. Finally, the bumper went straight to the shipping dock just in time for scheduled shipment.

So, basically the production lines were made to flow continuously without stoppage to wait for the completion of the batch.

*Batch & Queue process*

*Continuous Flow in an Automobile Assembly Line*

| WAS | | | IS |
|---|---|---|---|
| **LC** | **No.** | **Navigator** | **Substantiation for the Extracting** |
| + | 01 | Change in the aggregate state of an object | Transition of the assembly process from "batching" to "continuous" |
| ++ | 02 | Preliminary action | Prepare the production line in advanced so that each part can be transferred to next process individually. |
| + | 07 | Dynamization | Make the "queue" material to "move" |
| + | 11 | Inverse action | Instead of having the assembly process in batch, the process can be performed in continuous one. |
| ++ | 40 | Uninterrupted useful function | Eliminate idle running and interruptions. |

**Extracting-2**

**Standard Contradiction:** Mass production with "batch" process use less power for transportation between station, but the parts needed to wait for completion of the whole batch before they can be moved to next station.

**Radical Contradiction:** The product stock should be exist so that the part would not be transported individually between stations, but it should also not exist so it would not cause waiting time.

**FIM** (shortly): [ production with less waiting time ].

**fig. 11.45.** Extracting: Continuous Flow Process

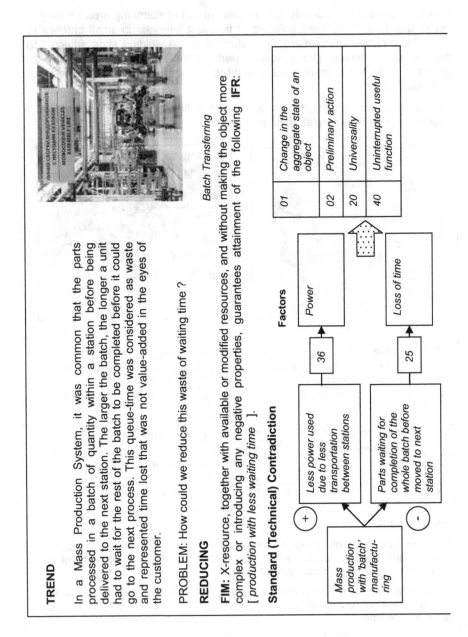

## TREND

In a Mass Production System, it was common that the parts processed in a batch of quantity within a station before being delivered to the next station. The larger the batch, the longer a unit had to wait for the rest of the batch to be completed before it could go to the next process. This queue-time was considered as waste and represented time lost that was not value-added in the eyes of the customer.

PROBLEM: How could we reduce this waste of waiting time ?

## REDUCING

FIM: X-resource, together with available or modified resources, and without making the object more complex or introducing any negative properties, guarantees attainment of the following IFR:

[ production with less waiting time ].

## Standard (Technical) Contradiction

*Batch Transferring*

| | | |
|---|---|---|
| 01 | Change in the aggregate state of an object | |
| 02 | Preliminary action | |
| 20 | Universality | |
| 40 | Uninterrupted useful function | |

**Factors**

Power

Loss of time

Mass production with 'batch' manufacturing

(+) Less power used due to less transportation between stations — 36 → Power

(-) Parts waiting for completion of the whole batch before moved to next station — 25 → Loss of time

**fig. 11.46 - beg.** Reinventing: Continuous Flow Process

## Radical (Physical) Contradiction

| The product stock | ⬆ | **should be exist**, so that the part would not be transported individually between stations | **&** | **should not be exist**, so it would not cause waiting time |

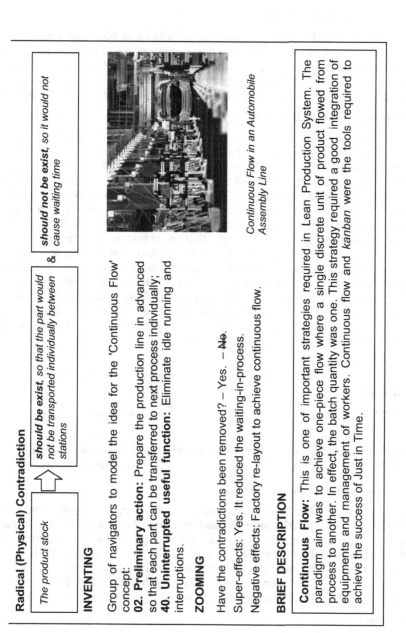

*Continuous Flow in an Automobile Assembly Line*

### INVENTING

Group of navigators to model the idea for the 'Continuous Flow' concept:

**02. Preliminary action:** Prepare the production line in advanced so that each part can be transferred to next process individually;

**40. Uninterrupted useful function:** Eliminate idle running and interruptions.

### ZOOMING

Have the contradictions been removed? – Yes. – ~~No~~.

Super-effects: Yes. It reduced the waiting-in-process.

Negative effects: Factory re-layout to achieve continuous flow.

### BRIEF DESCRIPTION

**Continuous Flow:** This is one of important strategies required in Lean Production System. The paradigm aim was to achieve one-piece flow where a single discrete unit of product flowed from process to another. In effect, the batch quantity was one. This strategy required a good integration of equipments and management of workers. Continuous flow and *kanban* were the tools required to achieve the success of Just in Time.

**fig. 11.46 - end.** Reinventing: Continuous Flow Process

### 11.2.3 Production Leveling

Processing in large batch is one characteristics of Mass Production that is considered necessary to reduce the changeover time and by that increase the effective production time and output. The batch process itself gives consequence of batch scheduling, in which large batch of certain type of product is produced within one period of time before the line is changed to another type of products.

### The Prototype – Batch Scheduling

Most of Bumper Works' customers were usually ordering massive batches – one-month lots to be delivered by the last day of the month. This caused Bumper Works to create its production schedule in large batches. For example, if there were orders for the end of the month for 2,000 of Bumper A, 2,000 of Bumper B and 4,000 of Bumper C – with 20 days of working per month – Bumper Works will produce Bumper A within the first 5 days of the month, Bumper B in the next 5 days, and Bumper C in the last ten days. They did that in order to avoid frequent changeovers that could take sixteen hours and consumed the production time. This batch scheduling – according to Toyota – gives some disadvantages. It would not suit to Toyota's production system that processed many variations of product type in small batches in short period of time, in order to avoid inventory of many kinds of raw materials and finished goods. The scheduling would also not be flexible in fulfilling the Toyota's after sales service division that frequently need a specific bumper to be delivered immediately during repairing customers' car.

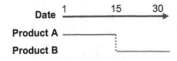

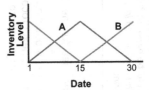

**fig. 11.47.** Batch production scheduling in Mass Production

**fig. 11.48.** Finished goods inventory level in Mass Production

### The Result of Contradiction Analysis

The "plus-state" of the "batch scheduling" mentality can be determined as in As-Matrix as **number 09 (ease of manufacture)**, because it did not require frequent changeover of the machines and change of the raw material and parts. The "minus-state" of the "batch scheduling" can be determined as in As-Matrix as **number 26 (quantity of material)**, because it had weakness of high inventory of one kind of product and shortage of the other one. This made the company could not be flexible in adapting the changes in customer demand.

### Inventing Idea

Combining the plus and minus state factors above, the results will show several points in As-navigators. The target is to enable the company to have balance in its ability to provide products demanded by customer. The new alternative will give the company possibility to change its production line more frequent without consuming long time for changeover. After combining the $A_S$-Matrix 09 and 26, the idea navigators resulted :

- **Navigator 03 (segmentation)** – according to this navigator, it is necessary to "disassemble" the batching scheduling into smaller ones.

- **Navigator 06 (periodic action)** – according to this navigator, the continuous batching scheduling need to be transformed into several periodic scheduling.

### The Result-artifact (Solution) – Mixed-model Production Leveling

Current market condition forces companies to produce different models continuously and limits the fluctuations in scheduled production requirements. This can be done by leveling the volume and mix the product types at the assembly line that needs support of the ability to produce small batches in more frequent changeovers and smaller WIP (work-in-process) inventories.

Production leveling is one of the prerequisites for having good Lean Production. The intent of production leveling is to maintain the availability of many types of products in order to have smooth fulfillment of customer orders and sequence them over time (Art of Lean, Inc., 2011).

The principle of mixed-model production leveling is by producing the batch volume of each product types in a shorter period of time, i.e. everyday. This will need more frequent changeovers, but with implementation of SMED, this will give no problems.

Production leveling for mixed products has advantages in eliminating work-in-process and finished goods accumulation by producing small batches in small period of time. Moreover, it can respond rapidly to the fluctuation in customer demand without having need to put huge amount of finished goods inventories and facilitates planning by letting the workers know at the beginning of the process what the average load will be.

Starting in 1992 – under Toyota's guidance – Bumper Works decided to prepare its daily schedule using Toyota's production leveling techniques. Shahid Khan's production managers would take the orders for the next month, let's say 2,000 units of Bumper A, 2,000 units of Bumper B, and 4,000 units of Bumper C. They would add them up (8,000 units) and divide by the number of working days in a month (say, twenty) to discover that Bumper Works would need each day to make 100 of Bumper A, 100 of Bumper B, and 200 of Bumper C each day. This would require three changeovers of the blanking and stamping machines everyday.

Because of its previous capability of quick changeover and continuous flow, it became possible for Bumper Works to implement this mixed-model production leveling in its assembly line. The new system gave Bumper Works the ability to balance its ability in dealing with fluctuation in customer demand in short notice without having to provide huge inventories of finished goods and raw materials. Thus, the inventory cost and area could also be reduced.

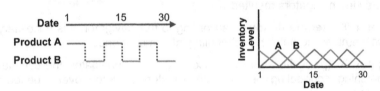

**fig. 11.49.** Production scheduling in TPS

**fig. 11.50.** Finished goods inventory level in TPS

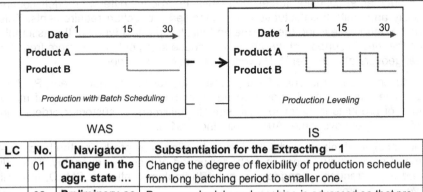

| LC | No. | Navigator | Substantiation for the Extracting – 1 |
|---|---|---|---|
| + | 01 | Change in the aggr. state ... | Change the degree of flexibility of production schedule from long batching period to smaller one. |
| + | 02 | Preliminary action | Prepare schedule and machine in advanced so that production can cover all types of products in small batching period. |
| ++ | 03 | Segmentation | "Disassemble" the batch scheduling into smaller ones. |
| ++ | 08 | Periodic action | Transition from a continuous batching schedule into several periodic ones. |
| + | 11 | Inverse action | Instead of producing one type of product within long period, produce many types of products within short period. |

**Extracting-2**

**Standard Contradiction:** Mass production with "batch" scheduling did not require frequent change over and change of material, but it produced high inventory of one type of product and shortage of the other one.

**Radical Contradiction:** The batch scheduling should be exist so that the change over time could be minimum, but it should also not be exist so that it could be flexible to customer change of demand.

**FIM:** X-resource, together with available or modified resources, and without making the object more complex or introducing any negative properties, guarantees attainment of the following IFR: [ even workload and inventory].

**fig. 11.51.** Extracting: Mixed-model Production Leveling

**TREND**

In a Mass Production System, it was common that the production was arranged to finish demand for one type of product (i.e. product A), before it moved to another product (i.e. product B). It was also common to have overtime to create stock enough to deal with future customer peak demand. Lean Production System saw this as another type of waste: *Mura* (unevenness) and *Muri* (overburden). The inventory for Product A would increase while the one of product B would decrease. Thus, the company could not have flexibility to adapt with customer change of demand. Furthermore, it would force the workers and the machines to do more than they were capable of / designed to.

PROBLEM: How could we reduce the unevenness of workload and inventory?

**REDUCING**

**FIM:** X-resource, together with available or modified resources, and without making the object more complex or introducing any negative properties, guarantees attainment of the following **IFR**: [ *even workload and inventory* ].

**Standard (Technical) Contradiction**

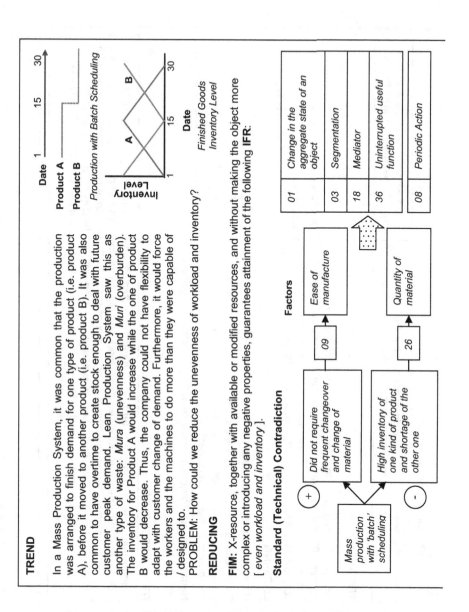

fig. 11.52 - beg. Reinventing: Mixed-model Production Leveling

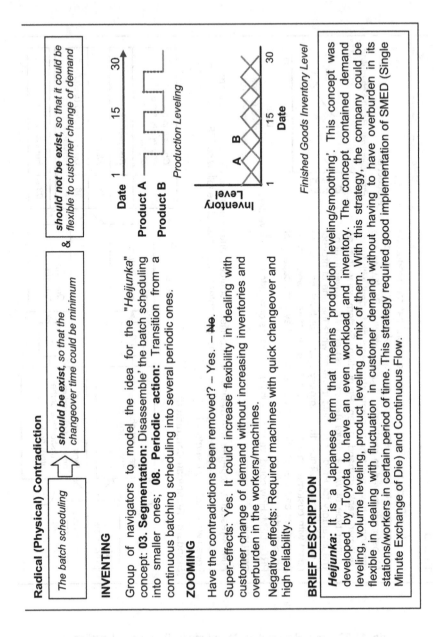

**Radical (Physical) Contradiction**

The batch scheduling ⟹ **should be exist**, so that the changeover time could be minimum & **should not be exist**, so that it could be flexible to customer change of demand

Date 1  15  30
Product A
Product B

*Production Leveling*

Inventory Level
A
B
Date 1  15  30

*Finished Goods Inventory Level*

**INVENTING**

Group of navigators to model the idea for the "*Heijunka*" concept: **03. Segmentation:** Disassemble' the batch scheduling into smaller ones; **08. Periodic action:** Transition from a continuous batching scheduling into several periodic ones.

**ZOOMING**

Have the contradictions been removed? – Yes. – No.

Super-effects: Yes. It could increase flexibility in dealing with customer change of demand without increasing inventories and overburden in the workers/machines.

Negative effects: Required machines with quick changeover and high reliability.

**BRIEF DESCRIPTION**

*Heijunka:* It is a Japanese term that means 'production leveling/smoothing'. This concept was developed by Toyota to have an even workload and inventory. The concept contained demand leveling, volume leveling, product leveling or mix of them. With this strategy, the company could be flexible in dealing with fluctuation in customer demand without having to overburden in its stations/workers in certain period of time. This strategy required good implementation of SMED (Single Minute Exchange of Die) and Continuous Flow.

**fig. 11.52 - end.** Reinventing: Mixed-model Production Leveling

### 11.2.4 Poka-yoke

In a Mass Production System, one of the characteristics is the high output rate of production. Supported with the moving assembly line and division of labors, the line can produce output in high speed.

#### The Prototype – Inspection and Rework

With output rate as its main concern, it become something in common for a company that adopt Mass Production to focus on the quantity, and put aside the concern of the quality of the products.

During its peak production in 1910s, there were hardly any inspections of finished automobiles in Ford's Model T assembly line at Highland Park. There was no engine checking until the vehicle was finished, and no products were ever road-tested. The company did not deliver the product at the highest quality, mostly because there was no urgency for it at that time. The cosmetic failures such as gaps in the fender panels, stumbled engines or electrical problems did not bother the customers.

However, as the market competition became tighter, the demand of the quality of the vehicles was increased and automobile companies usually put Final Inspection Station at the end of the assembly line to check for quality problems. This was because the companies wanted to keep the production line running to achieve the output target without having to stop for errors.

**fig. 11.53.** Reworking activity at an automobile assembly line

The defects then multiplied and accumulated from station to station and cause burden for the workers in the Final Inspection and Rework Stations. Because the errors did not tried to be fixed, the errors keep continued and created defects in subsequent products. Defective products needed to be inspected, reworked or even scrapped and it required allocation of man-hour and product replacement, which meant additional waste of cost.

#### The Result of Contradiction Analysis

The "plus-state" of the "inspection and rework" concept can be determined as in $A_S$-Matrix as **number 22 (speed)**, because each station would be able to run in high speed and did not need to stop for inspecting products for defects. The "minus-state" of the "inspection and rework" can be determined as in $A_S$-Matrix as **number 05 (precision of manufacture)**, because the defects produced in each station would be accumulated and give burden to the Final Inspection Station and consume a lot of effort and cost for rework.

### Inventing Idea

Combining the plus and minus state factors above, the results will show several points in $A_S$-navigators. The target is to make it possible to inspect and monitor the quality of the products in each process so that the defective product will accumulated in the Final Inspection and give burden to the workers in the Rework Station. The new alternative will give possibility to the system to do monitoring without sacrificing the productivity. After combining the $A_S$-Matrix 22 and 05, the idea navigators resulted:

- **Navigator 02 (preliminary action)** – according to this navigator, the machine need to be prepared by implementing partial change in it so that it can detect any defects or non-conformances.

- **Navigator 29 (self-service)** – according to this navigator, the machine need to serve itself with auxiliary to detect defect / non-conformance and stop automatically.

### The Result-artifact (Solution) – Poka-yoke System

Lean Production System considers quality as one of its goal. In order to achieve highest quality of products (zero defect), lean company shall realize that defects must be prevented from occurring. To do so, inspections need to be done – not merely to find defects after it happened – but it must prevent defects.

For a complete elimination of defects, 100% inspection must be adopted. Sampling inspection is not enough, because it can not guarantee product quality. However, the traditional 100% inspection in the Final Inspection Station is not suitable to be implemented since it will cause accumulation of products in that station, increase the waiting time and throughput time of a product.

Inspection by workers also not in line with TPS' principle of *jidoka* (autonomation / pre-automation / automation with human touch), where workers need to be separated from machines through the use of mechanisms to detect production abnormalities.

There are said to be twenty-three stages from purely manual work to full automation (s. Shingo). To be fully automated, a machine must be able to detect and correct its own operating problems. Since it is very difficult and not feasible to build an equipment that can correct its error by itself, then it is preferable to limit the machine capability to pre-automation stage, where it can detect its error automatically by using a simple mistake proofing device (*poka-yoke*).

Ninety percent of the result of full automation can be achieved at relatively low cost if machines are designed to merely detect problems and leave the correction of problems to the workers.

Successive, self, and source inspection can all be achieved through the use of *poka-yoke*. *Poka yoke* can achieve 100% inspection through mechanical or physical control.

There are two ways in which *poka-yoke* can be used to correct mistakes:

- Warning type: when the *poka-yoke* is activated, a buzzer sounds or a lamp flashes to alert the worker.

- Control type: when the *poka-yoke* is activated, the machine shuts down.

The warning *poka-yoke* still have weakness since it allows error to continue if workers do not respond to the warning. The control *poka-yoke* is the strongest corrective device because it shuts down the process until the error has been corrected.

Arawaka Shatai supplied car doors for Toyota. One of the production processes was to put a board covered with leather as a back lining plate for the door, which was attached by 20 retainers. Workers sometimes forgot to attach one or two retainers that caused defects on the products. They were advised to be more careful and the rate of defects dropped for a while but then returned to previous level. Since this appeared to be a recurring problem, twenty proximity switches were installed in the equipment as warning *poka-yoke*. If a retainer was left off, the equipment would stop and a buzzer sounds to alert the workers about the problem. With the use of the *poka-yoke*, the problem could be eliminated and the defect dropped to zero. In this case, a contact type *poka-yoke* performs 100% successive inspection.

**fig. 11.54.** Proximity switch: a sample of poka-yoke device

**fig. 11.55.** Barcode Reader, a poka-yoke system used in automobile industry to check completeness of the components need to be mounted to the passenger seats (belt buckle, side air bag, passenger check module, seat weight sensor and seat position sensor)

**fig. 11.56.** Toyota 4 Runner: award recipient for 2011 Most Dependable Vehicles for Midsize Crossover/SUV), based on J.D. Power surveys

**fig. 11.57.** A result of jidoka (build-in quality) concept by Toyota: Lexus GS, award recipient for 2010 Initial Quality Study for Midsize Premium Car, based on J.D. Power surveys to provide feedback on quality during the first 90 days of new-vehicle ownership

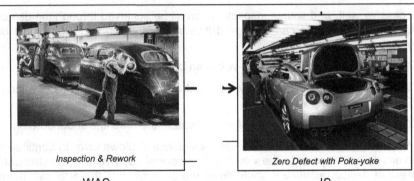

| | | WAS | IS |
|---|---|---|---|
| | | *Inspection & Rework* | *Zero Defect with Poka-yoke* |

| LC | No. | Navigator | Substantiation for the Extracting |
|---|---|---|---|
| ++ | 02 | **Preliminary action** | Prepare the machine by implementing partial change in it so that it can detect any defect or non-conformance. |
| + | 04 | **Replacement of mechanical matter** | Use of devices with electrical / magnetic / electro-magnetic / optical principle as sensors. |
| + | 12 | **Local property** | Change the structure of the machines that there is one part of the machine that can detect defect automatically. |
| + | 18 | **Mediator** | Use another object to transmit an action of detecting defective parts or non-conformance process. |
| + | 21 | **Transform damage into use** | Use the "damaging factor" of stopping machine (stop production) for a useful effect (eliminate the cause of the problem). |
| + | 28 | **Prev. installed cushion** | Install an additional object in the machine that can detect failure of process/defect on products. |
| ++ | 29 | **Self-servicing** | The machine service itself with auxiliary to detect defects/non-conformance and stop automatically. |

**Extracting-2**

**Standard Contradiction:** Inspection and Rework practice in the Final Station made it possible for the assembly line to run fast and achieved high productivity, but this concept would produce defects that could not be caught during process.

**Radical Contradiction:** The assembly line should run continuously to achieve high output, but should also be stop periodically so that the process and pro-duct can be monitored.

**FIM** (shortly): [ automation of inspection ].

**fig. 11.58.** Extracting: Poka-yoke System

## TREND

In a Mass Production System, it was a common sense that the goal was to achieve high output of products. The process run in high speed and the workers did not have time and concern to check the quality of the products. Defects could only be caught in the Final Inspection Station, which caused many products needed to be reworked or scrapped. This was a kind of waste of money and material that needed to be eliminated by Lean Production System.

*Rework activity in car assembly line*

PROBLEM: How could we catch the defect before Final Inspection Station without stopping the machines to perform periodic inspection in every station ?

## REDUCING

FIM: X-resource, together with available or modified resources, and without making the object more complex or introducing any negative properties, guarantees attainment of the following IFR: [ *automation of inspection* ].

## Standard (Technical) Contradiction

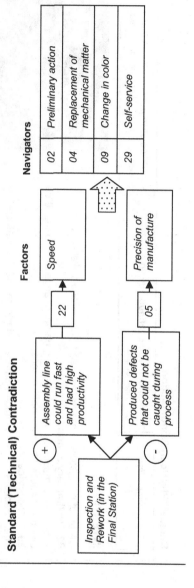

fig. 11.59 - beg. Reinventing: Poka-yoke System

## Radical (Physical) Contradiction

| The assembly line | ⟰ | **should run continuously**, to achieve high output | & | **should be stop periodically**, so that the process and product can be monitored |

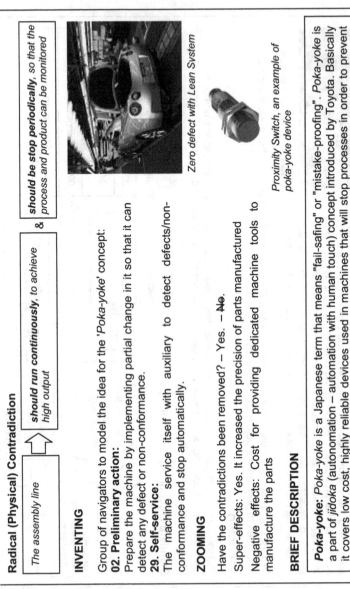

Zero defect with Lean System

Proximity Switch, an example of poka-yoke device

## INVENTING

Group of navigators to model the idea for the 'Poka-yoke' concept:

**02. Preliminary action:**
Prepare the machine by implementing partial change in it so that it can detect any defect or non-conformance.

**29. Self-service:**
The machine service itself with auxiliary to detect defects/non-conformance and stop automatically.

## ZOOMING

Have the contradictions been removed? – Yes. – No.

Super-effects: Yes. It increased the precision of parts manufactured

Negative effects: Cost for providing dedicated machine tools to manufacture the parts

## BRIEF DESCRIPTION

**Poka-yoke:** *Poka-yoke* is a Japanese term that means "fail-safing" or "mistake-proofing". *Poka-yoke* is a part of *jidoka* (autonomation – automation with human touch) concept introduced by Toyota. Basically it covers low cost, highly reliable devices used in machines that will stop processes in order to prevent the production of defective parts. One example of the *poka-yoke* device is proximity switch that used as sensor to detect presence of material / parts during production process.

**fig. 11.59 - end.** Reinventing: Poka-yoke System

### 11.2.5 Andon System

Henry Ford: *We expect the men to do what they are told. The organization is so highly specialized and one part is so dependent upon another that we could not for a moment consider allowing men to have their own way. Without the most rigid discipline we would have the utmost confusion.*[189]

**The Prototype – Passing-on Defect.** As has been mentioned earlier in section 5.2.4.1, in a Mass Production system there was a tendency to pass on errors to keep the line running that caused errors to multiple endlessly. Every worker assumed that errors would be caught at the end of assembly line. Even if the worker wanted to stop the machine due to errors, he had no access or tools to inform it to the responsible line managers, who had the authority to stop the line. This kind of discipline was assured as the managers at Ford posted the production output of each man on production board. The figures[190] were "posted hourly, and the records of those who equal or better quota set are written down in the colored crayon". This standard measure was taken to stir up "competition among workers, who performed the same operation".

So, rather than leaving his station and losing productivity, the operators preferred to pass on the defective product to next station and continue to process subsequent products. Hence, the so called "passing-on defect mentality", was commonly practiced that avoided line stoppages at any price and emphasized repair in the end of assembly line.

**fig. 11.60.** An example of Final inspection area in an Austin Automobile assembly line (1950s). Cars received an inspection and final polishing before went to test drive.

**The Result of Contradiction Analysis.** The "plus-state" of the "passing-on defect" mentality can be determined as in $A_S$-Matrix as **number 25 (loss of time)**, because the production process could take place continuously that would increase the productivity. The "minus-state" of the "passing-on defect" can be determined as in $A_S$-Matrix as **number 12 (loss of information)**, because the operators did not have chance and access to inform the defect occurred in the line. It also caused the defects multiply from station to station and created enormous amount of rework effort and cost to fix them.

**Inventing Idea.** Combining the plus and minus state factors above, the results will show several points in $A_S$-navigators. The target is to enable the workers to stop the process that produce defect as immediately as possible so that the defective product will not be pass on to the next station and accumulated in the Final Inspection and Rework Station. The new alternative will give possibility to

---

[189] McNairn, W. and McNairn, M. (1978) *Quotations from the Unusual Henry Ford.* – Quotamus Press, Redono Beach, CA
[190] Porter, H. F. (1917) *Four Big Lessons from Ford's Factory.* – System, 31 (June 1917)

stop the machine and inform the responsible persons only when the defect is occurred so that it will not sacrifice the productivity and manpower effort to do machine monitoring all the time. After combining the A$_S$-Matrix 25 and 12, the idea navigators resulted:

- **Navigator 09 (change in color)** – according to this navigator, it is advised that the machine that creates problem can "change its color".

- **Navigator 18 (mediator)** – according to this navigator, it is needed to use an object to transfer information about problem in certain machine/station.

- **Navigator 35 (unite)** – according to this navigator, instead of leaving the problem to the operator in a problematic machine, the whole responsible team need to come and work on the problem.

## The Result Artifact (Solution) – Andon System

At TPS, *jidoka* also means build-in quality. It means achieving highest quality during processing the materials[191]. One concept of *jidoka* – separating man's and machine's work has been mentioned earlier by the use of *poka-yoke*. The other concept of *jidoka* is to stop the work if found abnormality. According to TPS principle, passing on defective products to the end of assembly line is also one type of waste, since there will be human effort and cost involved to rework or scrap the product.

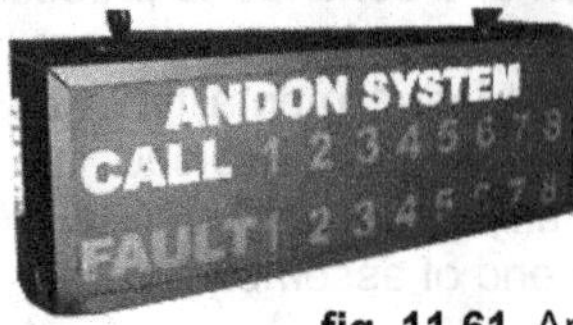

Moreover, since there is no corrective action performed to the problematic process or machine, similar defect will be produced from time to time. To avoid these wastes, it is necessary to stop the operation or machine if there is trouble with machine operation.

**fig. 11.61.** Andon system to communicate problems in machines

When problem occurs, visual control or *andon* (indicator lights) show the workers, supervisors, technicians and other related personnel where the trouble is. The troubles are then being able to be immediately communicated to everyone.

One of the important mentality to support this tool is the urgency to stop the production line and fix the problem. Top management need to commit itself to halt the machines or production lines when there is trouble.

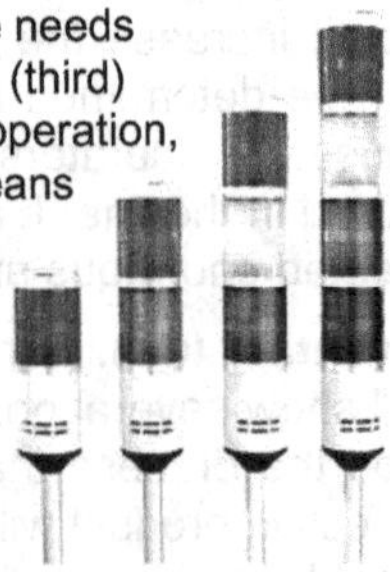

**fig. 11.62.** A type of andon lamp. Red colour (top) means machine error, yellow (second from top) means machine needs attention, green (third) means normal operation, blue (fourth) means short of material.

This is the key that makes *jidoka* concept with *poka-yoke* and *andon* tools become famous as autonomation – automation with a human touch.

---

[191] Liker, J. K. (2004) *The Toyota Way*. – McGraw-Hill, New York

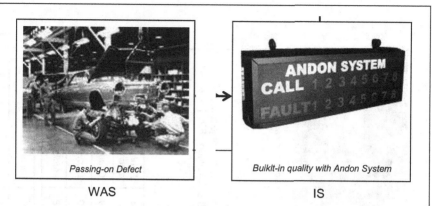

| WAS | IS |
|-----|-----|
| *Passing-on Defect* | *Buiklt-in quality with Andon System* |

| LC | No. | Navigator | Substantiation for the Extracting |
|-----|-----|-----------|-----------------------------------|
| + | 01 | Change in the aggr. state ... | Change the degree of flexibility to stop the production line whenever there is problem. |
| + | 02 | Preliminary action | A device is provided in advances o that the worker can inform problem in the production line. |
| ++ | 09 | Change in color | Change the color of the machine /station that give problem. |
| + | 11 | Inverse action | Instead of implementing "passing-on defect" mentality, the company needs to encourage the workers to stop and inform for any kind of problems. |
| + | 12 | Local property | Change the structure of the working station to add a device that act as information board for management. |
| ++ | 18 | Mediator | Use an object to transfer information about problem in certain machine / station. |
| + | 21 | Transform damage into use | Use the "damaging factor" of stopping machine (stop production) for a useful effect (eliminate the cause of the problem). |

## Extracting-2

**Standard Contradiction:** Passing-on defect mentality made the machine can work with low stoppage time, but it gave no access and tools to inform the defect that cause high rework and scrap rate at the Final Station.

**Radical Contradiction:** The assembly line should run continuously to achieve high output, but should also stop periodically so that the error could be fixed.

**FIM** (shortly): [ low rework rate ].

**fig. 11.63.** Extracting: Andon System

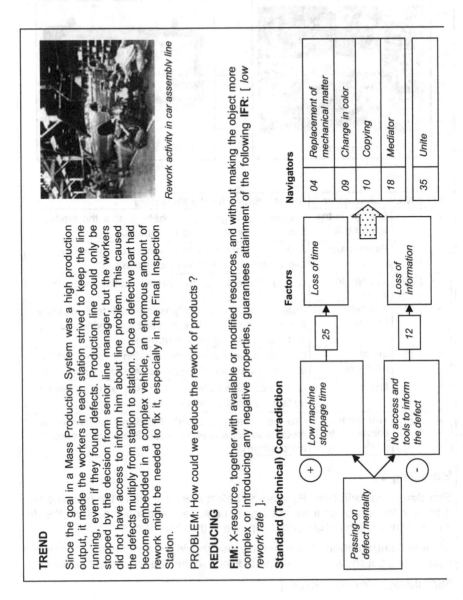

*Rework activity in car assembly line*

## TREND

Since the goal in a Mass Production System was a high production output, it made the workers in each station strived to keep the line running, even if they found defects. Production line could only be stopped by the decision from senior line manager, but the workers did not have access to inform him about line problem. This caused the defects multiply from station to station. Once a defective part had become embedded in a complex vehicle, an enormous amount of rework might be needed to fix it, especially in the Final Inspection Station.

PROBLEM: How could we reduce the rework of products ?

## REDUCING

**FIM:** X-resource, together with available or modified resources, and without making the object more complex or introducing any negative properties, guarantees attainment of the following **IFR**: [ *low rework rate* ].

## Standard (Technical) Contradiction

| Factors | Navigators | |
|---|---|---|
| Loss of time | 04 | Replacement of mechanical matter |
| | 09 | Change in color |
| | 10 | Copying |
| Loss of information | 18 | Mediator |
| | 35 | Unite |

Low machine stoppage time — 25 → Loss of time (+)

No access and tools to inform the defect — 12 → Loss of information (−)

Passing-on defect mentality

**fig. 11.64 - beg.** Reinventing: Andon System

## Radical (Physical) Contradiction

| *The assembly line* ⇧ | **should run continuously**, *to achieve high output* | & | **should be stopped periodically**, *so that the error could be fixed* |

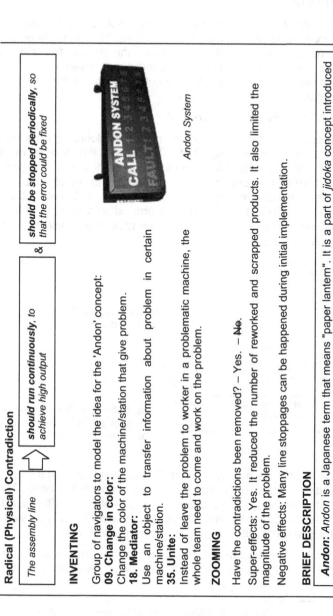

*Andon System*

### INVENTING

Group of navigators to model the idea for the 'Andon' concept:

**09. Change in color:**
Change the color of the machine/station that give problem.

**18. Mediator:**
Use an object to transfer information about problem in certain machine/station.

**35. Unite:**
Instead of leave the problem to worker in a problematic machine, the whole team need to come and work on the problem.

### ZOOMING

Have the contradictions been removed? – Yes. – ~~No~~.

Super-effects: Yes. It reduced the number of reworked and scrapped products. It also limited the magnitude of the problem.

Negative effects: Many line stoppages can be happened during initial implementation.

### BRIEF DESCRIPTION

**Andon:** *Andon* is a Japanese term that means "paper lantern". It is a part of *jidoka* concept introduced by Toyota. *Andon* is a system to notify management, maintenance, and other workers of a quality or process problem. The tool used can be a signboard incorporating signal lights/alarm sound that can be activated manually by workers or automatically by the equipment itself. *Andon* gives the worker the ability to stop production when a defect is found and immediately call for assistance. Therefore, it can stop further defective products being produced and reduce time and cost to rework the products.

**fig. 11.64 - end.** Reinventing: Andon System

### 11.2.6  Just-in-Time (JIT)

In Mass Production that adopted economic of scale principle, it is necessary to maximize output in order to reduce the production cost. This principle emphasis the maximum usage of the machines and "push" the material to the end of assembly line and encourages buildup of finished goods inventories. These inventories were considered as assets for the company, rather than accumulated waste of capital.

### The Prototype – "Push" Production Control

The production process[192] with "push" concept can be analogous by the water flow in the river with many rocks (fig. 12.65). The river is the material movement, the depth of water represent the inventory and Work-in-Process (WIP), and the rocks represent the problems encountered in assembly line. The "push" concept use high inventory (high level of water) as safety buffer to deal with quality problems (defective products), long production time (due to batch processes and machine problems), long set-up time (due to complex machine), or long delivery time from suppliers.

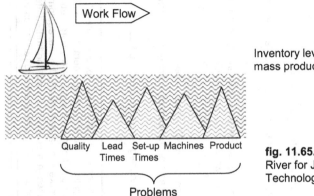

Inventory level with mass production

**fig. 11.65.** The Rocks in the River for JIT (after Tangram Technology, Ltd. 2005)

When Eiji Toyoda[193] and his managers from Toyota went to Ford's plant in 1950, to their surprised, they found that the company had many flaws. They saw the discrete process steps were based on large volumes, with interruptions between these steps causing large amounts of material to sit and wait in WIP-inventory. They saw the massive and expensive equipments.

The company tried to maintain efficiency in reducing the cost per piece, with workers keeping busy by keeping the equipment busy, resulting in a lot of

---

192 Silver, E. A., Pyke, D. F. and Peterson, R. (1998) *Inventory Management and Production Planning and Scheduling* (pp. 592 – 623, 631 – 653). – John Wiley & Sons, Inc., New York

193 Eiji Toyoda (b. 1913) – a Japanese engineer, cousin of Kiichiro Toyoda (founder of Toyota Motor Corp.). He became the President (1967 – 1981) and Chairman (1981 – 1994) of Toyota

overproduction and an uneven flow, with defects hidden in large batches that could go undiscovered for weeks.

Entire workplaces were disorganized and out of control. With big trucks moving materials everywhere, the factories looked more like warehouse than an automobile plant.

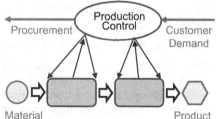

**fig. 11.66.** Push Production Control Principle usually adopted by Mass Production System

fig. 11.67. Unsold Dodge SUVs sat at the Atlantic Marine Terminal at the port of Baltimore, Maryland. More than 57,000 cars waiting for buyers during Q1-2009. Maryland paid USD 5.26 million for almost 10 hectares of additional car storage space near the port

## The Result of Contradiction Analysis

The "plus-state" of the "push production control" can be determined as in $A_S$-Matrix as **number 09 (ease of manufacture)**, because each station needed to think only of their own production pace. The "minus-state" of the "push production control" can be determined as in $A_S$-Matrix as **number 02 (universality, adaptability)**, because this concept was not flexible in dealing with the fluctuation of the customer demand. Accumulation of inventories between stations or in the final storage could be expected.

## Inventing Idea

Combining the plus and minus state factors above, the results will show several points in $A_S$-navigators. The target is to avoid accumulation of inventories between stations and in the final storage that consume cost and area. The new alternative will give possibility to have low stock of products without sacrificing the fulfillment of customer demands. After combining the $A_S$-Matrix 09 and 02, the idea navigators resulted :

- **Navigator 11 (inverse action)** – according to this navigator, instead of using "push" concept, the production can be controlled by 'pull' concept;

- **Navigator 02 (preliminary action)** – according to this navigator, the production line can be prepared in advanced so that they can adapt the change in customer demand without loss of time;

- **Navigator 18 (mediator)** – according to this navigator, a mediator is required to transfer information from customer / downstream stations to upstream stations and supplier in order to have a quick response without accumulating stocks.

### The Result-artifact (Solution) – Just-in-Time

Just-in-Time (JIT) is a production planning system that first introduced by Toyota Motor Company during 1960s. It was created as Toyota's answer for the inventory management and control system that were commonly used by major automotive industries at that time, especially in the United States (s. Womack).

The JIT was said to be inspired by supermarket concept. The supermarket customers may go to the shelves and buy what they want, when they need it. The shelves then are refilled as products are sold. This concept gives an easy system for suppliers to see how many products have been taken and avoid overstocks. Thus, the most important feature of a supermarket system is that stocking is triggered and maintained by actual demand. Toyota has used this concept to create a flexible production system that is characterized as the "pull" system of order-based production.

There are two kinds of known definitions of the Just-in-Time concept:

1. JIT in production (production when it is called for): produce parts just in time; all production activities are performed when the customer requires it.

2. JIT in delivery (stockless inventory): All external supplies are delivered exactly at the moment they are required, at the correct quantity with the correct type. It can be said that the inventory is taking place on the road.

So, the objective of JIT is to produce and deliver right quantity of products with the highest quality at the right time. During this process, it is necessary to gain it with minimizing inventory and lead time, and suppress failure and defects (s. Silver). In other words, JIT system tries to pursue zero inventories, zero transaction, and zero disturbances.[194]

In connection with "river and rocks" analogy mentioned in fig. 11.65 (s. Silver), as JIT requires low inventory (low level of water, figure 11.68), the problems (rocks) will be revealed and draw attention to be solved, otherwise the production can be stopped.

This concept is in contrast with "push" system implemented by Mass Production which implement high inventory (high level of water) to cover the problems.

---

[194] Vollman, T. E., Berry. W. L., Whybark, D. C. and Jacobs F. R. (2005) *Manufacturing Planning and Control for Supply Chain Management.* – McGrawHill, New York

JIT system gives several important advantages (s. Silver):

- Reduction of inventory space and cost
- Less material movements in/out of storage
- Reduced transactions
- Simplified manufacturing and planning control systems.
- Reduction of production throughput time
- Greater responsiveness to market demands
- Improvement of quality and reduction of quality cost.

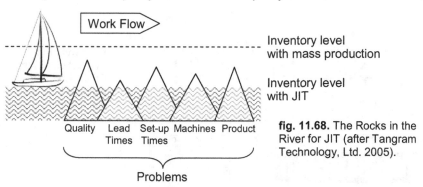

**fig. 11.68.** The Rocks in the River for JIT (after Tangram Technology, Ltd. 2005).

As JIT is a "pull" system, it uses a unique mechanism of information flow which is called *kanban* (Japanese term for "card") (s. Shingo). When TPS started receiving international attention in the seventies, many people were wrongly understood JIT as the "*kanban* method".

*Kanban* is actually tools used in JIT to help implement the "pull" principles of TPS. A *kanban* can be a variety of things, most commonly it is a card, but sometimes it is a cart, while other times it is just a marked space. In all cases, its purpose is to facilitate flow, bring about pull, and limit inventory. It is one of the key tools in the battle to reduce overproduction.[195]

One important *kanban* rule requires that all materials and products be accompanied by a *kanban* card. Thus, *kanban* connected the material and information flow between working station. The downstream station gives signal for required parts to its upstream center using *kanban* cards.

The *kanban* used in the TPS serves three main functions (s. Shingo):

- identification tag: indicates type of the product;
- job instruction tag: indicates process type, quantity and delivery time required for the product;

---

[195] Wilson, L (2010) *How to Implement Lean Manufacturing.* – McGraw-Hill, New York

- transfer tag: indicates from where and to where the product should be transported.

Basically *kanban* system can be applied in plants involved in repetitive production. *Kanban* system are not applicable in one-of-a-kind production based on infrequent and unpredictable orders.

With development in technology, the manual information system using *kanban* cards now is started to be improved by integrating the computer network system called *e-kanban*. This system gives simpler, more accurate, and more responsive flow of information between customer and suppliers, and within the organizations themselves.[196]

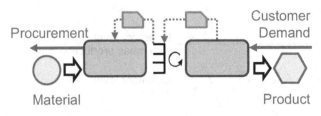

**fig. 11.69.** Pull Production Control Principle adapted by Lean Production System

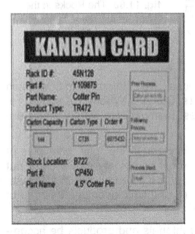

**fig. 11.70.** A conventional kanban card

**fig. 11.71.** Kanban card supplemented with barcode for simpler checking

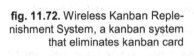

**fig. 11.72.** Wireless Kanban Replenishment System, a kanban system that eliminates kanban card

---

[196] Drickhamer, D. (2005) *The kanban e-volution.* – Material Handling Management (March), 24-26

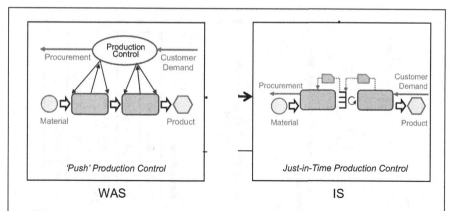

| LC | No. | Navigator | Substantiation for the Extracting |
|---|---|---|---|
| ++ | 02 | **Preliminary action** | Prepare the production line in advanced so that they can adapt the change of customer demand without loss of time. |
| + | 05 | **Separation** | Separate the 'buffer stocks' unneeded by customer by producing only the required ones. |
| ++ | 11 | **Inverse action** | Instead of using 'push' concept, the production can be controlled by 'pull' concept. |
| ++ | 18 | **Mediator** | Need a mediator to transfer information from customer / downstream station. |
| + | 36 | **Feedback** | Use 'feedback' (order) from customer to define the quantity of products need to be produced. |

**Extracting-2**

**Standard Contradiction:** Mass production with 'push' concept made each station only need to think of their own pace of production, but it was not flexible in dealing with fluctuation of customer demand / downstream station condition.

**Radical Contradiction:** The product stock should be exist to serve customer order, but should not be exist so it would not cause problems (holding cost, etc.)

**FIM:** X-resource, together with available or modified resources, and without making the object more complex or introducing any negative properties, guarantees attainment of the following **IFR:** [ production control with less inventories ].

**fig. 11.73.** Extracting: Just-in-Time

## TREND

In a Mass Production System, every station in assembly line had to "push" each product to next station as quickly as possible. This "make-to-stock" concept was not flexible in adapting the customer demand fluctuation and condition of the downstream stations. This caused many semi-finished/finished products waited between stations / in the final storage as inventories. These inventories were considered as problems since they were restrained capital, consumed area and holding cost and they could be worn-out / damaged over time.

PROBLEM: How could we cope with problems created by "push" control?

## REDUCING

**FIM:** X-resource, together with available or modified resources, and without making the object more complex or introducing any negative properties, guarantees attainment of the following **IFR**:

[ *production control with less inventories* ].

**Standard (Technical) Contradiction**

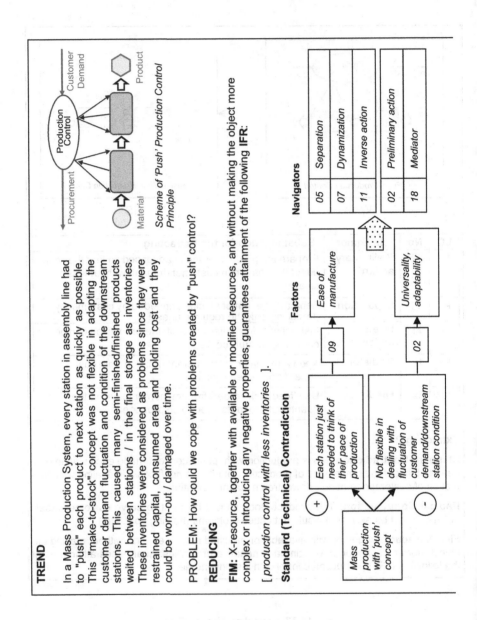

*Scheme of 'Push' Production Control Principle*

| Navigators | |
|----|----|
| 05 | Separation |
| 07 | Dynamization |
| 11 | Inverse action |
| 02 | Preliminary action |
| 18 | Mediator |

**fig. 11.74 - beg.** Reinventing: Just-in-Time

## Radical (Physical) Contradiction

| The product stock | ⬆ | **should be exist**, to serve customer order | & | **should not be exist**, so it would not cause problems (holding cost, etc.) |

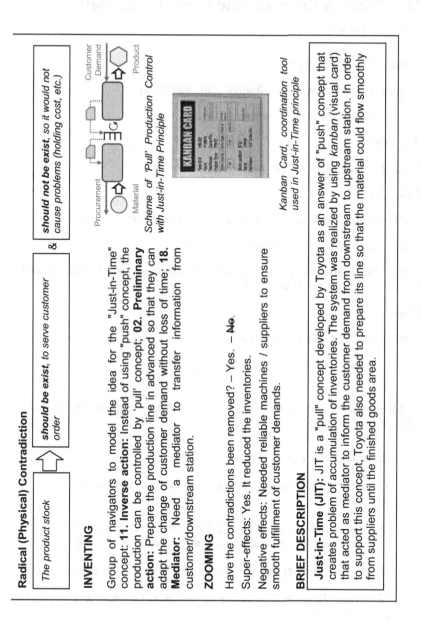

Scheme of 'Pull' Production Control with Just-in-Time Principle

Kanban Card, coordination tool used in Just-in-Time principle

### INVENTING

Group of navigators to model the idea for the "Just-in-Time" concept: **11. Inverse action:** Instead of using "push" concept, the production can be controlled by 'pull' concept; **02. Preliminary action:** Prepare the production line in advanced so that they can adapt the change of customer demand without loss of time; **18. Mediator:** Need a mediator to transfer information from customer/downstream station.

### ZOOMING

Have the contradictions been removed? – Yes. – ~~No~~.

Super-effects: Yes. It reduced the inventories.

Negative effects: Needed reliable machines / suppliers to ensure smooth fulfillment of customer demands.

### BRIEF DESCRIPTION

**Just-in-Time (JIT):** JIT is a "pull" concept developed by Toyota as an answer of "push" concept that creates problem of accumulation of inventories. The system was realized by using kanban (visual card) that acted as mediator to inform the customer demand from downstream to upstream station. In order to support this concept, Toyota also needed to prepare its line so that the material could flow smoothly from suppliers until the finished goods area.

**fig. 11.74 - end.** Reinventing: Just-in-Time

# All It Takes is Creativity and a Bit of Courage[197]

*Author's Afterword*

*Creativity and courage.*

*Resoluteness to take the initiative and leadership.*

*This is necessary to innovate and invent.*

*And labor.*

*Persistently. Relentlessly. Attentively. Responsibly.*

*Everyday.*

*Until a complete victory over yourself and over the problem!*

*If, of course, the goal is worth your effort.*

*You must overcome yourself, your weaknesses and fatigue in order to defeat the problems that attack you.*

*There is no other way.*

*This is the only way to mastery.*

• • •

And to complete this book not so instructive, we recall some stories and parables.

### 1. I hope this is not about us...

> After the first success in cutting trees, a young woodcutter began to cut less and less.
>
> *"I must be losing my strength"*, the woodcutter thought. He went to the boss and said that he couldn't understand what was going on.
>
> *"When was the last time you sharpened your axe?"* the boss asked.
>
> *"Sharpen? I had no time to sharpen my axe. I have been very busy trying to cut trees..."*

---

[197] You have already met these important words in this book. Do you remember where?

## 2. *Mr. President says…*

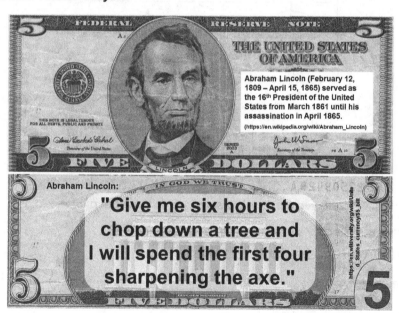

Herbert Paul Brooks Jr.
(1937 – 2003)

### 3. The great coach says…

He was an American ice hockey[198] player and coach. His most notable achievement came in 1980 as head coach of the gold medal-winning U.S. Olympic hockey team at Lake Placid, USA.

GATHERING THE TEAM[199]:

**"You think you will win with only your talent... Gentlemen, you don't have enough talent to win on talent alone."**

TRAINING WITH "HERBIE" METHOD:

*"Run to the blue line, back. Far blue line, back. Far red line, back. And you have 45 seconds to do it. **Get used to this drill.** You'll be doing it "a lot". Why? Because the legs feed the wolf, gentlemen. **I can't promise you we'll be the best team at Lake Placid next February. But we will be the best conditioned.** That I can promise you."*

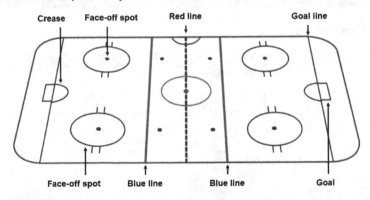

THE VICTORY:

*"We were a fast, CREATIVE team
that played extremely disciplined without the puck."*

**THUS THE VICTORY WAS MADE BEFORE THE GAME!**

**AND ONLY THEN – DURING THE GAME!**

*Michael Orloff*                                                    *Berlin, October, 2019*

---

[198] https://en.wikipedia.org/wiki/Miracle_on_Ice
[199] www.herbbrooksfoundation.com

# Part IV

# Reference Materials

## Important personal qualities to invent and innovate according[200] to Altshuller's *"Theory of Creative Person Development"*

| | |
|---|---|
| **Worthy Objective:** | the individual must have a novel, previously unattained, significant, socially useful, worthy objective (or system of objectives); |
| **Plans:** | the individual must have a plan (or set of plans) designed to assure attainment of his/her objective(s) and monitor implementation progress; |
| **Capacity to Work:** | the individual must be willing to perform, and actually perform, a huge volume of work to implement his/her plans; |
| **Problem-Solving Method:** | the individual must have a method that can be used to solve problems he/she may encounter on the way to his/her final destination; |
| **Resilience:** | the individual must be capable of defending his/her ideas, dealing with public censure and incomprehension, "sticking to his/her guns", remaining true to his/her ideals; |
| **Effectiveness:** | the results attained (or their scope) must be commensurate with the original objective(s). |

*Genrikh Altshuller*

---

[200] Translated by author from Altshuller, G.S., Vertkin, I.M., *How to Become a Genius. Life Strategy of a Creative Personality.* – Minsk: Belarus, 1994. – (Альтшуллер Г.С., Верткин И.М. *Как стать гением. Жизненная стратегия творческой личности.* – Минск: Беларусь, 1994)

# R1. A-Matrix*

## R1.1. List of 39 Plus- and Minus-factors

| # | Factor |
|---|--------|
| 01 | productivity |
| 02 | universality, adaptability |
| 03 | level of automation |
| 04 | reliability |
| 05 | precision of manufacture |
| 06 | precision of measurement |
| 07 | complexity of construction |
| 08 | complexity of inspection and measurement |
| 09 | ease of manufacture |
| 10 | ease of use |
| 11 | ease of repair |
| 12 | loss of information |
| 13 | external damaging factors |
| 14 | internal damaging factors |
| 15 | length of the moveable object |
| 16 | length of the fixed object |
| 17 | surface of the moveable object |
| 18 | surface of the fixed object |
| 19 | volume of the moveable object |
| 20 | volume of the fixed object |
| 21 | shape |
| 22 | speed |
| 23 | functional time of the moveable object |
| 24 | functional time of the fixed object |
| 25 | loss of time |
| 26 | quantity of material |
| 27 | loss of material |
| 28 | strength |
| 29 | stabile structure of the object |
| 30 | force |
| 31 | tension, pressure |
| 32 | weight of the moveable object |
| 33 | weight of the fixed object |
| 34 | temperature |
| 35 | brightness of the lighting |
| 36 | power |
| 37 | energy use of the moveable object |
| 38 | energy use of the fixed object |
| 39 | loss of energy |

---

* the tables are for general reference only in a format that may not be convenient enough for continuous work; please use materials from the web site www.mtriz.com

© Springer Nature Switzerland AG 2020
M. A. Orloff, *Modern TRIZ Modeling in Master Programs*,
https://doi.org/10.1007/978-3-030-37417-4

# R1.2. Table of A-matrix

Problem-factor (minus-factor) ➡

Trend-factor (plus-factor) ⬇

| Problem-factor (minus) / Trend-factor (plus) | # | 01 productivity | 02 universality, adaptability | 03 level of automation | 04 reliability | 05 precision of manufacture | 06 precision of measurement | 07 complexity of construction | 08 complexity of inspection and meas. | 09 ease of manufacture | 10 ease of use | 11 ease of repair | 12 loss of information | 13 external damaging factors |
|---|---|---|---|---|---|---|---|---|---|---|---|---|---|---|
| productivity | 01 | | 03.01 04.27 | 35.37 01.10 | 03.01 02.30 | 09.03 06.02 | 03.02 15.04 | 37.19 04.18 | 01.06 13.05 | 01.04 05.18 | 03.04 34.08 | 03.09 02.29 | 11.07 36 | 21.01 11.18 |
| universality, adaptability | 02 | 01.04 20.27 | | 13.15 01 | 01.11 32.18 | 27.12 | 01.35 27.04 | 07.14 27.04 | 03 | 03.11 31 | 07.15 03.16 | 03.16 34.24 | 27.11 | 01.28 09.31 |
| level of automation | 03 | 35.37 01.10 | 13.24 03.01 | | 28.13 09 | 04.10 06.36 | 04.10 03.02 | 07.18 29 | 15.13 11 | 03.10 11 | 03.37 17 | 03.01 11 | 01.38 | 05.38 |
| reliability | 04 | 03.01 01.10 | 11.01 32.18 | 28.11 13 | | 28.09 03 | 09.12 28.36 | 11.01 03 | 13.17 04 | 04.02 24.17 | 13.19 17 | 03.28 | 02.04 | 13.01 05.17 |
| precision of manufacture | 05 | 02.06 09.23 | 01.11 18 | 10.04 06.36 | 28.09 03 | | 10.04 18.11 | 10.05 06 | 02.36 | 18.03 | 03.09 01.36 | 29.02 | 18.34 29 | 10.04 02.26 |
| precision of measurement | 06 | 02.15 04.09 | 11.01 05 | 04.05 02.15 | 35.28 03.36 | 10.09 27 | | 13.01 02.15 | 10.18 09.04 | 20.01 29.06 | 03.11 19.15 | 03.09 11.28 | 34.10 01.21 | 04.18 21.10 |
| complexity of construction | 07 | 37.19 04.27 | 14.07 04.27 | 07.03 18 | 11.01 03 | 10.18 09 | 05.10 02.15 | | 07.02 27.04 | 13.10 03.11 | 13.39 10.18 | 03.11 | 34.08 09.03 | 21.08 14.17 |
| complexity of inspection and meas. | 08 | 01.06 | 03.07 | 15.33 | 13.17 04.32 | 27.34 24.09 | 10.18 09.04 | 07.02 27.04 | | 35.04 28.14 | 05.35 | 37.10 | 01.38 13.21 | 21.08 14.04 |
| ease of manufacture | 09 | 01.03 02.04 | 05.11 07 | 32.04 03 | 39.04 13.38 | 37.20 11.07 | 03.01 04.06 | 13.10 28.03 | 20.04 | | 05.35 11.16 | 01.03 28.39 | 09.18 06.16 | 18.05 |
| ease of use | 10 | 07.03 04 | 07.15 37.12 | 03.15 32.17 | 19.13 01.36 | 03.09 | 29.11 15.19 | 09.29 37.19 | 03.02 | 05.35 37 | | 37.10 03.09 | 24.02 13.21 | 05.29 04.23 |
| ease of repair | 11 | 03.09 24.16 | 34.03 34.11 | 15.01 | 28.02 03.16 | 29.02 | 02.05 11.28 | 01.03 19.11 | 01.24 28.02 | 03.01 10 | 03.37 | | 12.39 11.10 | 05.16 |
| loss of information | 12 | 11.36 07 | 18.35 39.17 | 01 | 02.04 36 | 27.24 09.34 | 09.03 | 20.11 18.24 | 01.38 | 09 | 13.21 | 05.02 | | 21.02 03 |
| external damaging factors | 13 | 21.01 11.18 | 01.28 21.31 | 38.12 15 | 13.18 05.17 | 10.04 02.06 | 04.38 36.10 | 21.08 14.17 | 21.08 14.17 | 18.01 05 | 05.29 04.23 | 01.02 05 | 21.02 05 | |
| internal damaging factors | 14 | 21.01 06.23 | 07.19 24.18 | 05 | 18.05 17.23 | 24.19 15.10 | 12.38 10 | 08.03 31 | 05.33 13.03 | 24.17 18.34 | 24.18 22.20 | 03.18 | 02.33 14 | 01.18 34 |
| length of the moveable object | 15 | 22.24 04.14 | 22.07 03.16 | 19.18 10.16 | 02.22 14.17 | 02.04 14.27 | 04.09 24 | 03.08 10.18 | 01.03 10.18 | 03.14 19 | 07.14 01.24 | 03.04 02 | 03.18 | 03.07 19.18 |
| length of the fixed object | 16 | 25.22 34.10 | 03.01 | 02.11 | 07.14 04 | 05.09 02 | 09.04 12 | 03.10 | 10 | 07.19 13 | 05.29 | 12 | 18.10 | 03.06 |
| surface of the moveable object | 17 | 02.10 15.05 | 07.25 | 22.25 04.36 | 14.39 | 05.09 | 10.04 09.12 | 22.03 11 | 05.26 11.06 | 11.03 10.18 | 07.19 11.16 | 07.11 02.03 | 25.10 | 21.38 04.03 |
| surface of the fixed object | 18 | 02.07 19.34 | 07.16 | 36 | | 09.01 17.24 | 05.14 06.26 | 10.04 09.12 | 05.01 26 | 05.01 25.06 | 17.16 | 16.24 | 16 | 25.16 13.05 23.01 |
| volume of the moveable object | 19 | 02.20 05.15 | 07.14 | 01.15 16.18 | 22.03 17.28 | 29.04 05.16 | 19.10 04 | 10.03 | 14.10 24 | 14.03 17 | 07.11 25.37 | 02 | 05.21 | 21.33 13.01 |
| volume of the fixed object | 20 | 01.27 02.05 | 31.20 11.09 | 02.03 11.18 | 05.01 16 | 01.02 29 | 19.10 09.31 | 03.31 | 05.19 10 | 01 | 34.10 18.19 | 03 | 09.22 18 | 15.23 08.13 |
| shape | 21 | 19.10 15.02 | 03.07 14 | 07.03 09 | 02.17 16 | 09.25 17 | 04.09 03 | 16.14 03.04 | 07.11 23 | 03.09 19.04 | 09.07 10 | 05.11 03 | 19.34 07.09 | 21.03 05.01 |
| speed | 22 | 04.02 11 | 07.02 10 | 02.06 | 28.01 13.04 | 02.04 09.29 | 04.09 03.18 | 02.04 24.15 | 12.15 13.16 | 01.11 32.03 | 09.04 11.37 | 15.05 04.13 | 11.10 | 03.04 01.36 |
| functional time of the moveable object | 23 | 01.19 22.08 | 03.01 11 | 20.02 | 28.05 11 | 12.13 16.17 | 12 | 02.24 14.07 | 08.14 23.01 | 13.03 24 | 37.13 | 14.02 13 | 02 | 21.07 38.04 |
| functional time of the fixed object | 24 | 40.02 16.30 | 05 | 03 | 15.13 20.17 | 02.10 18 | 02.10 18 | 35.02 24 | 29.15 20.01 | 01.02 | 03 | 03 | 02 | 19.03 17.38 |
| loss of time | 25 | 02.18 24.11 | 01.04 | 18.04 01.25 | 02.25 24 | 18.10 04.06 | 18.15 04.09 | 20.14 | 06.04 09.02 | 01.04 15.24 | 24.04 02.15 | 09.03 02 | 18.04 04.09 | 01.06 15 |
| quantity of material | 26 | 11.14 12.13 | 07.12 14 | 32.01 | 06.12 04.17 | 38.25 | 12.05 04 | 12.11 13.02 | 12.13 14.06 | 14.03 01.13 | 01.14 02.29 | 05.09 02.29 | 18.04 01 | 14.31 |
| loss of material | 27 | 04.01 02.36 | 07.02 05 | 01.02 06 | 02.14 23.01 | 01.02 18.31 | 16.15 31.04 | 01.02 04.18 | 01.06 02.11 | 07.15 38 | 09.04 05.18 | 05.01 15.13 | 09.03 02.12 | 38.21 25.17 |
| strength | 28 | 14.01 02.22 | 07.12 09 | 07 | 28.12 | 12.13 | | 12.13 16 | 05.11 04 | 13.12 07.17 | 28.12 02.09 | 09.17 04.05 | 13.28 12 | 06.01 27.03 |
| stabile structure of the object | 29 | 36.01 17.12 | 01.25 15.05 | 03.32 01 | 01.17 18 | 06 | 11 | 05.01 21.10 | 01.21 23.36 | 01.08 | 09.01 25 | 05.01 02.16 | 18.02 09 | 01.18 06.25 |
| force | 30 | 12.04 01.27 | 07.19 06.40 | 05.01 | 12.01 11.33 | 04.14 27.26 | 01.02 36.18 | 10.01 02.06 | 26.27 02.08 | 07.27 06.03 | 03.04 12.29 | 07.03 28 | 09.34 02 | 03.01 17.06 |
| tension, pressure | 31 | 02.22 01.27 | 01 | 01.18 | 02.11 08.01 | 12.01 | 20.04 03 | 08.03 27 | 05.26 16 | 03.01 16 | 28 | 05 | 04.05 34.18 | 21.05 27 |
| weight of the moveable object | 32 | 01.12 18.27 | 14.35 07.32 | 10.01 06.08 | 12.28 03.13 | 04.01 10.06 | 04.13 01.10 | 10.25 26.15 | 04.14 10.09 | 13.04 03.26 | 01.12 05.18 | 05.13 04.28 | 02.18 01 | 21.33 06.13 |
| weight of the fixed object | 33 | 03.04 07.01 | 08.07 14 | 05.10 01 | 02.04 32.12 | 02.03 01.19 | 06.10 04 | 03.02 10.23 | 29.04 19.07 | 04.03 09 | 20.11 03.09 | 05.13 04.28 | 02.07 01 | 05.08 21.27 |
| temperature | 34 | 07.04 01 | 05.06 13 | 10.05 08.16 | 08.01 12.02 | 18 | 09.08 18 | 05.19 16 | 12.13 01.31 | 10.13 | 10.13 | 24.02 16 | 39.18 | 21.38 01.05 |
| brightness of the lighting | 35 | 05.29 16 | 07.03 08 | 05.10 02 | 01.03 | 12.09 | 28.07 09 | 20.09 11 | 09.07 | 08.01 04.10 | 04.10 08 | 07.19 11.16 | 03.20 | 07.08 |
| power | 36 | 04.01 15 | 08.19 15 | 04.05 19 | 08.18 10.31 | 09.05 | 09.07 05 | 40.08 25.15 | 08.01 16 | 10.02 15 | 10.01 02 | 01.05 02.15 | 02.08 | 08.21 31.05 |
| energy use of the moveable object | 37 | 37.04 01 | 07.19 11.16 | 09.05 | 08.33 28.13 | 12.35 04.32 | 12.03 09 | 05.14 13.04 | 01.30 | 04.10 25 | 08.01 | 03.07 19.04 | 11.18 34.21 | 03.01 20.13 |
| energy use of the fixed object | 38 | 03.20 | 01.11 | 02.05 | 02.26 36 | 04.24 06 | 04.01 | 04.24 | 08.01 16.29 | 03.24 | 11.18 | 01.19 | 05.02 34 | 02.05 21.27 |
| loss of energy | 39 | 04.02 14.01 | 07.22 11.31 | 05 | 28.02 01 | 18.10 09.04 | 09 | 34.36 | 01.12 07.36 | 02.01 | 01.09 03 | 05.08 | 08.02 | 33.21 01.05 |

| Problem-factor (minus-factor) → Trend-factor (plus-factor) ↓ | # | internal damaging factors 14 | length of the moveable object 15 | length of the fixed object 16 | surface of the moveable object 17 | surface of the fixed object 18 | volume of the moveable object 19 | volume of the fixed object 20 | shape 21 | speed 22 | functional time of the moveable object 23 | functional time of the fixed object 24 | loss of time 25 | quantity of material 26 |
|---|---|---|---|---|---|---|---|---|---|---|---|---|---|---|
| productivity | 01 | 01.21/06.23 | 06.24/04.30 | 25.34/22.10 | 02.10/15.31 | 02.01/19.34 | 05.20/15.02 | 01.27/02.05 | 22.02/15.17 | 01.12/18 | 01.02/05.06 | 40.02/16.30 | 12.15 | 01.30 |
| universality, adaptability | 02 | 18.24/28 | 01.03/14.05 | 03.01/16 | 01.25 | 07.16 | 07.01/14 | 07.31/16.12 | 07.27/03.32 | 01.02/22 | 11.03/01 | | 05.16 | 01.04 / 12.01/07 |
| level of automation | 03 | 05 | 22.11/04.19 | 36 | 19.22/11 | 11.10/24 | 01.11/16 | 10.11/18.31 | 07.09/03.11 | 04.02 | 20.39 | 02.16/11 | 18.04/01.25 | 01.11 |
| reliability | 04 | 01.05/17.10 | 07.39/22.24 | 07.14/04.28 | 19.02/22.16 | 09.01/17.24 | 12.02/22.18 | 05.01/18 | 01.03/16.28 | 33.01/28.04 | 05.01/12.29 | 15.13/20.17 | 02.25/24 | 33.04/17.12 |
| precision of manufacture | 05 | 24.19/15.10 | 02.04/14.27 | 05.09/02 | 04.38/14.09 | 05.14/06.26 | 09.04/05 | 29.02/01 | 09.25/17 | 02.04/09 | 12.13/17 | 18.04/15 | 09.10/04.06 | 09.25 |
| precision of measurement | 06 | 12.38/23.02 | 04.10/35.16 | 09.04/12.16 | 10.04/09.12 | 10.04/09.12 | 09.11/20 | 10.18/11.31 | 20.04/09 | 04.11/09.18 | 04.20/09 | 02.10/18 | 18.15/04.09 | 05.20/09 |
| complexity of construction | 07 | 08.03 | 03.08/10.18 | 10 | 22.03/11.16 | 20.26 | 15.10/20 | 03.16 | 14.11/04.07 | 15.02/04 | 02.24/04.07 | 02.11 | 20.14 | 11.12/13.02 |
| complexity of inspection and meas. | 08 | 05.33 | 16.19/10.18 | 10 | 05.11/06.19 | 05.23/25.16 | 14.03/24.16 | 05.06/10.31 | 13.11/03.23 | 12.24/16.01 | 08.14/29.23 | 29.15/20.01 | 06.04/09.39 | 12.13/14.06 |
| ease of manufacture | 09 | 33.14/23 | 03.14/11.19 | 07.19/13 | 11.03/10.37 | 16.17 | 11.14/03.17 | 01 | 03.04/11.13 | 01.11/32.03 | 13.03/24 | 01.16 | 01.04/15.24 | 01.36/03.18 |
| ease of use | 10 | 31.18 | 03.19/11.37 | 03.24 | 03.19/16.23 | 06.16/07.23 | 03.16/01.07 | 24.06/23.31 | 07.15/14.04 | 06.11/15 | 14.12/32.29 | 03.16/29 | 24.04/02.15 | 37.01 |
| ease of repair | 11 | 07.37/04 | 03.04/02.29 | 12.06/31 | 07.11/09 | 16.29 | 29.05/01.28 | 03 | 03.11/05.24 | 15.39 | 28.14/04.13 | 03 | 09.03/02.29 | 05.04/02.29 |
| loss of information | 12 | 02.33/21 | 03.10 | 10 | 25.10 | 25.16 | 34.31 | 05.21 | 24.09 | 10.09 | 02 | 02 | 18.10/04.09 | 18.04/01 |
| external damaging factors | 13 | 11.18/19.24 | 19.03/23.24 | 03.06 | 21.03/38.04 | 13.05/23.01 | 21.36/27.01 | 15.23/08.13 | 21.03/12.01 | 33.21/01.04 | 21.07/38.04 | 19.03/17.38 | 01.06/15 | 01.38/14.31 |
| internal damaging factors | 14 | | 19.07/16.21 | 22.18 | 19.05/06.23 | 21.03/17 | 19.05/17 | 25.06/01.24 | 01.03 | 01.04/12.36 | 07.21/38.31 | 33.23/24 | 03.21 | 12.18/23.03 |
| length of the moveable object | 15 | 19.07 | | | 19.11/25 | 07.19/24 | 34.07/03.24 | 34.19/04.01 | 19.31/08.24 | 03.32/02.14 | 11.24/32 | 08 | 02.01/08 | 07.05/14 / 14.01 |
| length of the fixed object | 16 | 01.31 | 12.03/24 | | 24.19/01 | 19.34/02.17 | 25.34/07 | 01.32/05.22 | 11.22/07.34 | 22.24/31 | 01.14/31 | 03.17/22 | 25.14 | 24.31/29.22 |
| surface of the moveable object | 17 | 19.05/06.23 | 22.07/06.24 | 22.19/24.11 | | | 03.24/12.18 | 34.22/19.24 | 22.11 | 35.15/14.24 | 14.25/24.15 | 20.12 | 03.12/08 | 10.24 / 14.25/20.11 |
| surface of the fixed object | 18 | 21.03/17 | 12.03 | 10.34/39.23 | 24.31/34 | | | 22.34/11 | 10.11/24 | 24.34/10 | 10.35 | 11.01 | 05.02/08.25 | 02.01/24.06 / 05.06/17.24 |
| volume of the moveable object | 19 | 19.05/17.03 | 03.34/01.04 | 34.07/24 | 03.34/24.19 | 22.34/31 | | 01.22/18 | 03.07/14.24 | 14.20/30.15 | 20.01/24 | 25.11/03 | 05.20/15.02 | 14.25/34 |
| volume of the fixed object | 20 | 25.06/01.24 | 08.22 | 01.32/05.22 | 22.24/25.11 | 34.11/07 | 11.04/05.34 | | 34.05/01 | 17.05/04 | 08.03/07.15 | 25.11/30 | 01.16/09.06 | 01.12 |
| shape | 21 | 01.03 | 14.15/35.24 | 11.22/02.34 | 19.07/25 | 22.19/24.02 | 19.22/09.24 | 22.24/07.21 | 34.05/01 | | 01.07/15.06 | 22.10/39.29 | 25.11/35.21 | 22.02/15.19 / 26.21 |
| speed | 22 | 05.18/01.33 | 11.22/32 | 19.07/25 | 14.25/15 | 22.19/24 | 34.14/15 | | 04.34 | | 01.07/06.15 | 12.08/01.35 | 12.11/11.02 | 02.08/14.30 |
| functional time of the moveable object | 23 | 33.23/16.21 | 05.08/39 | 37.39/08 | 01.06/08 | 19.37/08 | 02.05/08.25 | 02.25/37 | 22.10/04.29 | 12.01/35 | | 02.40/24 | 40.02/04.06 | 12.01/02.17 |
| functional time of the fixed object | 24 | 21 | 19.17/08 | 03.17/01 | 01.06/08.22 | 01.19/12.34 | 01.12/11.24 | 01.15/30 | 19.38/11.34 | 14.24/22.11 | 18.04 | | 04.40/02.16 | 12.01/31 |
| loss of time | 25 | 01.21/06.23 | 07.05/14 | 25.18/22.35 | 10.24/35.16 | 02.01/19.24 | 05.35/15.02 | 01.16/09.06 | 24.02/15.19 | 04.10/02.24 | 40.02/04.06 | 02.16 | | 01.30/06.16 |
| quantity of material | 26 | 12.01/17.23 | 14.22/01.06 | 07.31/01.24 | 07.22/14 | 05.06/26.04 | 07.40/14 | 01.30/31.03 | 01.22 | 01.14/15.04 | 12.01/02.17 | 01.30/31 | 01.30 | 06.16/16.06 |
| loss of material | 27 | 02.03/15.14 | 22.14/02.23 | 02.04/18 | 02.31 | 02.06/23.31 | 03.14/25.26 | 02.07/06.31 | 22.05/12.35 | 14.01/04.30 | 02.11/12.06 | 04.13/06.30 | 07.06/01.02 | 20.12/02.18 |
| strength | 28 | 07.01/21.05 | 03.07/32.01 | 07.22/04.10 | 12.15/17.14 | 39.17/04 | 02.07/22.34 | 03.22/19.07 | 02.25/01.17 | 32.11/10.22 | 13.12/10 | 18.14/24 | 14.12/04.02 | 14.02/13 |
| stabile structure of the object | 29 | 01.17/13.23 | 11.07/03.04 | 27 | 05.28/07 | 23 | 04.02/08.23 | 15.04/01.17 | 21.03/06.24 | 38.07/04.06 | 11.13/02.01 | 23.12/01.36 | 01.13 | 07.09/01 |
| force | 30 | 11.12/26.18 | 19.08/39.26 | 04.02 | 08.02/07 | 03.06/26.27 | 07.39/37.27 | 05.26/06.27 | 02.01/17.15 | 11.04/07.37 | 08.05 | 05.02/10 | 02.27/26 | 22.14/26.26 |
| tension, pressure | 31 | 05.38/13.06 | 01.02/26 | 01.03/22.16 | 02.07/26.04 | 02.07/26.27 | 20.01/02 | | 01.18 | 01.24/20 | 08.12/01.26 | 22.01/13 | 27.26/19.05 | 02.22/24 / 02.22/26 |
| weight of the moveable object | 32 | 21.01/31.23 | 07.32/14.15 | 07.19/37.14 | 14.19/30.15 | 04.03/31.24 | 14.05/17.04 | 17.05/24.34 | 02.22/01.17 | 05.32/07.30 | 35.15/31.01 | 02.04 | 02.01/40.04 | 12.10/06.31 |
| weight of the fixed object | 33 | 01.21/03.23 | 19.24 | 02.03/14.01 | 34.01 | 01.25/11.05 | 22.11/12 | 35.01/24 | 11.02/14.22 | 01.19/25 | 01.02/37 | 05.13/08.20 | 02.40/01.10 | 08.20/06.10 |
| temperature | 34 | 21.01/05.18 | 07.08/39 | 07.08/37 | 12.01/17 | 01.30 | 15.23/17.06 | 01.20/24 | 22.21/08.09 | 05.04/26.25 | 08.11/23 | 08.06/26.17 | 01.04/33.06 | 12.19/25.23 |
| brightness of the lighting | 35 | 01.08/09.23 | 08.09/16 | 22.01 | 08.09/10 | 19.24/03 | 05.11/02 | 22.11/02 | 09.25 | 02.11/08 | 05.08/20 | 20.02/04 | 08.03/10.19 | 03.08 |
| power | 36 | 05.01/06 | 03.02/01.27 | 03.01/24 | 08.30 | | 19.09/11.30 | 01.20/30 | 25.20/29 | 14.22/17.15 | 07.01/05 | 08.01/02 | 16 | 01.40/02.20 / 24.15/08 |
| energy use of the moveable object | 37 | 05.01/20 | 37.04 | 05.08/07.04 | 07.08/29 | 01.05/19.07 | 01.11/06 | 01.30/01.34 | 12.15/14 | 37.05/01 | 32.07/20.06 | 04.01/02.08 | 04.11/08.06 | 01.30/15.36 / 16.06 |
| energy use of the fixed object | 38 | 08.21/06 | 19.24/37 | 19.39/16 | 11.37/18 | 19.16/08 | 11.08/04 | | 01.06 | 34.18 | 03.04 | 08.24/04 | 19.04/10 | 02.17/34.39 / 12.01/31 |
| loss of energy | 39 | 33.01/05.21 | 34.05/20.11 | 20.30/34 | 07.10/19.25 | 19.34/25.06 | 34.06/36 | | 34 | 24.18 | 16.01/30 | 33.01 | 19.31 | 02.06/09.34 / 34.06/29 |

| Problem-factor (minus-factor) → / Trend-factor (plus-factor) ↓ | # | loss of material | strength | stable structure of the object | force | tension, pressure | weight of the moveable object | weight of the fixed object | temperature | brightness of the lighting | power | energy use of the moveable object | energy use of the fixed object | loss of energy |
|---|---|---|---|---|---|---|---|---|---|---|---|---|---|---|
| | # | 27 | 28 | 29 | 30 | 31 | 32 | 33 | 34 | 35 | 36 | 37 | 38 | 39 |
| productivity | 01 | 04.02 01.36 | 14.04 02.06 | 01.12 21.23 | 04.07 02.26 | 02.27 22 | 01.10 18.27 | 04.13 07.12 | 01.33 04.02 | 10.19 08.03 | 01.40 02 | 01.02 30.08 | 03 | 04.02 14.01 |
| universality, adaptability | 02 | 07.02 05.11 | 01.12 09.20 | 01.25 22 | 07.19 40 | 01.16 | 03.20 07.32 | 08.07 14.16 | 13.05 12.01 | 20.21 10.03 | 08.03 14 | 08.01 14.11 | 16.03 37 | 06.07 03 |
| level of automation | 03 | 01.02 06.35 | 29.11 | 06.03 | 05.01 | 11.01 | 04.10 06.01 | 04.10 01.02 | 10.05 08 | 32.09 08 | 04.05 13 | 05.09 11 | 01.03 05.11 | 36.04 |
| reliability | 04 | 02.01 14.23 | 28.04 | 12.03 18.05 | 32.04 02.12 | 02.18 01.08 | 12.32 02.17 | 12.02 32.04 | 12.01 02 | 28.09 08 | 33.28 11 | 33.28 10.31 | 26.36 | 02.28 01 |
| precision of manufacture | 05 | 01.31 02.18 | 12.13 | 25.06 | 04.08 15.26 | 12.01 | 04.09 11.06 | 10.01 13.39 | 08.10 | 12.09 | 09.05 | 09.05 | 02.18 35.40 | 11.09 05 |
| precision of measurement | 06 | 02.16 31.04 | 04.20 09 | 09.01 11 | 09.05 | 20.04 09 | 09.01 10.04 | 04.01 29.10 | 20.08 04.18 | 20.03 09 | 12.20 09 | 12.20 09 | 18.16 24 | 10.09 13 |
| complexity of construction | 07 | 01.02 04.14 | 05.11 04 | 05.21 19.08 | 10.16 | 08.03 01 | 10.25 15.26 | 05.10 01.23 | 05.19 11 | 18.19 11 | 40.08 25.15 | 13.05 14.04 | 04.02 11 | 02.01 11.05 |
| complexity of inspection and meas. | 08 | 03.06 12.07 | 13.12 04 | 28.21 23.25 | 26.04 17.08 | 01.26 27.09 | 13.10 04.01 | 20.11 04.03 | 12.13 01.16 | 05.18 10 | 08.03 16.02 | 01.30 | 08.01 27 | 01.12 07.08 |
| ease of manufacture | 09 | 07.15 02.09 | 03.12 03 | 28.11 | 01.37 | | 01.08 03.27 | 04.14 07.16 | 03.13 26.11 | 13.10 06 | 04.18 13.03 | 13.03 37.18 | 04.10 03.24 | 08.01 |
| ease of use | 10 | 04.09 05.18 | 09.17 12.04 | 09.01 25 | 04.11 01 | 05.09 37 | 29.05 11.07 | 20.11 03.29 | 10.13 11 | 11.19 13 | 01.15 05.02 | 03.11 18 | 18.37 11 | 05.08 11 |
| ease of repair | 11 | 05.01 15.13 | 03.28 05.39 | 05.01 | 03.28 02 | 11 | 05.13 01.28 | 05.13 01.28 | 24.02 | 07.03 11 | 07.02 09.05 | 07.03 04.16 | 03.07 11.16 | 07.03 09.08 |
| loss of information | 12 | 34.19 12.11 | 18.31 22.17 | 25.10 18.35 | 11.03 | 18.21 | 02.18 01 | 02.01 35 | 21.03 18 | 08 | 02.08 | 03.40 08 | 18 | 08.02 |
| external damaging factors | 13 | 38.21 08.17 | 06.01 27.03 | 01.18 25.06 | 11.01 23.06 | 21.05 27 | 21.33 23.13 | 05.21 11.18 | 21.38 01.05 | 03.08 09.11 | 08.21 31.05 | 03.18 20.13 | 02.05 21.27 | 33.21 01.05 |
| internal damaging factors | 14 | 02.03 15 | 07.01 21.05 | 01.17 13.23 | 01.04 03.17 | 05.38 13.06 | 08.21 07.23 | 01.21 03.23 | 21.01 05.13 | 01.18 09 | 08.18 06 | 05.01 20 | 08.21 06 | 33.01 05.21 |
| length of the moveable object | 15 | 24.14 36.02 | 32.01 14.15 | 03.32 07.15 | 19.02 24 | 03.32 01 | 32.07 14.15 | 03.19 07.24 | 02.07 09 | 09 | 03.01 | 32.01 18 | 24.03 07 | 34.05 01.23 |
| length of the fixed object | 16 | 02.04 18.01 | 07.22 04.10 | 23.27 01 | 04.02 | 03.22 01 | 25.31 32.17 | 01.04 17.14 | 12.01 30.06 | 12.29 | 37.32 | 01.18 33.11 | 25.31 37.11 | 20.04 |
| surface of the moveable object | 17 | 02.01 05.23 | 12.07 17.22 | 28.05 11.23 | 08.25 01.05 | 02.07 26.04 | 05.19 14.04 | 12.31 05 | 05.07 16 | 07.09 08.11 | 08.02 09.06 | 08.01 | 19.08 35 | 07.19 25.10 |
| surface of the fixed object | 18 | 02.22 06.23 | 17 | 05.30 | | 03.06 01.26 | 02.07 26.27 | 22.31 24.11 | 25.05 22.06 | 01.23 30 | 03.18 01.09 | 19.09 | 08.11 04 | 19.34 25 |
| volume of the moveable object | 19 | 26.23 15.02 | 39.22 07.34 | 04.02 03.23 | 07.01 26.27 | 20.01 26.27 | 05.10 14.17 | 31.17 10 | 15.23 02.06 | 02.11 05 | 01.26 11.06 | 01 | 30.38 08 | 34.07 11.16 |
| volume of the fixed object | 20 | 02.23 01.15 | 39.22 19.07 | 18.04 01.17 | 05.06 27 | 18.01 | 31.25 24.08 | 01.02 08.22 | 01.20 24 | 02.11 04.34 | 25.20 | 02.08 31.11 | 01.17 | 01.17 |
| shape | 21 | 01.14 12.35 | 25.22 02.17 | 38.03 06.24 | 01.02 27.17 | 15.07 02.22 | 32.02 14.17 | 32.07 10.12 | 21.22 08.09 | 11.07 09 | 24.20 05 | 05.20 15.22 | 22.07 31 | 22 |
| speed | 22 | 02.11 04.30 | 32.12 10.22 | 04.38 03.06 | 11.04 07.08 | 20.06 30.17 | 05.04 11.30 | 03.11 | 24.25 26.05 | 02.11 08 | 08.01 30.05 | 32.07 01.30 | 01.08 | 22.40 08.01 |
| functional time of the moveable object | 23 | 04.13 12.06 | 13.12 02 | 11.12 01 | 08.05 16 | 08.12 13 | 08.35 15.31 | 31.15 | 08.01 23 | 03.08 24.01 | 08.02 01.30 | 04.20 01.06 | 20.06 | 02.18 01 |
| functional time of the fixed object | 24 | 13.16 06.30 | 01.39 | 23.12 01.36 | 19.17 39 | 19.24 17 | 31.24 07 | 20.13 08.16 | 08.06 26.17 | 17.18 .34 | 16 | 11.17 18 | 01.17 | 02.17 15.18 |
| loss of time | 25 | 01.06 02.23 | 14.12 04.06 | 01.12 21.35 | 02.27 26.35 | 27.26 24 | 02.40 27.01 | 02.40 10.35 | 01.14 33.06 | 03.08 10.19 | 01.40 02.20 | 01.30 08.06 | 03 | 02.35 06.09 |
| quantity of material | 26 | 20.12 02.18 | 22.01 15.02 | 07.05 19.17 | 01.22 12 | 26.28 22.12 | 01.20 06.31 | 13.10 06.01 | 12.19 23 | 01.04 25.31 | 01 | 15.14 16.06 | 12.01 31 | 34.06 29 |
| loss of material | 27 | | 01.04 31.17 | 11.19 01 | 22.07 06.17 | 12.26 27.02 | 01.20 36.17 | 01.20 21.09 | 33.26 23.31 | 23.31 11 | 04.13 18 | 01.06 18.35 | 04.13 37.31 | 01.13 05.31 |
| strength | 28 | 01.04 31.17 | | 11.19 01 | 02.06 12.22 | 02.12 06.17 | 03.32 17.07 | 17.10 13.03 | 25.02 17 | 01.08 | 02.10 01.04 | 08.01 02 | 01 | 01 |
| stable structure of the object | 29 | 05.22 25.17 | 19.39 07 | | 02.01 33.16 | 05.01 17 | 33.01 05.23 | 10.23 03.17 | 01.03 09 | 19.12 13.07 | 09.01 13.31 | 11.08 | 03.16 14.06 | 22.05 23.20 |
| force | 30 | 32.01 17.35 | 01.02 22.13 | 01.02 33 | | 06.33 28 | 32.03 27.06 | 06.11 03.04 | 01.02 33 | 11.08 01.18 | 03.01 06.27 | 08.19 02 | 03.16 26.27 | 22.07 |
| tension, pressure | 31 | 02.26 12.27 | 39.06 12.17 | 01.38 05.17 | 26.01 33 | | 02.26 27.17 | 11.14 27.17 | 01.23 08.05 | 18.37 21.01 | 02.01 22 | 22.18 02.27 | 19.24 37.18 | 05.26 29 |
| weight of the moveable object | 32 | 35.01 12.31 | 04.13 06.17 | 03.01 08.23 | 32.02 06.27 | 02.26 27.17 | | 08.01 | 20.14 24.30 | 08.03 09 | 37.26 06.31 | 01.37 15.31 | 01.03 04 | 20.05 15.08 |
| weight of the fixed object | 33 | 35.32 11.25 | 04.05 02.13 | 10.23 03.17 | 32.02 08.01 | 11.14 02.06 | 17.31 03 | | 04.08 09.21 | 01 | 07.08 06.21 | 12.19 | 06.08 04.03 | 06.08 04.07 |
| temperature | 34 | 33.26 14.31 | 02.25 21.17 | 03.01 09 | 01.02 09 | 01.23 12.33 | 26.21 20.30 | 21.01 09 | | 09.25 33.16 | 05.22 19.29 | 08.07 12.19 | 01.12 | 33.19 01.30 |
| brightness of the lighting | 35 | 11.03 | 01.08 | 09.12 13 | 10.08 20 | 25.01 37 | 08.03 09 | 05.01 09 | 09.01 08 | | 09 | 09.03 08 | 09.01 03.07 | 08.16 03.20 |
| power | 36 | 04.13 06.30 | 10.02 04 | 01.09 07.31 | 10.05 26.01 | 21.02 01 | 32.26 19.13 | 08.10 19.13 | 05.22 19.29 | 16.20 08 | | 16.20 08.27 | 08.07 16.03 | 02.01 30 |
| energy use of the moveable object | 37 | 01.18 06.35 | 35.08 39.01 | 08.11 19.18 | 16.10 13.05 | 36.22 29 | 37.06 04.31 | 04.37 31.11 | 08.18 12.22 | 05.07 08 | 20.08 27.06 | | 07.11 | 37.21 07.18 |
| energy use of the fixed object | 38 | 04.13 06.31 | 01 | 13.24 14.06 | 26.27 | 19.39 01.17 | 04.11 32.08 | 08.39 20.13 | 01.08 33.26 | 08.05 | 35.08 11.01 | 05.08 11.35 | | 01.31 03.18 |
| loss of energy | 39 | 01.13 05.27 | 10 | 22.05 23.20 | 26.30 | 24.11 | 07.20 08.04 | 08.20 06.39 | 08.30 34 | 03.11 09.07 | 12.30 | 01.08 12 | 01.08 24 | |

# R2. As-Catalog

## R2.1. List of 40 navigators (specialized transformation models)

| # | Navigator |
|----|-----------|
| 01 | Change in the aggregate state of an object |
| 02 | Preliminary action |
| 03 | Segmentation |
| 04 | Replacement of mechanical matter |
| 05 | Separation |
| 06 | Use of mechanical oscillations |
| 07 | Dynamization |
| 08 | Periodic action |
| 09 | Change in color |
| 10 | Copying |
| 11 | Inverse action |
| 12 | Local property |
| 13 | Inexpensive short-life object as a replacement for expensive long-life one |
| 14 | Use of pneumatic or hydraulic constructions |
| 15 | Discard and renewal of parts |
| 16 | Partial or excess effect |
| 17 | Use of composite materials |
| 18 | Mediator |
| 19 | Transition into another dimension |
| 20 | Universality |
| 21 | Transform damage into use |
| 22 | Spherical-shape |
| 23 | Use of inert media |
| 24 | Asymmetry |
| 25 | Use of flexible covers and thin films |
| 26 | Phase transitions |
| 27 | Use of thermal expansion |
| 28 | Previously installed cushions |
| 29 | Self-servicing |
| 30 | Use of strong oxidants |
| 31 | Use of porous materials |
| 32 | Counter-weight |
| 33 | Quick jump |
| 34 | Matryoshka (nested doll) |
| 35 | Unite |
| 36 | Feedback |
| 37 | Equipotentiality |
| 38 | Homogeneity |
| 39 | Preliminary counter-action |
| 40 | Uninterrupted useful function |

## R2.2. Table of As-Catalog

| 01. Change in the aggregate state of an object | a) this includes transitions into "pseudo-states" ("pseudo-liquid") and into transitional states such as the use of the elastic properties of solid objects as well as simple transitions such as from a solid to a liquid state;<br>b) changes in concentration or consistency, in the degree of flexibility, in temperature, etc. |
|---|---|
| 02. Preliminary action | a) previous necessary (partial or complete) change of an object;<br>b) prepare objects in advance so that they can be put to work from the best position and are available without loss of time. |
| 03. Segmentation | a) disassemble an object into individual parts;<br>b) make it possible to disassemble an object;<br>c) raise the degree of disassembly (reduction into parts) of an object. |
| 04. Replacement of mechanical matter | a) replace mechanical schemes with optical, acoustic, or olfactory schemes;<br>b) use of electrical, magnetic, or electromagnetic fields for the interaction of objects;<br>c) replacement of static fields with dynamic ones, from temporally fixed to flexible fields, from unstructured fields to fields with a specific structure;<br>d) use of fields in connection with ferric-magnetic particles. |
| 05. Separation | separate the "incompatible part" ("incompatible property") from the object or - turned completely around - include the only really necessary part (necessary property) into the object. |
| 06. Use of mechanical oscillations | a) cause an object to vibrate;<br>b) raise the frequency of the vibrations up to and including ultra-high frequencies if the object is already in motion;<br>c) use of the resonating frequency, application of quartz vibrators;<br>d) use of ultra-sound vibrations in connection with electromagnetic fields. |
| 07. Dynamization | a) the characteristics of an object or an environment are changed to optimize every work procedure;<br>b) disassemble an object into parts that are moveable among each other;<br>c) make an object moveable that is otherwise fixed. |
| 08. Periodic action | a) transition from a continuous function to a periodic one (impulse);<br>b) change the periods if the function already runs that way;<br>c) use the breaks between impulses for other functions. |
| 09. Change in color | a) change the color of an object or its environment;<br>b) change the level of the transparency of an object or its environment;<br>c) use color supplements to observe objects or processes that are difficult to see;<br>d) add lighting if this kind of supplements is already in use. |
| 10. Copying | a) use a simplified and inexpensive copy instead of an inaccessible, complicated, expensive, inappropriate, or fragile object;<br>b) replace an object or a system of objects with optical copies; use here a change in measure (blow-up or reduce the copy);<br>c) if visible copies are used, they can be replaced with infra-red or ultra-violet copies. |

| 11. Inverse action | a) instead of an action prescribed by the conditions of an assignment, complete a reverse action (heat an object instead of cooling it); <br> b) make a moveable part of an object or the environment fixed or a fixed part moveable; <br> c) turn an object "upside down" or around. |
|---|---|
| 12. Local property | a) change the structure of the object (the external environment, external influences) from the same to a different one; <br> b) different parts of an object have different functions; <br> c) every object should exist under conditions that correspond best to its functions. |
| 13. Inexpensive short-life object as a replacement for expensive long-life one | replace an expensive object with a group of inexpensive objects without certain properties, for example, long life. |
| 14. Use of pneumatic or hydraulic constructions | use gaseous or fluid parts instead of fixed parts in an object: parts that can be blown up or filled with hydraulic fluid, air-cushions, hydrostatic or hydro-reactive parts. |
| 15. Discard and renewal of parts | a) parts that have fulfilled their task and are no longer part of an object should be disposed of (dissolved, evaporated, etc.); <br> b) used parts of an object should be immediately replaced during work. |
| 16. Partial or excess effect | when it is difficult to achieve the desired effect completely, we should try to achieve a bit less or a bit more. This can make the task much easier. |
| 17. Use of composite materials | move from homogeneous materials to combinations. |
| 18. Mediator | a) use another object to transfer or transmit an action; <br> b) temporarily connect an object with another (easily separable) object. |
| 19. Transition into another dimension | a) an object is shaped so that it can move or is placed not only in a linear fashion, but also in two dimensions, meaning on a surface. It is also possible to improve the transition from a surface to a three-dimensional space; <br> b) do construction on several floors; tip or turn the object on its side; use the back of the space in question; <br> c) optical rays that strike a neighboring space or the back of the present space. |
| 20. Universality | an object has several simultaneous functions so that other objects are not needed. |
| 21. Transform damage into use | a) use damaging factors, especially damaging influences from the environment to achieve a useful effect; <br> b) eliminate a negative factor by combining it with other negative factors; <br> c) support the damaging factor until it is no longer causes damage. |
| 22. Spherical-shape | a) change from linear parts of the objects to curved ones, from flat surfaces to spherical ones; from parts shaped like cubes or parallelepipeds to round structures; <br> b) use rollers, balls, and springs; <br> c) change to turning movements by using centrifugal force. |
| 23. Use of inert media | a) replace a normal medium with an inert one; <br> b) complete a process in a vacuum. |

| 24. Asymmetry | a) move from a symmetrical shape of an object to an asymmetrical one; <br> b) increase the degree if the object is already asymmetrical. |
|---|---|
| 25. Use of flexible covers and thin films | a) flexible covers and thin layers are used in place of the usual constructions; <br> b) isolate objects from the external world with flexible covers or thin layers. |
| 26. Phase transitions | full use of phenomena that occur during phase transitions such as a change in volume, radiation or absorption of warmth, etc. |
| 27. Full use of thermal expansion | a) full use of the expansion (or reduction) of materials when heating them; <br> b) use of materials with different coefficients of heat expansion. |
| 28. Previously installed cushion | increase the relatively low security of an object with safety measures in advance. |
| 29. Self-servicing | a) the object services itself with auxiliary and repair functions; <br> b) reuse waste (energy, material). |
| 30. Use of strong oxidants | a) replace normal air with an enhanced stream; <br> b) replace an enhanced stream with oxygen; <br> c) influence air or oxygen with ionizing rays; <br> d) use of oxygen with ozone; <br> e) replace ionized or ozone-oxygen with ozone. |
| 31. Use of porous materials | a) make an object porous or use supplementary porous elements (inserts, coverings, etc.); <br> b) if the object already consists of a porous material, the pores can be filled with some kind of material in advance. |
| 32. Counter-weight | a) compensate for the weight of an object with its connection to another object with lifting power; <br> b) compensate for the weight of an object using interaction with the external environment (for example, with aerodynamic or hydrodynamic forces). |
| 33. Quick jump | complete a process or some of its (damaging or dangerous) stages at high speed. |
| 34. Matryoshka (nested doll) | a) an object is inside another object that is also inside another, etc.; <br> b) an object runs through a hollow space in another object. |
| 35. Unite | a) unite similar objects or objects for neighboring operations; <br> b) temporarily unite similar objects or objects for neighboring operations. |
| 36. Feedback | a) create a retroactive influence; <br> b) change a retroactive influence that already exists. |
| 37. Equipotentiality | change work conditions so that it is not necessary to lift or lower an object. |
| 38. Homogeneity | objects that interact with the object in question must be made from the same material (or from one with similar properties). |
| 39. Preliminary counter-action | if the conditions of a task require an action, then a opposite action should be taken in advance. |
| 40. Uninterrupted useful function | a) complete a job without interruptions where all parts work continuously at full capacity; <br> b) eliminate idle running and interruptions. |

# R3. Afs-Catalog

| Fundamental transfor-mation | Relationship to As-navigators |
|---|---|
| **Separation of conflicting properties in space:** One part of system space has property **A**, while an-other part of system space has property **not-A**. | **05 separation:** remove the disruptive part, emphasize the part needed. **10 copy:** use of simplified and inexpensive copies. **19 transition to another state:** increase the freedom of an object, use construction in several layers, use lateral and other surfaces. **22 spherical-shape:** transition to curved surfaces and shapes, use of wheels, balls, or springs. **24 asymmetry:** transition to asymmetrical shapes, increase asymmetry. **25 use of flexible covers and thin layers:** use flexible covers and thin layers instead of normal constructions. **34 matryoshka:** store an object in another one in stages, place an object in the hollow space of another one. |
| **Separation of conflicting properties in time:** During one time interval, the system has property **A**; during an-other time interval, it has property **not-A**. | **02 preliminary action:** run the necessary effect partially or com-pletely; arrange the objects so that they can go to work faster. **07 dynamization:** make an object (or its parts) moveable, optimize the characteristics of the process (of an object) in every stage. **08 periodic action:** a) transition from a continuous functioning to a periodic one (impulse); b) change the periods if the functioning al-ready runs that way; c) use the breaks between impulses for other functions.. **18b mediator:** connect an object with another (easily removable) one for a specific time. **28 previously installed cushion:** consider possible disruptions in advance. **33 quick jump:** accelerate a process strongly so that damages don't even appear. **35b unite:** temporally connect similar or neighboring operations with each other. **39 preliminary counter-action:** an opposite effect must be com-pleted in advance to run the primary effect. **40 uninterrupted useful function:** eliminate idle time and interrup-tions so that all parts of an object function at full capacity. |
| **Separation of conflicting properties in structure:** Some ele-ments of the system have property **A**, while other | **03 segmentation:** disassemble an object into its parts, increase the degree of "disassembly". **04c replacement of mechanical matter**: replacement of static fields with dynamic ones, from temporally fixed to flexible fields, from unstructured fields to fields with a specific structure. **11 inverse action:** take the opposite action from the one that is ap-parently given by the conditions at hand. **12 local property:** transition from a similar to a different structure so |

| | |
|---|---|
| elements or the system as a whole have (has) property **not-A**. | that every part can complete its function under the best condition.<br><br>**15 disposal and regeneration of parts:** a used-up part can be disposed of or regenerated during its function.<br><br>**18 mediator:** a) use another object to transfer or transmit an action; b) temporarily connect an object with another (easily separable) object.<br><br>**32a counter-weight**: compensate for the weight of an object with its connection to another object with lifting power.<br><br>**35a unite:** connection of similar objects or objects for neighboring operations with each other.<br><br>**36 feedback**: a) create a retroactive influence;  b) change a retroactive influence that already exists. |
| **Separation of conflicting properties in material / energy:**<br>For one purpose, the material has property **A**; for another purpose, it has property **not-A**. | **01 change of the aggregate state of an object:** change the concentration or consistency, full use of properties like the elasticity of materials, etc.<br><br>**14 use of pneumatic or hydraulic constructions:** use gaseous or fluid parts instead of fixed parts in an object: parts that can be blown up or filled with hydraulic fluid, air-cushions, hydrostatic or hydroreactive parts.<br><br>**17 use of materials that consist of several components:** transition from similar materials to those consisting of several components.<br><br>**15 discard and renewal of parts:** a) parts that have fulfilled their task and are no longer part of an object should be disposed of (dissolved, evaporated, etc.); b) used parts of an object should be immediately replaced during work.<br><br>**16 partial or excess effect:**  when it is difficult to achieve the desired effect completely, we should try to achieve a bit less or a bit more.<br><br>**23 use of inert media:** replace a medium with something inert. Let processes run in a vacuum.<br><br>**26 use of phase transitions:** full use of phenomena that occur during phase transitions: changes in volume, in the radiation or absorption of heat.<br><br>**27 use of heat  expansion:**  full use of the heat expansion of materials, the use of materials with different heat expansion.<br><br>**29b self-servicing:** use of the waste of material and energy.<br><br>**30 use of strong oxidants:** replace air with oxygen, influence air with ion beams, use of ozone.<br><br>**31 use of porous materials:** shape an object in a porous way, fill porous parts with some kind of material.<br><br>**38 homogeneity:** manufacture objects that influence each other from one and the same material. |

# Main References

## To part 8. Modeling the Green Automotive Innovations:

8.1 New Fuel Alternative
(http://www.ec.gc.ca/cleanair-airpur/Electricity-WSDC4D330A-1_En.htm) - Electricity Generation" from website Environment Canada
http://en.wikipedia.org/wiki/Fossil_fuel
http://www.biodiesel.org/
http://journeytoforever.org/biodiesel.html
http://www.darvill.clara.net/altenerg/fossil.htm
8.2 New Alternative Fuel Improvement
http://www.technologyreview.com/biztech/20319/?a=f
http://sustainabledesignupdate.com/?p=950
http://algaefuel.org/
http://www.oilgae.com/algae/oil/biod/cult/cult.html
8.3 Electrical Energy Supply Improvement
http://electronics.howstuffworks.com/lithium-ion-battery.htm
http://en.wikipedia.org/wiki/Lithium-ion_battery
http://www.automotto.org/entry/body-work-powered-cars-are-the-future-of-electric-hybrid-cars/
http://www.euinfrastructure.com/news/car-powered-by-body-work/
http://www.euinfrastructure.com/
8.4 Ocean Transport Energy Alternative
http://thetravelersnotebook.com/how-to/how-to-travel-by-cargo-ship/
http://www.ships-info.info/
http://www.tsuneishi.co.jp/english/horie/about.html
http://www.grist.org/article/shipping2/
http://www.2wglobal.com/www/environment/orcelleGreenFlagship/index.jsp
http://www.batteryuniversity.com/partone-5A.htm
http://blog.modernmechanix.com/2007/12/13/sea-waves-to-drive-ocean-liner/
8.5 Solar Technology Improvement
http://www.solarelectricalvehicles.com/
http://news.nationalgeographic.com/news/2005/01/0114_050114_solarplastic.html
http://green.venturebeat.com/2009/08/26/innovalight-says-spray-on-solar-ink-is-coming-soon/
8.6 Energy Supply for Rail Transport Improvement
http://www.jrtr.net/jrtr17/f40_technology.html
http://www.kaist.ac.kr/english/01_about/06_news_01.php?req_P=bv&req_BIDX=10&req_BNM=ed_news&pt=17&req_VI=2207
http://www.iav.com/en/7_press/press_releases.php?we_objectID=15760
http://www.bombardier.com/en/transportation/sustainability/technology/primove-catenary-free-operation
8.7 New Technology of Engine Performance
http://www.consumerenergycenter.org/transportation/fuelcell/index.html
http://www.afdc.energy.gov/afdc/fuels/hydrogen.html
http://en.wikipedia.org/wiki/Hydrogen_vehicle
http://www.mdi.lu/
http://news.bbc.co.uk/2/hi/science/nature/7243247.stm
http://www.aaat.com/green-car-technologies.cfm
8.8 Public Transportation Improvement – Bus
www.bvg.de
http://en.wikipedia.org/wiki/QR_Code
http://www.tudelft.nl/live/pagina.jsp?id=f51868f8-2bad-4f43-b22a-d864e926b389 - Technical University of Delft, Netherland project of "Superbus", lead by Prof. Wubbo Ockels.
http://www.superbusproject.com/
8.9 Rail and Road Transport Alternative
http://www.ktmb.com.my/

© Springer Nature Switzerland AG 2020
M. A. Orloff, *Modern TRIZ Modeling in Master Programs*,
https://doi.org/10.1007/978-3-030-37417-4

http://stringtransport.com/
http://www.gizmag.com/unitsky-string-transport-rail-
suspended/15300/http://en.wikipedia.org/wiki/Electric_multiple_unit

**To part 9.   Managing Value Across Industries: Modern TRIZ's  Modeling of Jerome Lemelson's Inventions:**

*References for Text*

- Altschuller, G 1996, 'And Suddenly the Inventor Appeared: TRIZ, the Theory of Inventive Problem Solving'; 2nd edition, Technical Innovation Center, Inc., Massachusetts
- Aluminium Can Beverage (2012): Pages of Time Company,
- http://www.gono.com/history/marktest.htm
- Batterman et al. (1992): Eric P. Batterman et al., 1992, Multiple resolution machine readable symbols, U.S. Patent 5,153,418. Oct. 6
- Ben-Dor et al (1981): Effraim Ben-Dor et al, 1981, Roller skate, U.S. Patent 4,273,345. Jun. 16
- Chain Link (2012): Youtube Chain Link Fence Manufacturing, http://www.youtube.com/watch?v=E3LnjVXTfXc&feature=related
- Colstoun et al., (1986): Francois P. D. Brown de Colstoun et al, 1986, Method for detecting a source of heat, more particularly a forest fire in a watched area, and system for carrying out said method, U.S. Patent 4,567,367. Jan. 28
- Compact Disc (2012): Council on Library and Information Resources, http://www.clir.org/pubs/reports/pub121/sec3.html
- Dyke Gasoiline Engine (1920): Dyke's automobile and gasoline engine encyclopedia, http://archive.org/details/dykesautomobile00dykegoog
- Forbes Amazon (2010): Forbes How Amazon Maintains Its Edge, http://www.forbes.com/2010/09/13/amazon-innovation-change-management-leadership-managing-human-capital-10-disruption.html
- Jerome H. Lemelson et al. – pointed in text
- Kunz et al.(1971): Oskar Kunz et al., 1971, Water Pistol, U.S. Patent 3,575,318. Apr. 20
- Pedersen et al. (2005): Robert D. Pedersen et al, 2005, Robotic manufacturing and assembly with relative radio positioning using Radio based position determination, U.S. Patent 6,898,484. May 24
- Reuters Smartphone War (2011): Reuters: Samsung, Apple to end Nokia's smartphone reign, http://www.reuters.com/article/2011/06/13/us-nokia-smartphones-idUSTRE75C18O20110613
- Rickards, T 1997, 'Creativity and Problem Solving at Work', Gower Publishing, Hampshire
- Robert D. Pedersen et al., (1998): Robert D. Pedersen & Jerome H. Lemelson, 1998, Fire
- Detection Systems and Methods, U.S. Patent 5,832,187. Nov.3
- Roller Skate (2012): Encyclopedia, http://www.encyclopedia.com/topic/roller_skating.aspx#1-1E1:rollersk-full
- Smithsonian (2012): List of Patents, http://invention.smithsonian.org/about/about_patents.aspx
- Tape Recorder (2012): e how, http://www.ehow.com/about_5040701_history-tape- record-ers.html#ixzz1pJ0q86qc
- Water Gun (2001): Water Gun Museum, http://www.sinasnet.nl/Watergun.HTML
- Wildfire (2012): National Geographic Environmental Natural-disasters-wildfires, http://environment.nationalgeographic.com/environment/natural-disasters/wildfires/
- Wire Recorder (2011): Video Interchange, http://www.videointerchange.com/wire_recorder1.htm

*References for Figures*

- Bombay Harbor (2012): A chain link fence, http://www.bombayharbor.com/productImage/0146866001320822186/Chain_Link_Fence.jpg

- Colour box (2011): A Beverage Can, http://www.colourbox.com/preview/2735748-794811-can-of-cola-with-straw-isolated-over- white-background.jpg
- Dawanda (2012): Retro reflective thread, http://en.dawanda.com/product/20780061-Retro-reflective-machine-embroidery-THREAD
- Dimensions info (2009): An Audio Cassette with tape, http://www.dimensionsinfo.com/wp-content/uploads/2009/11/Audio-Cassette.jpg
- Ecouterre (2012): A suit with reflective stickers, http://www.ecouterre.com/reversible-tron-inspired-jacket-reflects-light-for-safer-night- biking
- Environmental Graffiti (2012): Regions of Forest fire occurrence in Africa, http://static.environmentalgraffiti.com/sites/default/files/images/800px-2002_african_fires_nasa.img_assist_custom-600x410.png
- Flickr (2011): A Pyramidal Reflector, http://www.flickr.com/photos/sschie/6106484130/
- Forgotten futures (1900): A Transmitter in operation at the 'New York Herald'
- Office, http://www.forgottenfutures.co.uk/fax/fax_6.gif
- Free patent online (2005): A Magnetic Strip and picture recording strip, http://www.freepatentsonline.com/6947306-0-large.jpg
- Free patents online (2005): Strip, http://www.freepatentsonline.com/6947306-0- large.jpg
- Gadgether (2010): Macegun, http://gadgether.com/wp- content/uploads/2010/09/macegun.jpg
- Gothamist (2005): A water gun, http://gothamist.com/attachments/arts_jen/2005_07_artswatergun.jpg
- Honda Tuning (2012): A manual wheel removal in progress, http://www.hondatuningmagazine.com/tech/htup_0512_honda_civic_hatchback_wilwood_big_brake_kit/photo_02.html
- Kids Army (2006): A toy grenade, http://www.kids- army.com/product_images/i/837/toy-hand-grenade-1_59123_zoom.jpg
- Kuka Robotics (2012): An Assembly Robot, http://www.kuka- robotics.com/NR/rdonlyres/EC6ED75B-8EEE-49FE-993D-73488D7CD3C4/0/PR_KUKA_Industrial_Robot_KR5_Sixx_R850_01.jpg
- Manufacturer (2012): Extruded Fence Nets, http://www.manufacturer.com/cimages/buyLeads/www.alibaba.com/1229/l/Inquiry_For_Extruded_Fence_Nets.jpg
- Miller Mc Cune (2010): An image of the landscape from the infrared camera, with yellow regions signifying high heat signature, http://www.miller-mccune.com/wp- content/uploads/2010/01/mmw_wildfires1.jpg
- My lot (2006): A contact lens, http://images.mylot.com/userImages/images/postphotos/1652430.jpg
- Nellis AF (2012): A towed watercraft, http://www.nellis.af.mil/shared/media/photodb/photos/110817-F-GQ230-264.jpg
- Night Gear (2012): A sweat shirt made from reflective thread, http://www.night-gear.com/stores/n/nightgear/catalog/9000_1_photo.jpg
- Nova Diamant (2012): Microscopic View of Valve surface coated with synthetic Diamond, http://www.novadiamant.com/res/Default/eggb.jpg
- Science photo (2012): A Facsimile Apparatus, http://www.sciencephoto.com/image/352469/530wm/T5500008-Fax_machine_transmitting_a_document-SPL.jpg
- Shopping shadow (2012): An Eyeglass, http://di1-4.shoppingshadow.com/images/pi/32/3e/1d/104326539-260x260-0-0_prada%2Bprada%2Bpr06nv%2B1ab1o1%2Beyeglasses%2Bgloss%2Bblack%2Bf.jpg
- Shree enterprisers (2012): Surgical Instruments, http://shreeenterprisers.com/images/Surgical%2520instruments-big.jpg
- University Products (2012): A Foam Sheet, https://www.universityproducts.com/secure/images/categories/main_45.jpg
- Uploads (2012): A poppet Valve, http://174.123.135.195/uploads05/8/Y/ZQ111128715334.jpg

- White house museum (1902): A Telegraphic Office, http://www.whitehousemuseum.org/west-wing/old- eob/telegraph-office-c1902.jpg
- Wikimedia (2012): An Inline Skate, http://upload.wikimedia.org/wikipedia/commons/6/66/Inline_skate_wheels.jpg

**To part 10.   Modeling the Great Inventions of Carl and Werner von Siemens with Modern TRIZ:**

*References for Text*

- Wikipedia: Problem. http://en.wikipedia.org/wiki/Problem
- Glenn Mazur (1995): Theory of Inventive Problem Solving (TRIZ); http://www.mazur.net/triz
- I-TRIZ: Introduction to Basic I-TRIZ, What is an Inventive Problem; http://www.ideationtriz.com/source/12_Inventive_Problem.htm
- Katie Barry, Ellen Domb and Michael S. Slocum: TRIZ – What is TRIZ. http://www.triz-journal.com/archives/what_is_triz
- Kelly McEntire/DDR (2010): Improving Innovation Through TRIZ; http://www.slideshare.net/QRCE/triz-product-design-development-presentation
- Toru Nakagawa: TRIZ Home Page in Japan. http://www.osaka-gu.ac.jp/php/nakagawa/TRIZ/eTRIZ
- Wikipedia: Productive Thinking Model; http://en.wikipedia.org/wiki/Productive_Thinking_Model
- Widipedia: Lateral thinking. http://en.wikipedia.org/wiki/Lateral_thinking
- Wikipedia: Method of focal objects; http://en.wikipedia.org/wiki/Method_of_focal_objects
- Wikipedia: TRIZ. http://en.wikipedia.org/wiki/TRIZ
- Modern TRIZ Academy International (2010-2011): TRIZ. http://www.modern-triz-academy.com/triz.htm
- Wikipedia: Genrich Altshuller. http://en.wikipedia.org/wiki/Genrich_Altshuller
- http://www.teachingenuity.com/2011/03/13/genrikh-altshullerIdeation International Inc. (2006-2011): History of TRIZ and I-TRIZ. http://www.ideationtriz.com/history.asp
- Valeri Souchkov (2008): A Brief History of TRIZ; http://www.xtriz.com/BriefHistoryOfTRIZ.pdf
- Michael Orloff (2006): Inventive Thinking through TRIZ. A Practical Guide.
- David Conley: TRIZ-Definitions and Characterization; http://www.innomationllc.com/Files/whatstriz.pdf
- Trevor Mendham: TRIZ. http://www.wyrdology.com/mind/creativity/TRIZ.html
- B.Rossi & V.Muzi: Benefits of TRIZ. http://www.know-it-project.eu/docs/TRIZ_introduction_en.pdf
- Valeri Souchkov, ICG T&G (2007): Accelerate Innovation with TRIZ; http://www.xtriz.com/publications/AccelerateInnovationWithTRIZ.pdf
- True North Innovation (2007): Introduction to TRIZ; http://www.truenorthinnovation.co.uk/downloads/introduction_to_triz.pdf
- The Altshuller Institute for TRIZ Studies (2011): What is TRIZ; http://www.aitriz.org/index.php?option=com_content&task=view&id=18&Itemid=32
- Modern TRIZ Academy International (2010-2011): Introduction. http://www.modern-triz-academy.com/publications-intro.htm
- Darrel Mann: The Four Pillars of TRIZ. http://www.systematic-innovation.com/Articles/99,%2000,%2001/Mar00-The%20Four%20Pillars%20of%20TRIZ.pdf
- Wikipedia: Siemens. http://en.wikipedia.org/wiki/Siemens
- Siemens AG: Travel back through the history of Siemens; http://www.siemens.com/history/en/history/index.htm
- Wikipedia: Aktiengesellschaft. http://en.wikipedia.org/wiki/Aktiengesellschaft
- Siemens (2007): Werner von Siemens (1816-1892); http://www.siemens.com/history/pool/perseunlichkeiten/gruendergeneration/werner_von_siemens_en.pdf
- Siemens AG: Worldwide Presence. http://www.siemens.com/about/en/worldwide.htm

- Dr. Florian Kiuntke (2011): 1870 – Around the world in 28 minutes; http://www.siemens.com/history/en/news/1069_indo-european_telegraph-line.htm
- Montreal Environment: A conversation with M.Roland Aurich CEO Siemens Canada. http://www.montrealenvironment.ca/a-conversation-with-m-roland-aurich-ceo-siemens-canada/
- Wikipedia: The Siemens Regenerative Furnace; http://en.wikipedia.org/wiki/Siemens_regenerative_furnace
- Encyclopedia: Sir William Siemens. http://www.1902encyclopedia.com/S/SIE/william-siemens.html
- Bessemer process. http://en.wikipedia.org/wiki/Bessemer_process
- ENotes.com, Inc. (2011): History, What is Bessemer Process; http://www.enotes.com/history/q-and-a/what-bessemer-process-286670
- Wikipedia: Open hearth furnace. http://www.wisegeek.com/what-is-an-open-hearth-furnace.htm
- New World Encyclopedia: Thermometer; http://www.newworldencyclopedia.org/entry/Thermometer
- Wikipeida: Resistance thermometer; http://en.wikipedia.org/wiki/Resistance_thermometer
- Vincent Grosjean (2007): Email in the 18th Century: the optical telegraph; http://www.lowtechmagazine.com/2007/12/email-in-the-18.html
- Scripophily.com: Siemens Electric Business Inc- 1901-1913; http://scripophily.net/noname19.html
- Siemens AG: Information & Communications; http://www.siemens.com/history/en/innovations/information_and_communications.htm
- Wikipedia: Dynamo. http://en.wikipedia.org/wiki/Dynamo#cite_ref-3
- Wikipedia: Electric generator. http://en.wikipedia.org/wiki/Electric_generator, visited on 12.19.2011
- Wikipedia: Rail transport. http://en.wikipedia.org/wiki/Rail_transport
- Wikipedia: Diesel locomotive. http://en.wikipedia.org/wiki/Diesel_locomotive
- Wikipedia: Electric locomotive. http://en.wikipedia.org/wiki/Electric_locomotive
- Wikipedia: Railway electrification system; http://en.wikipedia.org/wiki/Railway_electrification_system
- Mary Bellis: History of the Elevator; http://inventors.about.com/od/estartinventions/a/Elevator.htm
- IEEE: The Electric Elevator; http://www.ieeeghn.org/wiki/index.php/The_Electric_Elevator
- Dr. Florian Kiuntke (2010): 1880 – Werner von Siemens present the world's first electric elevator. http://www.siemens.com/history/en/news/1043_elevator.htm
- ReferenceAnswers: elevator. http://www.answers.com/topic/elevator
- Correios De Macau: Morse Telegraph; http://macao.communications.museum/eng/exhibition/secondfloor/MoreInfo/2_5_2_MorseTelegraph.html

*References for Figures*

- Johann Geory Halske (b. July 30, 1814, Hamburg – d. March 18, 1890, Berlin). http://matidavid.com/pioneer_files/siemens.htm
- Sir William Siemens in 1850. http://www.siemens.com/history/en/history/index.htm
- Siemens' first cable plant in Woolwich in 1863. http://www.siemens.com/history/en/history/index.htm
- Shares of Siemens. http://www.siemens.com/history/en/history/index.htm
- Berlin Siemensstadt in 1914. http://www.siemens.com/history/en/history/index.htm
- Osram GmbH KG. http://www.siemens.com/history/en/history/index.htm
- Fusi Denke Seizo KK. http://www.siemens.com/history/en/history/index.htm
- Hermann von Siemens. http://www.siemens.com/history/en/history/index.htm
- Shares of Siemens AG in 1966. http://www.siemens.com/history/en/history/index.htm
- New location in Munich-Perlach. http://www.siemens.com/history/en/history/index.htm

- Carl Wilhelm Siemens. http://en.wikipedia.org/wiki/Carl_Wilhelm_Siemens
- Ernst Werner von Siemens in 1880. http://www.ssplprints.com/image/89967/unattributed-ernst-werner-von-siemens-german-electrical-engineer-and-inventor-c-1880
- Siemens' house in Lenthe where Werner von Siemens was born. http://matidavid.com/pioneer_files/siemens.htm
- A monument in honour of Werner von Siemens in the village where he was born. http://matidavid.com/pioneer_files/siemens.htm
- Werner von Siemens a few months before his death. http://matidavid.com/pioneer_files/siemens.htm
- Steven Roberts, History of the Atlantic Cable & Undersea Communications. The route of the Indo-European Telegraph Line. http://atlantic-cable.com/CableCos/Indo-Eur/index.htm
- Blueprint of Siemens regenerative furnace. http://www.siemens.com/history/en/history/index.htm
- Regenerative furnace produced in Guangdong, China. http://www.fslangdun.com/en/displayproduct.php?id=27&cid=76
- Bessemer converter, Kelham Island Museum, Sheffield, England. http://en.m.wikipedia.org/wiki/File:Bessemer_5180.JPG
- Bessemer converter component. http://en.wikipedia.org/wiki/Bessemer_process
- Oral clinical mercury thermometer. http://www.tradevv.com/chinasuppliers/opersonvila_p_110981/china-oral-clinical-thermometer-mercury-thermometer.html
- Thermocouples. http://www.kryothermusa.com/indexc18c.html?tid=87
- Platinum Resistance Thermometers. http://gb-international.com/educational-60-4-10.html
- Email in the 18th century: the optical telegraph. Optical Telegraph, France. http://www.lowtechmagazine.com/2007/12/email-in-the-18.html
- Early letter pointer telegraph instrument at Hamar museum. http://members.ozemail.com.au/~telica/Norway_Hamar_Railway_Museum_Main_Exhibition_Hall.html
- Ernst Werner von Siemens.Pointer Telegraph in 1847. http://matidavid.com/pioneer_files/siemens.htm
- Battery-free telegraph. http://www.siemens.com/history/en/history/index.htm
- Faraday disk. http://en.wikipedia.org/wiki/File:Faraday_disk_generator.jpg
- Energy use in the US. Pixxi's dynamo, http://www.timetoast.com/timelines/energy-use-in-the-us--41
- Siemens' dynamo. http://matidavid.com/pioneer_files/siemens.htm
- Trevithick's locomotive, 1804; http://en.wikipedia.org/wiki/File:Locomotive_trevithick.jpg
- A steam locomotive at the Gare du Nord, Paris, 1930; http://en.wikipedia.org/wiki/File:Steam_Locomotive.jpg
- http://en.wikipedia.org/wiki/Electric_locomotive
- Map of electrified train in Northern Europe. http://en.wikipedia.org/wiki/File:Map_of_electrified_railways_in_Northern_Europe.jpg
- Spanish modern electric locomotive (AVE Class 102 type train); http://en.wikipedia.org/wiki/File:Talgo_350.jpg
- Swiss Electric locomotive at Brig, Switzerland; http://en.wikipedia.org/wiki/File:SwissMGB.jpg
- Otis demonstrating safety elevator 1854 NYC; http://archhistdaily.wordpress.com/2012/01/23/january-23-all-safe/
- Werner von Siemens, Electric elevator, 1880 Mannheim; http://www.siemens.com/history/en/news/1043_elevator.htm

**To part 11. Reinventing of Automobile Production Systems Evolution:**

*References for Text*

- AITRIZ – The Altshuller Institute for TRIZ Studies (2011): What is TRIZ?. www.aitriz.org/index.php?option=com_content&task=view&id=18&Itemid=32
- Altshuller, G. S. (1979): Creation as an Exact Science. Sow. Radio Publishers, Moscow.

- American Decades (2001): The Automobile; www.encyclopedia.com/topic/The_Automobile.aspx
- Art of Lean, Inc. (2011): Toyota Production System Basic Handbook; artoflean.com/files/Basic_TPS_Handbook_v1.pdf
- Barry, K., Domb, E. and Slocum, M. S. (2010): Triz – What is Triz. The Triz Journal; www.triz-journal.com/archives/what_is_triz
- Bellis, M. (2011): The History of the Automobile; inventors.about.com/library/weekly/aacarssteama.htm
- Biggs, L. (1996): The Rational Factory, Architecture Technology and Work in America's Age of Mass Production. The John Hopkins University Press, Baltimore, London.
- Bryan, F. R. (2009): The Birth of Ford Motor Company. Henry Ford Heritage Association. http://hfha.org/HenryFord.htm#Ford-Motor-Co
- Clarke, C. (2005) : Automotive Production System and Standardisation : From Ford to the Case of Mercedes Benz. Physica-Verlag, Heidelberg.
- Dennis, P. (2007) : Lean Production Simplified : A-Plain Language Guide to the World's Most Powerful Production System. Productivity Press, New York.
- Drucker, P. F. (1946): Concept of the Corporation. John Day Company, New York, NY.
- Eckermann, Erik (2001): World History of the Automobile. SAE Press, p.14.
- Encyclopedia of World Biography (2004): Adam Smith; http://www.encyclopedia.com/topic/Adam_Smith.aspx#4
- EPI – Earth Policy Institute (2011): World Bicycle and Passenger Car Production, 1950-2007. www.earth-policy.org/data_center/C26
- ESB – Encyclopedia of Small Business (2007): Brainstorming; www.encyclopedia.com/doc/1G2-2687200064.html
- EyeWitness to History (2005): Henry Ford Changes the World, 1908; www.eyewitnesstohistory.com/ford.htm
- Flink, J. J. (1988): The Automobile Age. MIT Press, Cambridge, MA.
- Ford, H. and Crowther, S. (1922): My Life and Work. Nevins and Hill, Ford TMC.
- FMC – Ford Motor Company (2011): Model T Facts; media.ford.com/article_display.cfm?article_id=858
- Gartman, D. (1986): Auto Slavery: The Labour Process in the American Automobile Industry, 1897 – 1950. Rutgers University Press, New Brunswick and London.
- Glasscock, C. B. (2011): Car History: A Vision Becomes Reality – Part 1: The Automobile Aristocracy of America. www.americanautohistory.com/Articles/Article001.htm
- Gorgano, G. N. (1985) : Cars : Early and Vintage, 1886-1930. Grange-Universal, London.
- Graham (2010): Automotive Industry: Trends and Reflections. International Labour Organization, Geneva.
- Harvey, D. (2004) : Lean, Agile. www.davethehat.com/articles/LeanAgile.pdf
- HFHA – Henry Ford Heritage Association (2009): The Ford Story; hfha.org/HenryFord.htm#Ford-Motor-Co
- Hines, P., Found, P., Griffiths, G. and Harrison, R. (2011): Staying Lean: Thriving, Not Just Surviving. Productivity Press, New York.
- Hines, P. and Rich, N.(1997) : The Seven Value Stream Mapping Tools. International Journal of Operations & Production Management, Vol. 17 No. 1, pp. 46-64. MCB University Press.
- Hodzic, M. (2008): Gottlieb Daimler, Wilhelm Maybach and the "Grandfather Clock"; www.benzinsider.com/2008/06/gottlieb-daimler-wilhelm-maybach-and-the-grandfather-clock
- ICG T&C – ICG Training and Consulting (2010): TRIZ Success Cases. www.xtriz.com/documents/TRIZSuccessCases.pdf
- IESS – International Encyclopedia of Social Sciences (2008): Automobile Industry. www.encyclopedia.com/doc/1G2-3045300143.html
- IMC – Invention Machine Corporation, (2011): Invention Machine. inventionmachine.com
- Keller, Paul A., as appears in Pyzdek, T. (2000) : The Handbook for Quality Management. QA Publishing, Tucson, Arizona.

- Krasnoslobodtsev, V. and Langevin, R. (2006): Applied TRIZ in High-Tech Industry. www.triz-journal.com/archives/2006/08/2006-08.pdf
- LAB – Leading American Business (2003): Ford, Henry; www.encyclopedia.com/topic/Henry_Ford.aspx
- Lean Enterprise Institute (2010) : Principles of Lean; www.lean.org/whatslean/principles.cfm
- Lewis, D. (1976): The Public Image of Henry Ford: An American Folk Hero and His Company. Wayne State University Press, Detroit, MI.
- Liker, J. K. (2004) : The Toyota Way. McGraw-Hill, New York.
- Marr, K. (2009): Toyota Passes General Motors As World's Largest Carmaker. Washington Post Article. www.washingtonpost.com/wp-dyn/content/article/2009/01/21/AR2009012101216.html
- Mazur, G (1995): Theory of Inventive Problem Solving (TRIZ). www.mazur.net/triz
- Mertins, K. (2010): Just in Time – JIT. Handout Material for Global Production Management. Global Production Engineering, Technische Universität Berlin, Berlin.
- McCalley, B. W. (1994): Model T Ford: The Car That Changed the World. Krause Publications, Iola, WI.
- McNairn, W. and McNairn, M. (1978): Quotations from the Unusual Henry Ford. Quotamus Press, Redono Beach, CA.
- Mg12.info (2008): Magnesium in AutomotiveCenter; www.mg12.info/metallurgy/magnesium-applications/magnesium-in-automotive.html, visited on 26.05.2011.
- Montgomery, D. (1989): The Fall of the House of Labour: The Workplace, the State, and American Labour Activism, 1865-1925. Cambridge University Press, Cambridge, UK.
- NEW – New World Encyclopedia (2008): Karl Benz; www.newworldencyclopedia.org/entry/Karl_Benz
- NEW – New World Encyclopedia (2008): Automobile; www.newworldencyclopedia.org/entry/Automobile
- Norman, J. (2011): The First Production Automobiles (1893 – 1894); www.historyofinformation.com
- OICA – Organisation Internationale des Constructeurs d'Automobiles (2007): 2007 Production Statistics. oica.net/category/production-statistics/2007-statistics
- Ohno, T. (1988): Toyota Production System: Beyond Large-Scale Production. Productivity Press, New York, NY.
- Orloff, M. A. (2006): Inventive Thinking through TRIZ: A Practical Guide. Springer-Verlag, Berlin.
- Orloff, M. A. (2010): Modern TRIZ: Textbook for program of higher education in MTRIZ. Modern TRIZ Academy International, Berlin.
- Orloff, M. A. (2011): ABC TRIZ: Basics of Inventive Thinking for the Modern Young People. Technopark Adlershof, Berlin.
- Pereira, R. (2010): Guide to Lean Manufacturing. LSS Academy.
- Pine II, B. J. (1993): Mass Customization: The New Frontier in Business Competition. Harvard Business School Press, Boston, MA.
- Poppendieck, M. and Poppendieck T. (2007): Implementing Lean Software Development: from Concept to Cash. Addison-Wesley, Boston, MA.
- Porter, H. F. (1917): Four Big Lessons from Ford's Factory. System: The Magazine of Business Vol. 31 (June 1917). pp. 639-646.
- Quivik, F. L. (2003): Historic American Engineering Record: The Ford Motor Company's Richmond Assembly Plant a.k.a The Richmond Tank Depot. National Historic Park, Richmond.
- Raff, D (1987) : Wage Determination Theory and the Five-Dollar Day at Ford. Ph.D. Dissertation, Massachusetts Institute of Technology.
- Rother, M. and Shook, J. (1999) : Learning to See : Value Stream Mapping to Create Value and Eliminate Muda. The Lean Enterprise Institue : Massachusetts.
- Meyer III, S. (1981): The Five Dollar Day, Labor Management and Social Control in the Ford Motor Company 1908 - 1921. State University of New York Press, Albany.

- Saifutdinov, A. F., Beketov, Q. E., Ladoushkin, V. S. and Nesterov, G. A. (2002): Distillation Technology for the 21st Century. Hydrocarbon Asia, November/December 2002. p.40-43.
- Satyanarayana, A. (2008): Automobile Engine Constructional Details: Materials used for the Engine Block; www.brighthub.com/diy/automotive/articles/6639.aspx#ixzz1OW5A4uWt
- Shingo, S. (1989) : A study of the Toyota Production System from an Industrial Engineering Viewpoint. Productivity Press, Cambridge, MA.
- Silver, E. A., Pyke, D. F. and Peterson, R. (1998) : Inventory Management and Production Planning and Scheduling (pp. 592 – 623, 631 – 653). John Wiley & Sons, Inc., New York.
- Sorensen, S. E. (1956): My Forty Years with Ford. Norton, New York.
- Souchkov, V (2007a): Accelerate Innovation with TRIZ; www.xtriz.com/publications/AccelerateInnovationWithTRIZ.pdf
- Souchkov, V. (2007b): Made with TRIZ.; www.xtriz.com/publications/MadeWithTRIZ.pdf
- Souchkov, V. (2007c): TRIZ Success Cases; www.xtriz.com/publications/TRIZSuccessCases.pdf
- Souchkov, V (2008): A Brief History of TRIZ; www.xtriz.com/BriefHistoryOfTRIZ.pdf
- Tebo, L. (2010): C. Harold Wills, Automotive Pioneer; www.automotivetraveler.com/index.php?option=com_content&view=article&id=1022:c-harold-wills-automotive-pioneer&catid=196:guest-blogs&Itemid=362
- The Henry Ford (1995): The Showroom of Automotive History: 1896 Quadricycle; http://www.hfmgv.org/exhibits/showroom/1896/quad.html
- TMC – Toyota Motor Corp. (1995): History of Toyota; www.toyota-global.com/company/history_of_toyota
- Trizsite.com (2005): Four Pillars of TRIZ; www.trizsite.com/trizresource/fourpillars.asp
- TTG – The TRIZ Group (2007): Genrikh Altshuller – The Creator of TRIZ; www.trizgroup.com/altshuller-bio.html
- USITC – U.S. International Trade Commission (2002): Industry and Trade Summary: Motor Vehicles. U.S. ITC Publication 3545, USITC, Washington, DC.
- Vollman, T. E., Berry. W. L., Whybark, D. C. and Jacobs F. R. (2005) : Manufacturing Planning and Control for Supply Chain Management. McGrawHill, New York.
- Wilson, L (2010) : How to Implement Lean Manufacturing. McGraw-Hill, New York.
- Womack, J. P. and Jones, D. T. (1996) : Lean Thinking : Banish Waste and Create Wealth in Your Corporation. Simon and Schuster, New York.
- Womack, J. P., Jones, D. T. and Roos, D. (1990) : The Machine that Changed the World. Rawson Associates, New York.
- Zlotin, B., Zusman, A., Altshuller, G. and Philatov, V. (1999): Tools of Classical TRIZ. Ideation International Inc.

*References for Figures*

- Mazur (1995): Theory of Inventive Problem Solving (TRIZ); www.mazur.net/triz
- dentist.net (2010): Crest Whitestrips Premium; www.dentist.net/crest-whitestrips-premium.asp, visited on 03.06.2011
- Net Resources International (2011): KC-767 Tanker Transport Aircraft, USA; www.airforce-technology.com/projects/kc767/kc7671.html
- Fruchter, M (2008): Sharing, Self Promotion Always a Two-Way Street; blog.louisgray.com/2008/11/sharing-self-promotion-always-two-way.htm
- 123RF (2011): Closed railway crossroad. jp.123rf.com/photo_6008120_opened-railway-crossroad.html
- Net Resources International (2011): Crossing Borders: Rail Travel Within the EU; www.railway-technology.com/features/feature116419/feature116419-1.html
- Searoader (2011): Searoder Ambulance; searoader.com
- engineblock.org (2011): Engine Block; engineblock.org/tips.htm
- Victorinox (2011): Traveller Lite; victorinox.com/product/1/100/1000/1.7905.AVT
- Glasscock, C. B. (2011): Car History - A Vision Becomes Reality. Part 1: The Automobile Aristocracy of America; www.americanautohistory.com/Articles/Article001.htm

- Christopher (2011): Mercedes-Benz Celebrates 125 Year; pursuitist.com/auto/mercedes-benz-celebrates-125-years
- Bert, P. (2010): Histoire de l'automobile; paul.bert.sens.pagesperso-orange.fr/activites/fetescie/2003/automobile.htm
- Ford Gallery (2008): Ford Gallery; www.rarecarrelics.com/gallery_of_Ford_photos.php
- Used Corolla (2011): Corolla History; www.usedcorolla.com/history
- Lean Enterprise Institute (2010) : Principles of Lean; www.lean.org/whatslean/principles.cfm
- Art of Lean, Inc. (2011) : Toyota Production System Basic Handbook; artoflean.com/files/Basic_TPS_Handbook_v1.pdf
- EyeWitness to History (2005): Henry Ford Changes the World, 1908; www.eyewitnesstohistory.com/ford.htm
- National Photo Company (1916): Auto Hacket Motorcar Co. Shop,
- Jackson, Michigan, 1916; blacksmithandmachineshop.com/Shop-Views-step-in-to-what-it-was-like-to-work-in-those-long-gone-faded-cold-dakr-pre-internet-grueling-years-of-your-grandfather.html
- Theobald, M. (2004): Fisher Body Company : 1908 – present; www.coachbuilt.com/bui/f/fisher/fisher.htm
- Huff Report (2011): The 1959 Chevrolet Assembly Line; www.huffreport.com/archives/literature/59chevies.htm
- National Geographic (1923): The Automobile Industry - An American Art that Has Revolution-ized Methods in Manufacturing and Transformed Transportation; www.flickr.com/photos/mytravelphotos/2155172271/in/set-72157603599425372
- Kleo (2008): 16 Most Innovative Car Inventions of All Times; allworld-cars.com/wordpress/?p=7305
- Bert, P. (2010): Histoire de l'automobile; paul.bert.sens.pagesperso-orange.fr/activites/fetescie/2003/automobile.htm
- Tebo (2010): C. Harold Wills: Automotive Pioneer; www.automotivetraveler.com/index.php?option=com_content&view=article&id=1022:c-harold-wills-automotive-pioneer&catid=196:guest-blogs&Itemid=362
- Ford Gallery (2008): A Model T Ford Farm Truck with a Load of Hay; www.rarecarrelics.com/gallery_of_Ford_photos.php
- Ford Gallery (2008): Ford Model T Assembly Line – Petrol Tank; www.rarecarrelics.com/gallery_of_Ford_photos.php
- Ford Gallery (2008): Factory Assembly Line; www.rarecarrelics.com/gallery_of_Ford_photos.php
- Baker, J. (2011): Austin Memories – Car Assembly; www.austinmemories.com/page24/page62/page62.html
- Theobald, M. (2004): Fisher Body Company : 1908 – present; www.coachbuilt.com/bui/f/fisher/fisher.htm
- Deaton, J. P. (2008): How Automotive Production Lines Work; au-to.howstuffworks.com/under-the-hood/auto-manufacturing/automotive-production-line.htm
- Greenless (2011): The Old Motor; theoldmotor.com/?m=201101
- Ford Gallery (1928): Body Assembly Room; www.rarecarrelics.com/gallery_of_Ford_photos.php
- Ford Gallery (1928): Assembly Station; www.rarecarrelics.com/gallery_of_Ford_photos.php
- Ford Gallery (1928): Early Model T Ford Assembly Line; www.rarecarrelics.com/gallery_of_Ford_photos.php
- Ford Gallery (2008): Moving Assembly Line; www.rarecarrelics.com/gallery_of_Ford_photos.php
- Giddy L. (2011): History of the Assembly Line; www.beaconschool.org/~lgiddy/AssemblyLine.htm
- Strieber, A. (2008): T Party: Ford celebrates 100th birthday of the Model T -- we take a spin in a Tin Lizzie;

www.motortrend.com/features/auto_news/2008/112_0807_ford_model_t_100_year_celebrati
on/index.html.

- Niebuhr, M. (2008): Photography: Considering Contemporary Issues of the Day. Photo by
  Charles Seeler; mniebuhr.com/2008/01/01/photography-considering-contemporary-issues-of-
  the-day
- National Geographic (1923): The Automobile Industry - An American Art that Has Revolution-
  ized Methods in Manufacturing and Transformed Transportation;
  www.flickr.com/photos/mytravelphotos/2155173759
- ArtofLean (2009): Set Up Reduction in Toyota; artoflean.com/blog1/?p=103
- Management and Development Center (2011): Single Minute Exchange of Dies SMED;
  mdcegypt.com/Pages/Management%20Approaches/Lean%20enterprise/SMED/Introduction.
  asp
- Kessel, D. (2011): Rows of Finished Jeeps Churned Out in Mass Production for War Effort as
  WWII Allies; www.allposters.com/-sp/Rows-of-Finished-Jeeps-Churned-Out-in-Mass-
  Production-for-War-Effort-as-WWII-Allies-Posters_i3781089_.htm,
- Russian Transport Daily Report (2011): Sollers-Far East Manufactures First Production
  Crossover SsangYong New Actyon; rtdr.org/news/automotive/sollers-
  far_east_manufactures_first_production_crossover_ssangyong_new_actyon.html
- Drukier, M. (2010): The Practical Entrepreneur: Purchasing and Negotiations, Part VIII;
  www.theepochtimes.com/n2/business/the-practical-entrepreneur-purchasing-and-
  negotiations-part-viii-38346.html
- Rabe, M. (2010): Lean Management, Handout material for Global Production Management.
  p.19. Global Production Engineering, Technische Universität Berlin, Berlin.
- Ibest Electrical Co. Ltd. (2011): Inductive Proximity Switch;
  www.supplierlist.com/product_view/jasmine84/25932/100526/Inductive_proximity_sensor.ht
  m.
- Keyence Corp. (2011): Automotive Industry Poka-yoke Guide Part 2;
  www.keyence.com/topics/barcode/bl/auto2.php#more
- J.D. Power (2010): Award Recipients: 2010 Initial Quality Study (Midsize Premium Car;
  www.jdpower.com/autos/car-photos/Initial-Quality/Award-Recipients/2010
- J.D. Power (2011): Award Recipients: 2011 Most Dependable Vehicles (Midsize Crosso-
  ver/SUV; www.jdpower.com/autos/car-photos/VDS/Most%20Dependable/2011
- Baker, J. (2011): Austin Memories – Car Assembly;
  www.austinmemories.com/page24/page62/page62.html
- Dollarpaisa.com (2010): M. K. Enterprises, Padam Nagar;
  www.dollarpaisa.com/Company/Sells/19617/19466/M._K._Enterprises_Padam_Nagar.htm
- Sergio (2008): Kaizen, gerencia visual y andon; www.sortega.com/blog/kaizen-gerencia-
  visual-y-andon
- Tangram Technology Ltd. (2005): The Manufacturing Strategy Series; www.tangram.co.uk
- Mertins, K. (2010): Just in Time – JIT. Handout material for Global Production Management.
  p.4. Global Production Engineering, Technische Universität Berlin, Berlin.
- Peskett (2009): Global Downturn Hits Hard. Photo by Chip Somodevilla/Getty Images;
  www.caradvice.com.au/26275/global-downturn-hits-hard-pictures
- Brady Worldwide, Inc. (2010): Productivity;
  www.bradyid.com/bradyid/cms/contentView.do/5906/gp.htm
- Manufactus (2010): Examples for Kanban Cards; www.manufactus.com/products/examples-
  for-kanban-cards/en
- Peskett (2009): Wireless Kanban Replenishment System';
  www.leansupermarket.com/servlet/Detail?no=340

The other references are given in the text. The author thanks all the authors of the illustrations and texts that he and his students used.

The author also apologizes for possible missing references and would be grateful for suggestions to improve the references.

Printed in the United States
By Bookmasters